Interdisciplinary Studies on Digital Transformation and Innovation:

Business, Education, and Medical Approaches

Ahmed Gomaa Radwan
Nile University, Egypt

Salwa K. Abd-El-Hafiz
Cairo University, Egypt

Islam Tharwat Abdel Halim
Nile University, Egypt

Yuhong Liu
Santa Clara University, USA

Meikang Qiu
Texas A&M University Commerce, USA

IGI Global
Scientific Publishing
Publishing Tomorrow's Research Today

Vice President of Editorial — Melissa Wagner
Managing Editor of Acquisitions — Mikaela Felty
Managing Editor of Book Development — Jocelynn Hessler
Production Manager — Mike Brehm
Cover Design — Phillip Shickler

Published in the United States of America by
IGI Global Scientific Publishing
701 East Chocolate Avenue
Hershey, PA, 17033, USA
Tel: 717-533-8845
Fax: 717-533-8661
Website: https://www.igi-global.com E-mail: cust@igi-global.com

Library of Congress Cataloging-in-Publication Data

Names: Radwan, Ahmed G. editor | Abd-El-Hafiz, Salwa K. editor | Tharwat Abdel-Halim, Islam, 1982- editor | Liu, Yuhong, 1982- editor | Qiu, Meikang editor
Title: Interdisciplinary studies on digital transformation and innovation : business, education, and medical approaches / edited by Ahmed Radwan, Salwa Abd-El-Hafiz, Islam Abdel-Halim, Yuhong Liu, Meikang Qiu.
Description: Hershey, PA : IGI Global Scientific Publishing, [2025] | Includes bibliographical references and index. | Summary: "The book endeavors to address the challenge of connecting theoretical research with real-world applications by offering pioneering research discoveries, inventive methodologies, and actionable initiatives. Through an exploration of the intersection between data analytics, security, and sustainability, this publication empowers readers with the necessary knowledge and resources to tackle global challenges, make well-informed choices, and actively contribute to the creation of a sustainable future"-- Provided by publisher.
Identifiers: LCCN 2024058258 (print) | LCCN 2024058259 (ebook) | ISBN 9798337311326 hardcover | ISBN 9798337311333 paperback | ISBN 9798337311340 ebook
Subjects: LCSH: Technological innovations--Case studies
Classification: LCC HD45 .I5545 2025 (print) | LCC HD45 (ebook) | DDC 658.5/14--dc23/eng/20250212
LC record available at https://lccn.loc.gov/2024058258
LC ebook record available at https://lccn.loc.gov/2024058259

British Cataloguing in Publication Data
A Cataloguing in Publication record for this book is available from the British Library.

Table of Contents

Detailed Table of Contents

In the rapidly evolving landscape of technology, innovation, and sustainability, integrating Artificial Intelligence (AI) in higher education presents significant opportunities to enhance learning, teaching and assessment, streamline administrative processes, and promote sustainable educational practices. Despite a vast amount of literature on AI in education, comprehensive use cases that can inform effective implementation of AI in higher education remain scarce. This chapter aims to fill this gap by exploring in-depth the AI applications in higher education, addressing key areas of AI integration such as curriculum design and content development, pedagogical strategies and learning environments, evaluation mechanisms and feedback systems, process streamlining and efficiency enhancement, learning analytics, academic research and ideation, and student support and services. As such, this chapter contributes to the broader goal of sustainable development in the context of education thus aligning with the themes of the SMART conference. This research may inform sustainable solutions, policy, and governance mechanisms in the context of AI-driven education.

Digital transformations often struggle due to employee resistance and low engagement. Gamification, incorporating game elements into non-game settings, tackles these issues. This study explored gamification's impact on a news organization's digital transformation. A gamification technique implemented with 164 journalists for five months. Results showed a significant increase in digital transformation metrics, indicating successful behavior change. Points, rankings, and badges were the most effective elements. Surveys confirmed this, with 80% of participants finding it appealing and 83.7% more interested in learning new digital methods. This research

advancements aim to improve the accuracy and relevance of innovation measurement, supporting SMEs in continuous improvement and strategic decision-making.

In today's globalized world, innovation is a key enabler that leads to economic growth, competitiveness, and sustainability. The gap between an innovative and any non-innovative SME in all regards is most probably widening, as much of small and medium business in Egypt are still to step into creative methodologies that many corporations have focused on to ensure their success in the market. By identifying the obstacles that prevent SMEs from being innovative, the research will attempt to determine what barriers hinder innovation in Egypt's non-innovative SMEs. This research seeks to identify the Sub-sectoral barriers in manufacturing sector specific to non-innovative SMEs to shed the light on these obstacles. This study is guided by this open ended question: What sub-sectoral barriers are hindering the innovative capacity of non-innovative SMEs in the Egyptian manufacturing sector? This study utilizes an analysis of existing data from the Community Innovation Survey (CIS).

This study explores the drivers and challenges of digital transformation in Non-Governmental Organizations (NGOs) in Egypt, aiming to provide insights and best practices for decision-makers to enhance the success of digital initiatives in this sector. Utilizing a quantitative research methodology, the online survey gathered responses from 187 participants to collect comprehensive data. Key internal drivers identified include data-driven decision-making (24%), operational efficiency (23%), and employee empowerment (14%). A significant external driver is the competitive environment, noted by 31% of respondents. Major challenges include limited financial resources (52%), inadequate technology infrastructure (43%), and resistance to change (35%). This research proposes a framework for digital transformation in NGOs, featuring five main pillars and 12 sub-pillars. Additionally, a Digital Readiness Score (DRS) tool has been introduced to assess the readiness of NGOs for digital initiatives, offering a structured approach to navigate the transformation process effectively.

Medication therapy management (MTM) is of importance for the assistance of healthcare providers to create medication plans that fit with each patient's specific requirements. MTM includes services provided by pharmacists or other qualified healthcare professionals to improve the treatment regimen for each patient. These services can involve assessment of the patients' health status, conduction of pharmaceutical care plan, and decrease the hospital stays.Artificial intelligence (AI) allows MTM to be conducted with higher efficiency and effectiveness in the applications of clinical pharmacy. AI has been indispensable in the medical field by assisting providers in reducing expenditures, enhancing patient outcomes, and increasing the overall efficacy of patient treatment through enhancing medication safety, predicting health risks and outcomes across diverse populations, reducing time and costs of treatment, monitoring medication adherence, and scheduling follow-up healthcare services. As a result, healthcare providers are currently obligated to implement AI to assist with MTM services

Recent advancements in human stem cell-derived brain organoids have the potential to replicate critical molecular and cellular aspects of learning, memory, and cognition in vitro. The term "organoid intelligence" (OI) is proposed to encapsulate these advancements, aiming to establish a multidisciplinary field dedicated to OI. Researchers can now manufacture standardized, 3D myelinated brain organoids with heightened cell density, increased levels of glial cells, and gene expression essential for learning. Advanced 3D microelectrode arrays allow precise spatiotemporal electrophysiological signaling and recording. This lets scientists study the molecular processes that make learning and memory possible. Emerging technologies enable

Preface

In the era of rapid technological advancements, the concept of digital transformation has permeated every aspect of modern life. From education and healthcare to business and beyond, innovation driven by technology is reshaping how we learn, work, and interact with the world around us. Digital transformation is not merely about adopting new tools; it is about rethinking processes, fostering collaboration, and embracing an interdisciplinary approach to solving complex challenges.

This book, Interdisciplinary Studies on Digital Transformation and Innovation: Business, Education, and Medical Approaches, serves as a valuable resource for understanding the multifaceted impact of digital transformation across key sectors. Through a collection of diverse chapters, the book explores how emerging technologies such as artificial intelligence, gamification, virtual reality, and machine learning are driving innovation while addressing critical challenges in education, business, and healthcare.

The chapters included in this book address pressing issues such as innovation measurement for SMEs, barriers to innovation in manufacturing, and digital transformation frameworks for NGOs. Healthcare innovation also takes center stage, with chapters exploring AI-driven medication therapy management, advancements in organoid intelligence for biocomputing, and the integration of explainable machine learning in medical diagnostics. Together, these contributions highlight how interdisciplinary collaboration and innovative thinking are essential for addressing global challenges and creating a sustainable future. Each chapter offers a unique perspective, rooted in cutting-edge research and practical applications, making this book an invaluable resource for a wide audience—including researchers, policymakers, industry professionals, and educators.

ORGANIZATION OF THE BOOK

In **Chapter 1**, the integration of artificial intelligence (AI) into higher education is explored, presenting opportunities to enhance learning, teaching, and administrative processes while fostering sustainable educational practices. Despite the abundance of literature on AI in education, there remains a lack of comprehensive case studies to guide its effective implementation in higher education. This chapter delves into key areas of AI integration, including curriculum design, pedagogical strategies, learning analytics, and student support systems. By addressing these facets, the research offers actionable insights for creating sustainable policies and governance mechanisms that leverage AI for transformative educational outcomes.

In **Chapter 2**, gamification is examined as a tool to overcome challenges in digital transformations, such as employee resistance and low engagement. Through a study involving 164 journalists over five months, the chapter highlights how gamification elements like points, rankings, and badges significantly improved digital transformation metrics within a news organization. The findings demonstrate the potential of gamification to drive positive behavior changes, enhance engagement, and reduce resistance to change, offering valuable guidance for organizations navigating digital transformation.

In **Chapter 3**, the intersection of technology and sustainable tourism is discussed, focusing on balancing environmental, economic, and social aspects for long-term viability. The chapter emphasizes the role of immersive technologies, particularly virtual reality, in addressing the deterioration of tourism resources. By enabling virtual tours of vulnerable destinations, these technologies contribute to resource rejuvenation and sustainable travel, aligning with the United Nations' Sustainable Development Goals for 2030.

In **Chapter 4**, the importance of innovation measurement for the success of small and medium enterprises (SMEs) is explored. This chapter reviews various methodologies for assessing innovation, highlighting their strengths and limitations in capturing inputs, activities, outputs, and external factors. The discussion includes challenges such as valuing intangible assets and addressing global disparities, with recommendations for advancing innovation measurement frameworks to better support SMEs in strategic decision-making and continuous improvement.

In **Chapter 5**, the barriers hindering innovation among non-innovative SMEs in Egypt's manufacturing sector are investigated. Using data from the Community Innovation Survey, the research identifies sub-sectoral obstacles that impede the adoption of innovative practices. This chapter sheds light on the critical challenges facing SMEs and provides insights into fostering innovation to bridge the gap between innovative and non-innovative enterprises, enhancing their competitiveness and sustainability.

In **Chapter 6**, the drivers and challenges of digital transformation in non-governmental organizations (NGOs) in Egypt are analyzed. Based on survey responses from 187 participants, the research identifies key internal drivers like data-driven decision-making and operational efficiency, alongside challenges such as limited financial resources and resistance to change. The chapter introduces a framework and a Digital Readiness Score (DRS) tool to assess and enhance NGOs' digital transformation efforts, offering practical strategies for overcoming obstacles in this sector.

In **Chapter 7**, the integration of artificial intelligence (AI) into medication therapy management (MTM) is explored, highlighting its role in optimizing healthcare delivery. AI-driven MTM applications improve patient outcomes by enhancing medication safety, predicting health risks, and monitoring adherence. This chapter underscores the growing necessity for healthcare providers to adopt AI in MTM, demonstrating its potential to reduce costs, improve treatment efficacy, and promote better health outcomes.

In **Chapter 8**, advancements in human stem cell-derived brain organoids are discussed under the emerging concept of "organoid intelligence" (OI). The chapter explores the development of standardized, 3D myelinated brain organoids with advanced gene expression and electrophysiological capabilities. These innovations enable the study of learning and memory processes, paving the way for biocomputing models and real-world applications through organoid-computer interfaces. The research outlines a visionary approach to integrating biological intelligence with technology.

In **Chapter 9**, the impacts of hearing loss on quality of life and its implications for social and economic sustainability are explored. The chapter reviews advancements in hearing aid technologies, including smartphone-based apps and machine learning for sound enhancement, alongside policies promoting accessibility. By addressing the needs of individuals with hearing loss, these innovations contribute to inclusive communication and improved well-being, particularly in low- and middle-income countries where the burden of hearing loss is most significant.

In **Chapter 10**, explainable machine learning (XML) techniques are investigated for their potential in early diagnosis of fatty liver disease (FLD) and Hepatitis B Virus (HBV) infections. By enhancing the interpretability of complex diagnostic models, XML improves clinician trust, patient outcomes, and healthcare adoption. This chapter emphasizes the ethical and practical importance of embedding explainability in medical diagnostics, laying a foundation for future research on the generalizability and real-time application of XML in healthcare.

We extend our deepest gratitude to the contributors whose dedication and expertise have made this book possible. Their efforts exemplify the spirit of collaboration and innovation that drives progress in addressing global challenges. We also acknowledge the support of the institutions and organizations that have championed these advancements, enabling the pursuit of research that bridges the gap between technology, sustainability, and societal well-being.

This book is an invaluable resource for researchers, practitioners, policymakers, educators, and anyone interested in the transformative potential of digital innovation. By bringing together perspectives from business, education, and medicine, it underscores the interconnectedness of these domains and the need for collaborative approaches to drive meaningful change.

As editors, we are honored to present this collection of insights contributed by distinguished experts from around the world. We hope this book inspires further exploration, fosters innovative practices, and contributes to the ongoing discourse on digital transformation and innovation.

Ahmed Gomaa Radwan
Nile University, Egypt

Salwa K. Abd-El-Hafiz
Cairo University, Egypt

Islam Tharwat Abdel Halim
Nile University, Egypt

Yuhong Liu
Santa Clara University, USA

Meikang Qiu
Augusta University, USA

Chapter 1

Integrating AI Into Higher Education: A Comprehensive Exploration

Laura Nicoleta Labib
https://orcid.org/0009-0005-1061-7631
Nile University, Egypt

ElHassan Anas ElSabry
https://orcid.org/0000-0001-6135-0280
Nile University, Egypt

ABSTRACT

In the rapidly evolving landscape of technology, innovation, and sustainability, integrating Artificial Intelligence (AI) in higher education presents significant opportunities to enhance learning, teaching and assessment, streamline administrative processes, and promote sustainable educational practices. Despite a vast amount of literature on AI in education, comprehensive use cases that can inform effective implementation of AI in higher education remain scarce. This chapter aims to fill this gap by exploring in-depth the AI applications in higher education, addressing key areas of AI integration such as curriculum design and content development, pedagogical strategies and learning environments, evaluation mechanisms and feedback systems, process streamlining and efficiency enhancement, learning analytics, academic research and ideation, and student support and services. As such, this chapter contributes to the broader goal of sustainable development in the context of education thus aligning with the themes of the SMART conference. This research may inform sustainable solutions, policy, and governance mechanisms in the context of AI-driven education.

DOI: 10.4018/979-8-3373-1132-6.ch001

INTRODUCTION

The integration of Artificial Intelligence (AI) in higher education is subject of intense scrutiny and debate, and the question on everyone's mind is: will this technology improve or worsen higher education outcomes? The Artificial Intelligence (AI) technology is seen both as a solution for the world's most pressing problems (Desai, 2020) and also as a significant risk or harm, with the capacity of disrupting societal systems (Hendrycks, Mazeika, & Woodside, 2023), thus a mix of feelings exists which are spanning from enthusiasm for the new opportunities it brings to fear over losing control over it.

Theoretical physicist Stephen Hawking stated that "AI is likely to be either the best or worst thing ever to happen to humans so there's huge value in getting it right" (Griffin, 2015). There is no doubt that AI is a vital component of the university of the future, namely a Smart AI-centered University (Babu & Ontario, 2023).

A Google Scholar search using the keywords "AI" and "higher education" at the time this chapter is written brings over 6 million results. Searching for "AI in higher education" brings over 5 million results. Despite this "ocean" of material, comprehensive use cases that can inform and educate the effective implementation of AI tools in higher education are scarce. Previous guidelines, such as UNESCO's "ChatGPT and Artificial Intelligence in Higher Education: Quick Start Guide" and the "Guidance for Generative AI in Education and Research" report (UNESCO, 2023), provide broad guidelines but lack comprehensive practical applications and step-by-step execution strategies. Similarly, Klutka, Ackerly, & Magda (2020) discuss various uses of AI in higher education, from student acquisition to institutional maximization, emphasizing the need for aligning AI with institutional goals, ensuring privacy, ethics, fairness, and transparency, and fostering a culture of innovation.

AI concepts and methods offer promising opportunities for enhancing the learning experience. Fuzzy Logic, for instance, accommodates the inherent uncertainty in student assessment, allowing for more nuanced grading that acknowledges the varying degrees of understanding (Khomeiny, Kusuma, Handayani, Wibawa, & Supadmi Irianti, 2020). Decision trees can help analyze academic data of students to understand the main factors affecting their academic performance in university courses (Al-Radaideh, Al-Shawakfa, & Al-Najjar, 2006). Bayesian networks are useful in higher education for creating detailed tests that adapt to each student's knowledge, helping teachers understand exactly what each student knows and needs to learn (Culbertson, 2015). They can also make online learning systems smarter by choosing questions that are just right for each student's level. However, using Bayesian networks widely in schools requires more research and understanding of how to set them up and use them effectively (Culbertson, 2015). Markov Models prove invaluable for understanding and predicting student behaviors and learning progress by analyzing sequences of data, these models can identify patterns and trends in students performance and learning habits (Gu & Deng, 2022). While the algorithms themselves are just tools, it is the mastery of these tools that allows educators, researchers, and technologists to apply them in creative ways, collectively contributing to making higher education more adaptable, personalized, and efficient, ultimately benefiting both educators and students.

The implications for theory, practice, and policy regarding AI or generative AI programs such as GPT technology in education highlight several key areas. These include the importance of adopting it in meaningful, ethical, and sustainable ways, the need to cultivate new literacies specific to AI/GPT technologies (UNESCO, 2023). Additionally, it is necessary for educators to critically comprehend their advantages and disadvantages before implementation, and the recognition that human educators play a pivotal role in education while technology should be seen as a supportive tool rather than an exaggerated replacement (Bozkurt, et al., 2023).

APPLICATIONS OF AI IN HIGHER EDUCATION

This literature review follows a thematic approach, organizing and analyzing research around the core areas of AI integration in higher education. Instead of adhering to a chronological order, the review explores key themes identified providing a focused examination of how AI is transforming the educational landscape. While the primary focus is on current advancements from 2022 to 2024, it also acknowledges foundational contributions, such as decision trees (Al-Radaideh et al., 2006), Bayesian networks (Culbertson, 2015), and intelligent tutoring systems like AutoTutor, which have shaped current innovations. By integrating these past insights, this review offers a comprehensive exploration of AI's evolution in higher education, highlighting how contemporary developments build on earlier research and technologies. Based on this thematic structure, the chapter examines seven key areas identified of AI application: curriculum design and content development, pedagogical strategies and learning environments, evaluation mechanisms and feedback systems, process streamlining and efficiency, learning analytics, academic research and ideation, and student support services.

In the broad scope of this review, it is also important to recognize the distinction between the use of commercial AI tools and more foundational algorithms such as machine learning, natural language processing (NLP), and sentiment analysis. While commercial tools are often widely adopted for practical purposes, they can lack a shared framework for comparison. On the other hand, fundamental algorithms provide a more structured, research-based approach that underpins many of the current advancements in AI-driven education. This thematic approach will explore how both commercial tools and foundational AI concepts contribute to the evolving educational landscape.

1. Curriculum Design and Content Development (the WHAT):

Numerous research articles have explored how AI-powered tools can be beneficial and integrated in the areas of curriculum design and content development in higher education. Most commonly research touches upon three areas:

a) the utilization of AI for curriculum structuring and planning

AI tools, such as ChatGPT, are increasingly being used as co-designers in higher education curriculum development. These tools assist educators in tasks like creating rubrics, improving curriculum accessibility, and generating course materials, including outlines, presentations, coding exercises, quizzes, and scientific papers (UNESCO, 2023; Qadir, 2022). Beyond content creation, AI can optimize curriculum delivery by analyzing student performance and feedback, helping educators tailor lessons and assessments. Additionally, AI can analyze job market trends to identify the skills and knowledge students will need, informing the development of future curricula (Igbokwe, 2023).

Fazlollahi, et al. (2023) in a cohort study, reportedly the first of its kind, examines the outcomes of applying an AI-enhanced curriculum for bimanual surgical skill acquisition in postgraduate medical training and it found both positive and negative effects; on the positive side safety indicators were improved but on the negative side, efficiency indicators declined. The surgical performance indicators for those who underwent training using an AI-enhanced curriculum were evaluated against those of a control group who received no feedback, and also compared with expert standards.

Ka Yuk Chan & Tsi (2023) research investigates the AI revolution in higher education, revealing that some teachers already utilize generative AI to design more engaging courses, generate scenarios tailored to students' majors, and create multiple-choice questions for interactive quizzes such as Kahoot! Games which engage students in problem-solving related to real-world issues, making their courses more interesting and engaging. Separately, Kahoot!, a platform for interactive quizzes where students of all ages, from primary to higher education, including business professionals can answer multiple-choice questions in a fun, competitive way, has incorporated an AI question generator into its system despite not being fully leveraging AI technology overall and relying more on gamification (Kahoot!, 2023). This feature, the AI question generator, can save time by automatically creating high-quality and varied questions based on a chosen topic, adjusting difficulty based on student performance, providing instant feedback with correct answers, and scaling to accommodate any class size, thereby enhancing student engagement and learning outcomes however, a prompt on their website warns users that this program might occasionally generate inaccurate or untrue responses, thus using it with caution is advised.

b) the application of AI for customizing curriculum to individual student's learning needs

Imagine one has a recipe for a dish and want to personalize it for different people with different tastes and dietary restrictions, using AI to personalize the curriculum is like adjusting the ingredients, measurements, and cooking instructions in the recipe based on each person's preferences and needs. For example, if someone prefers a spicier flavor, you can increase the amount of spices in their version of the dish, if someone else has dietary restrictions, you can modify the recipe to exclude certain ingredients so the focus here is on customizing the recipe itself to cater to each individual's specific requirements.

Therefore, AI models can play a significant role in curriculum design and content development in a customized way for neurodiverse learners as example by offering personalized learning paths and accommodating various modes of interaction, it can support students with different ways of accessing and processing information (Cardona, Rodriguez, & Ishmael, 2023). Moreover, contextualized learning tools like iTalk2Learn and AiDA have been developed to personalize the learning experiences of students by adapting the content, leveraging AI algorithms to enhance curriculum development through the provision of customized learning paths (Chaudhry & Kazim, 2021).

According to their website, iTalk2Learn has a specific feature called Adaptive Sequencer which involves using AI to dynamically sequence and personalize the curriculum content based on each individual student's needs, progress, and learning patterns, and here there is a "Lesson Recommender" function that takes into account all historical performance across an entire student base as well as the behavioral patterns of each individual student in order to adapt more intelligently to their needs (iTalk2Learn, 2023).

Relevant to curriculum customization, especially for self-study, are Intelligent Tutoring Systems (ITS). Intelligent Tutoring Systems (ITSs) educational tools that provide customized learning by adapting to students' needs and progress (Wang, et al., 2023). They track and respond to each student's knowledge, skills, and preferences, offering personalized feedback and tasks for optimal learning and while not all are AI-based recently AI has been incorporated to enhance personalized learning (Wang, et al., 2023). Originating in 1970, these systems aim to enhance student engagement and understanding (Lin, Huang, & Lu, 2023). While Intelligent Tutoring Systems will be discussed in the upcoming categories in the broader sense of AI application in education, it's worth nothing some common Intelligent Tutoring Systems (ITS) that help in customizing the curriculum to individual student's learning needs in higher education such as the AutoTutor, Why2-Atlas, COMET, VIPER and EER-Tutor (Chassignol, Khoroshavin, Klimova, & Bilyatdinova, 2018).

For the purpose of this chapter, it was attempted to try Why2-Atlas and AutoTutor, but despite efforts, a functional Why2-Atlas platform or comprehensive instructions could not be located. However, on a positive note AutoTutor was discovered and experimented with but it felt short on expectations in terms of the desired level of adaptation and personalization one would expect from an AI tool, even talking of analyzing performance or dynamically adjusting curriculum (AutoTutor, 2023).

In addition to supporting curriculum design, AI tools like ChatGPT can contribute to the development of gender and culturally inclusive learning materials and lessons by crawling global content in multiple languages, provided that educators specifically request its assistance in doing so (Bozkurt, et al., 2023). GPT 4 infrastructure seems to be popular among AI powered educational website, for example Khan Academy launched "Khanmigo", Khan Academy's AI-powered guide meant to be a tutor for learners and assistant for teachers, which also raises donations for its ChatGPT 4 (Khan Academy, 2023). Khanmigo suggests an attempt to incorporate AI-driven features for personalized guidance however, without detailed information on its functionalities, it is unclear how Khanmigo works in terms of customizing the curriculum to individual student's learning needs and further official information or clarification is recommended to examine the extend of its value proposition.

These instances serve as compelling evidence of the profound impact AI has or could have on curriculum design and content development and how by leveraging AI, educators can create adaptive and personalized learning materials that significantly enhance the overall learning experiences, foster higher levels of student engagement, inclusivity and deepen knowledge acquisition and despite not having any clear strong working examples in this area, the possibilities are rich.

2. Pedagogical Strategies and Learning Environment (the HOW):

Pedagogical strategies and learning environments have significantly evolved in recent years, with the integration of artificial intelligence (AI) playing a crucial role in transforming traditional teaching methods. AI-powered tools also aim to develop twenty-first-century skills beyond digital literacy, fostering curiosity, initiative, creativity, collaboration, and adaptability. Through the integration of AI, smart learning environments become more individualized, promoting personalized and adaptive teaching methods that cater to the specific needs, interests, and contexts of individual students taking education yet to a further level (Peng, Ma, & Spector, 2019).

This section explores the integration of AI in pedagogical strategies and learning environments and its mainly focused on three areas:

a) Integration of AI in personalized and adaptive teaching methods

Before delving into AI's role in personalized and adaptive learning, it is important to clarify the distinctions between these two approaches.

Personalized learning refers to educational experiences that are tailored to meet the individual needs, preferences, and interests of each student. This approach often allows students to have some control over the pace and content of their learning, with educational strategies designed to meet their unique needs. It is typically driven by data collected on students' learning styles, interests, and progress, enabling a more customized learning experience (Pane, Steiner, Baird, Hamilton, & Pane, 2017).

In contrast, adaptive learning uses technology and algorithms to adjust learning content in real-time based on a student's performance. Adaptive systems continuously assess a learner's knowledge and skills, modifying the difficulty level and content delivery dynamically to ensure that each student is provided with the appropriate level of difficulty and support (Johnson, et al., 2016). The key distinction between these approaches lies in the control and method: personalized learning offers more flexibility to the student in directing their learning path, while adaptive learning is system-driven, reacting to student performance and adapting the material accordingly. Both approaches are crucial in the integration of AI in education, as they work together to create a more responsive and individualized learning environment.

Imagine one is a chef teaching a cooking class and have a group of students with varying skill levels and learning styles. Integrating AI into personalized and adaptive teaching methods is like using technology to tailor the teaching approach to each student, it's like having AI-powered assistants in the kitchen that can provide individualized guidance to each student based on their progress and needs. For instance, if a student is struggling with a particular technique, the AI assistant can provide targeted feedback and suggest alternative approaches to help them improve. The focus here is on adapting the teaching methods using AI to accommodate the unique needs of each student (Labadze, Grigolia, & Machaidze, 2023).

One of the key aspects of integrating AI in pedagogical strategies is the use of Open Learner Models (OLMs) which are widely used in AI in Education (AIEd) to help learners, teachers, and parents understand and support the learning process (Chaudhry & Kazim, 2021). Online Learning Management (OLM) systems visually represent and track learner progress, providing insights into individual learning patterns, and enabling targeted interventions. By integrating with AIEd systems, they personalize learning experiences, adapt strategies, recommend content, and give tailored feedback, fostering a learner-centered approach and enhancing understanding in AIEd contexts.

At the same time, Learning Management Systems (LMSs), such as Moodle and Blackboard, are primarily designed to manage and deliver educational content. These systems support tasks such as course organization, content delivery, assessment management, and administrative functions (Georgieva, 2023). While OLMs and LMSs have different primary functions, they are often discussed together because they can work in tandem to improve the learning experience. When combined, OLMs provide deeper insights into student progress, while LMSs manage the content and structure of the courses. This partnership is particularly useful in visualizing student progress and enabling personalized learning pathways. To understand the distinction between OLMs (Open Learner Models) and LMS (Learning Management Systems), ChatGPT4 was asked to create a table comparing OLMs and LMSs:

Table 1. Comparison between Open Learner Models (OLMs) and Learning Management Systems (LMS) Note. Adapted from "Comparison between Open Learner Models (OLMs) and Learning Management Systems (LMS)" by Assistant, 2024. Copyright 2024 by OpenAI. Used with permission.

Aspect	OLMs (Open Learner Models)	LMS (Learning Management Systems)
Purpose	Provide insights into individual learner progress and learning patterns	Manage and deliver educational courses and training programs
Functionality	Visual representation of learner progress; personalized insights; learner-centered approaches	Course organization; content delivery; assessment management; administrative features
Integration with AI	Frequently integrated with AI in Education (AIEd) systems	Can integrate with AI technologies, but extent may vary
User Focus	Primarily focuses on individual learner experience	Cater to learners, instructors, administrators, and sometimes parents
Examples	Components of ITS like Cognitive Tutor, AutoTutor	Moodle, Blackboard, Canvas, Schoology

This comparison shows that although OLMs and LMSs serve different functions, their integration offers considerable benefits. OLMs focus on providing detailed, personalized insights into individual learners' progress, while LMSs handle the broader organizational and administrative aspects of education, such as content delivery and assessments. By combining these two systems, educators can gain a more holistic view of a student's progress, utilizing OLMs to visualize learning patterns and provide adaptive feedback, while the LMS manages course structure and content delivery. This integrated approach enables better personalization and responsiveness in teaching, ultimately enhancing both the learning and teaching experience (Ippakayala & El-Ocla, 2017).

As for widely used LMS platforms like Moodle, Blackboard, Canvas, and Schoology, these systems do not explicitly advertise the integration of AI features on their websites. However, the potential for AI-driven innovations within these platforms is substantial, and future developments could see the inclusion of AI-powered capabilities to enhance the user experience. Despite the lack of explicit AI features at present, these platforms remain widely adopted in education, and their potential for AI innovation is tremendous. AutoTutor was discussed earlier in this chapter. While the original Cognitive Tutor is not relevant in the context of this chapter as it is focusing on elementary students, the idea is still sound, and it saw further development. Initially Cognitive Tutor was developed by Carnegie Learning as a secondary mathematics curriculum, emphasizing on students' thinking and learning processes in mathematics and that curriculum could be implemented through textbooks, adaptive software, or a combination of both (U.S Department of Education, 2016). As of 2023, we can see different variations of this namely a cognitive tutor that uses the Davinci-003 API to create personalized questions for students, the tutor adjusts the difficulty level based on the student's understanding and connects the topic to different areas to enhance learning and by providing customized and adaptive questions, this approach aims to improve student learning outcomes (Walton, 2023).

As for other LMS systems, such as the well-known Moodle, Blackboard, and Canvas, which are extensively used in higher education, and the lesser-known Schoology, primarily used in K-12 education (Powerschool, 2024), it is not clear whether they currently use or plan to use any AI features. There is no explicit mention of AI technology on their respective websites. However, given their widespread global usage, the potential for innovation through AI integration is significant.

The comparison of OLMs and LMSs in this paper is not just about their differences, but about how their integration can significantly improve learning outcomes by leveraging AI. While LMSs organize and deliver educational content, OLMs provide real-time insights into student progress, enabling personalized learning experiences. When these two systems work together, supported by AI technologies, they create a more adaptive and responsive learning environment. This collaboration enhances both teaching and learning, making the comparison not only informative but essential for understanding the potential of AI in education.

A very important pillar in the context of personalized and adaptive teaching is also the Personalized Adaptive Learning (PAL). PAL is an advanced teaching method that goes beyond traditional online and classroom learning, leveraging smart technologies and it consists of two main parts namely personalized learning which is a learner-focused approach tailored to each student's needs to boost personal growth and adaptive learning which involves the use of technology to track students' progress and dynamically adjust teaching methods, enhancing the educational experience (Apoki, Hussein, Al-Chalabi, Badica, & Mocanu, 2022). PAL aims to cater to diverse learners in a digital environment, often integrating AI assuming the role of pedagogical agents for improved learning experiences.

One of the most prevalent applications of Personalized Adaptive Learning (PAL) in education is the use of Intelligent Pedagogical Agents (IPAs). Although not all pedagogical agents use AI, IPAs stand out for their ability to provide a more personalized, interactive, and effective learning experience. These AI-driven characters enhance engagement and learning outcomes by emulating, modeling, and augmenting human behavior. IPAs can assume roles such as conversational partners, empathetic supporters, or peer companions, thus improving distance education. By diminishing the sense of distance and boosting social presence, they create a more immersive learning process (Hammershaimb, 2020).

In addition to addressing the WHAT of educational learning, tools like iTalk2Learn and AiDA also effectively tackle the HOW. These tools exemplify this in theory by aiming to personalize the educational journey and offering feedback, which leads to more effective and engaging learning outcomes. Not only do they aim to work on the content of learning, but they also aim to focus on the manner of delivery by adjusting the learning process to accommodate individual students' specific needs, interests, and contexts and through providing feedback, they cultivate a more personalized, effective, and engaging learning environment (Chaudhry & Kazim, 2021).

In higher education, there is a growing number of AI-powered tools aimed at enhancing the learning experience and obtain better educational outcomes. These tools leverage artificial intelligence algorithms to analyze student data, provide personalized recommendations, and offer adaptive feedback. A comprehensive scan of available higher education platforms identified several key and popular examples such as Coursera, Top Hat, Gradescope, Echo360, and Smart Sparrow. These platforms leverage AI algorithms to analyze student data, provide personalized recommendations, and offer adaptive feedback, ultimately enabling a more individualized and effective learning experience in higher education and some will be discussed in other categories through the paper as they have various aims and applications in the context of higher education: Coursera, Top Hat, and Smart Sparrow contribute also to curriculum design, content development, pedagogical strategies, and learning environments (Coursera, 2023) (Top Hat, 2023) (Smart Sparrow, 2023), Gradescope aids in evaluation mechanisms and feedback systems (Gradescope, 2023), while Echo360 offers learning analytics and streamlining processes (Echo360, 2023). These tools also provide additional functionalities beyond their primary categories, and their applicability varies depending on specific institutional contexts.

These AI-driven platforms assist in tailoring the curriculum to individual students' needs, improving engagement, and optimizing learning outcomes and while this list is not exhaustive, it illustrates the increasing availability of AI-powered tools designed to personalize and enhance higher education instruction.

b) Incorporation of AI in interactive learning experiences

AI technologies enable the creation of interactive platforms that foster active participation, and immersive learning (Davé, 2023). These platforms could benefit from AI which could enhance traditional e-learning systems by leveraging intelligent virtual robots in virtual classrooms and smart campuses (Alam, 2022), simulated environments to provide students with opportunities to engage in problem-solving activities (Treceño-Fernández, et al., 2020), and adaptive learning systems, that allow students to receive personalized guidance thus enhancing problem-solving skills and understanding of complex concepts (Phobun & Vicheanpanya, 2010).

Intelligent Pedagogical Agents, Intelligent Tutoring Systems (ITS) and Open Learner Models (OLMs) discussed earlier are all very relevant in the context of an interactive learning experience. IPAs enhance the educational experience by adding a social and interactive dimension (Hammershaimb, 2020). ITS can adapt to individual learner needs and provide interactive and dynamic feedback, analyze learner responses, track progress, and provide targeted interventions or suggestions to enhance learning outcomes, thereby personalizing instruction to create an interactive learning environment that aligns with the individual needs and pace of each learner, making the learning experience more engaging and interactive (Wang, et al., 2023). A good example of AI powered ITS is DeepTutor, an advanced intelligent tutoring system in which students engage in quality interaction and instruction, fostering deep understanding of complex science topics by aligning with Learning Progressions (LPs) "to facilitate natural paths to mastery" (Rus, 2023). As early as 2015, in their paper "DeepTutor: An Effective, Online Intelligent Tutoring System That Promotes Deep Learning", the authors introduce DeepTutor, an innovative conversational intelligent tutoring system developed as a web application, which has shown promising results in large-scale experiments with high-school students, providing evidence of the feasibility of web-based conversational intelligent tutoring at scale (Rus, Niraula, & Banjade, 2015).

AI-powered OLMs provide an interactive interface that offers personalized insights into learners' strengths, weaknesses, and areas for improvement. Learners can engage with these models, using the AI-generated feedback to make informed decisions about their learning strategies. This interactivity, powered by AI, enhances the learning experience, making it more adaptive and personally tailored to the learner's progress and performance.

Firat (2023) explores the potential of AI, specifically ChatGPT, in revolutionizing higher education through his study which underlines how AI can create personalized and interactive learning experiences, transforming traditional learning systems and reshaping the roles of educators. It highlights that AI, through tools like ChatGPT, can foster an engaging, adaptive learning environment, enabling learners to interact with the content and receive instant, personalized feedback. However, it also addresses challenges, such as potential misinformation, data bias, and privacy issues, that need to be managed to fully leverage AI's potential in education.

ChatGPT can serve as collaboration coaches and study buddies which are forms of interactive learning powered by AI (UNESCO, 2023). Roles such as study buddies are meant to assist and support learners in their educational journey, provide personalized guidance, feedback, and resources to individual learners based on their specific needs and learning goals while collaboration coaches focus on facilitating collaboration among learners, encouraging teamwork, and fostering communication and problem-solving skills. Despite these roles do not explicitly exist in the program but one could assume it such roles and more and listed in the table below, a report from UNESCO released earlier in 2023 which aims to tackle ChatGPT and AI in higher education along with their various applications:

Table 2. AI Applications in Education, adapted from UNESCO's Quick Start Guide ChatGPT and AI in Higher Education 2023

Role	Description	Example of Implementation
Possibility engine	AI generates alternative ways of expressing an idea	Students write queries in ChatGPT and use the Regenerate response function to examine alternative responses.
Socratic opponent	AI acts as an opponent to develop an argument	Students enter prompts into ChatGPT following the structure of a conversation or debate. Teachers ask students to use ChatGPT to prepare for discussions.
Collaboration coach	AI helps groups to research and solve problems together	Working in groups, students use ChatGPT to find information to complete tasks and assignments.
Guide on the side	AI acts as a guide to navigate learning material and conceptual spaces	Teachers use ChatGPT to generate content for classes/courses (e.g., discussion topics) and advise on how to support students in learning specific concepts.
Personal tutor	AI tutors each student and gives immediate feedback on progress	ChatGPT provides personalized feedback to students based on information provided by students or teachers (e.g., tests).
Co-designer	AI assists throughout the design process	Teachers ask ChatGPT for ideas about designing or updating a curriculum (e.g., rubrics for assessment) and/or for topics to include in a syllabus, how to make the curriculum more accessible.
Exploratorium	AI provides tools to play with, explore and interpret data	Teachers provide problem formulation to students who write different queries in ChatGPT to find out more. ChatGPT can be used to support hypothesis learning.
Study buddy	AI helps the student reflect on learning material	Students explain their current level of understanding to ChatGPT and ChatGPT provides feedback. Teachers can use ChatGPT to help students reflect on and improve their learning.
Motivator	AI offers games and challenges to extend learning	Teachers or students ask ChatGPT for ideas about how to extend students' learning based on understanding of their current level of knowledge (e.g., quizzes).
Dynamic assessor	AI provides educators with a current picture of each student's current knowledge	Students interact with ChatGPT in a tutorial-type dialogue and then ask ChatGPT to produce a summary of their current state of knowledge to share with their teacher for assessment.

The existing literature is still limited, and further research on the perspectives of scholars and students is needed to contribute to the understanding of using ChatGPT in higher education.

c) Deployment of AI in facilitating and enhancing remote learning experiences

AI plays a crucial role in facilitating and enhancing remote learning experiences, particularly in the context of distance or online education as it helps overcome geographical barriers and enables students to access quality education from anywhere. Educational institutions must embrace this change, while educators need to adopt new roles to facilitate a successful transition. Generative AI has the potential to transform distance learning by reimagining the roles of educators and universities and as such both educators and universities must adapt, focusing on guiding and supporting students, leveraging learning analytics and educational technology, and integrating digital transformation strategies (Bozkurt & Sharma, 2023).

The benefits of generative AI in distance learning are significant, and with that new skills and literacies related to generative AI should be developed but overall, careful consideration and proactive measures are required to navigate the associated challenges and ensure equitable, transparent, and ethically sound implementation (UNESCO, 2023).

In a recent study, Qadir (2022) explores the role of generative AI such as ChatGPT in Engineering education employing quite a novel practice of asking ChatGPT itself about the future of education, particularly how universities should adapt as a result of these and to quote part of the response: " AI tools may enable universities to offer more online and distance learning options, making education more accessible to students who are unable to physically attend class". Seo, Tang, Roll, Fels, & Yoon (2021) discuss the AI potential to positively impact online learning, but there are important considerations and challenges that need to be addressed such as privacy, reliability, explainability, over-standardization and surveillance issues.

Modern ITSs offer dynamic and interactive learning experiences, including real-time feedback, adaptive learning paths, and customization for various learning styles and needs (Apoki, Hussein, Al-Chalabi, Badica, & Mocanu, 2022). This integration promotes an individualized and highly effective remote learning environment. However, concerns include the reliability and explainability of AI systems, the potential over-standardization of the learning process, and privacy issues related to the measurement of unconscious behavior (Bozkurt & Sharma, 2023; Cardona, Rodriguez, & Ishmael, 2023). Authors suggest solutions to these concerns, such as making AI systems explainable, involving human instructors in the learning process, and establishing clear data norms and agreements (Apoki, Hussein, Al-Chalabi, Badica, & Mocanu, 2022; Bozkurt & Sharma, 2023).

By embracing AI in pedagogical strategies and learning environments, education becomes more learner-centered, context-aware, and conducive to holistic skill development. The integration of AI in personalized teaching methods, interactive learning experiences, and remote learning facilitates a transformative educational landscape that empowers students and enhances their learning journey.

3. Evaluating Mechanisms:

a) Application of AI in grading and comprehensive assessment

AI has the potential to streamline the grading and assessment process in online learning by quickly analyzing student responses and providing consistent evaluations, saving time and reducing bias.

Chaudhry & Kazim (2021) discuss the potential of AI in revolutionizing the assessment system highlighting AIAssess, an intelligent assessment tool developed by researchers at UCL Knowledge Lab. AIAssess incorporates three models: a knowledge model, an analytics model, and a student model. The knowledge model stores information about each topic, the analytics model analyzes students' interactions, and the student model tracks their progress on specific topics. Chaudhry & Kazim also mentions the work of Samarakou et al., who developed an AI assessment tool aimed at reducing the workload of instructors by providing qualitative evaluations of student exercises. These assessment tools can be further enhanced through the application of machine learning techniques, such as semantic analysis, voice recognition, natural language processing, and reinforcement learning. These techniques contribute to improving the quality and effectiveness of assessments, as highlighted by Chaudhry & Kazim's paper.

Furthermore, Gradescope (2023) has revolutioned this aspect with having gradded over 700 milion questions from 2600 universities for 3.2 milion students. Gradescope is designed to support a wide range of subjects and disciplines, including mathematics, computer science, engineering, sciences, humanities, and social sciences. However, its effectiveness may vary depending on the nature of the subject and the specific requirements of the course, so instructors should assess whether it meets their needs. Many papers cover the aspect of AI's application in terms of evaluation and feedback mechanisms however, it is important to address concerns about AI grading, such as its potential for errors (Hsu, et al., 2021) by designing robust systems that are regularly checked by humans, ensuring transparency to students, and prioritizing accuracy, fairness, and clarity in the grading process and this has been coined into the term "AI explainability", such definition was obtained from

Veerappa & Rinzivillo (2023) and underscores the importance of transparency and understandability in the ethical use of AI.

b) Utilization of AI for predictive analysis in student performance

Mustapha (2023) in the article entitled "Predictive Analysis of Students' Learning Performance Using Data Mining Techniques: A Comparative Study of Feature Selection Methods" employed advanced AI technologies to predict and analyze students' academic performance by leveraging data mining techniques and machine learning algorithms he was able to accurately forecast how well students would perform in their studies. Through the analysis of large datasets encompassing student demographics, academic records, and learning behaviors, the research uncovered valuable patterns and indicators that contribute to academic success. This information enables educators to personalize interventions, tailor teaching strategies, and allocate resources effectively, ultimately promoting positive learning outcomes. The study's findings align with a broader landscape of research in the field of education, where data mining and AI techniques are being explored extensively.

Similarly, Igbokwe (2023) discusses how AI can analyze student data in order to predict their performance, identify at-risk students, and recommend interventions. Researchers across the globe are investigating the potential of AI-driven models to predict student performance and provide insights that can inform decision-making in education.

4. Streamlining Processes, and Enhancing Efficiency:

a) Automation of administrative tasks and routine academic tasks through AI

According to Siminto, Akib, Hasmirati, & Widianto (2023) AI significantly augments several aspects of educational management, and on the administrative side, AI optimizes resource management, improves the efficiency of operational processes such as scheduling and staff evaluations, and reduces the administrative burden, thereby allowing institutions to focus more on educational quality rather than procedural tasks.

Also, Igbokwe (2023) discusses how AI can significantly boost the efficiency of administrative tasks in educational settings by automating routine tasks like scheduling, record-keeping recruitment and admission support such as analyzing applicant data and identifying candidates more likely to succeed, automizing financial aid processes and thus reducing administrative burdens. Firat (2023) also discusses how AI chatbots could reduce university staff workload by answering frequently asked questions.

b) Application of AI in improving collaboration and communication efficiency between students, teachers, and administrators

Bozkurt (2023) presents the idea that generative AI has the potential of enhancing collaboration and cooperation during educational processes as generative AI-powered conversational educational agents represent an inevitable paradigm shift in the educational landscape by enabling seamless communication and collaboration among students and educators. However, Bozkurt's discussion remains at a high level without detailing the actionable steps on how this transformation might occur. Yet one of the most valuable insights Bozkurt offers is that prompt engineering emerges as the new digital literacy. This concept underscores the importance of developing skills to effectively interact with AI systems, as the ability to formulate precise and effective prompts becomes crucial in leveraging AI technologies for educational purposes.

Valenzuela-Fernández, Garro-Aburto, & Ocaña-Fernández (2019) highlight that AI-based educational formats promise substantial improvements by providing accurate personalization of learning according to individual student requirements. They argue that this personalization provided it integrates various forms of human interaction with information and communication technologies, could make communication within the educational environment more effective and coherent. This tailored approach addresses diverse student needs, facilitating better interaction between students, educators, and administrative staff, ultimately leading to a more efficient and responsive educational environment which implies overall improved communication and collaboration.

5. Learning Analytics:

The Society for Learning Analytics Research (2023) defines learning analytics as the process of collecting, analyzing, and interpreting data generated during the learning process to gain insights and improve educational outcomes. Igbokwe (2023) discusses how learning analytics based on AI can help educators analyze student data to identify trends and patterns, evaluate the effectiveness of teaching methods, and make data-driven decisions; AI algorithms can analyze data on student performance, attendance, and other factors to predict which students may be at risk of falling behind and intervene early. Collins (2023) underscores the value of learning analytics in education, data from learning management systems (LMS) can benefit students, instructors, administrators, researchers, and system designers. Students can track their progress, instructors can assess resource effectiveness, administrators can evaluate program performance, researchers can study learning outcomes, and system designers can enhance user experiences.

Also, a popular learning analytics platform is Echo360, used in educational settings to capture and analyze data related to student engagement and performance, it provides instructors with tools to record and upload lectures, track student attendance and participation, generate analytics reports, and facilitate interactive learning experiences (Echo360, 2023). By collecting and analyzing data on student behavior and performance, Echo360 helps educators gain insights into students' learning patterns and make data-informed decisions to improve teaching and learning outcomes. These innovative systems have the potential to drive analytic efforts forward significantly, ultimately, their aim being to create more engaging and data-driven learning experiences for both students and faculty.

Learning analytics promises to make education more data-driven and effective for all stakeholders. Sentiment analysis is an important component of learning analytics (Grimalt-Álvaro & Usart, 2023). Grimalt-Álvaro & Usart (2023) explore sentiment analysis in higher education assessment focusing mainly on the emotional climate and student performance stipulating that it can help teachers improve students' performance, predict grades, and identify at-risk students utilizing two major approaches namely NLP (Natural Language Processing) and ML (Machine Learning). Over the past decade, researchers have become increasingly interested in using sentiment analysis, which involves natural language processing (NLP), machine learning, and deep learning techniques, to understand how students feel, what they think, and how they behave in various educational settings.

Kastrati, Dalipi, Imran, Nuci, & Wani (2021) aimed to examine the existing literature by conducting a systematic mapping study, which involved selecting and analyzing 92 relevant research papers. The analysis considered different aspects, including the topics investigated in education, commonly used sources, research trends, tools, and techniques, as well as how data was represented in sentiment analysis. The study revealed that there has been a growing number of publications in recent years, especially in 2020, where deep learning (DL) techniques were prevalent. Researchers showed varied interests in topics such as the methods used for developing sentiment analysis systems, the metrics employed to evaluate these systems, and the types and sizes of datasets used. Despite the growth in this field, the study identified several challenges in applying sentiment analysis to understand students' feedback, including problems with fine-grained analysis, figurative language, generalization, complex language, and a lack of benchmark datasets, and recommended addressing them with structured datasets, more emotion detection, standardized metrics, and better contextualization or in simple words using structured data (data that is clear and organized), paying more attention to emotions, using standard ways to measure how well the systems work, and improving how the systems understand the meaning of words in context.

6. Academic Research and Ideation:

Academic research and ideation have experienced a significant transformation with the adoption of artificial intelligence (AI). The Ohio State University made a significant advancement in using AI for academic research in energy, where scientists have developed a deep learning model to identify hidden clean energy sources underground, focusing on hydrogen deposits and this model, combined with global satellite imagery, can detect surface signs of subsurface hydrogen, notably semicircular depressions associated with these deposits (Woodall, 2023). This innovative approach demonstrates AI's capability in enhancing clean energy exploration, showcasing its interdisciplinary application in technology and geoscience.

Another prominent application is the use of AI in conducting comprehensive literature reviews as they can efficiently sift through vast amounts of academic literature, making the process faster and more thorough and this not only saves researchers time but also ensures that they don't miss critical studies. One such example is SciSpace Literature Review, an "interactive literature review workstation" with a vast collection of over 270 million papers, specifically designed to provide researchers with a consolidated space where they can compare and contrast different papers, identify existing gaps in the literature, and establish a strong evidence foundation for their own studies (SciSpace, 2023). Also, research assistance comes from generative AI such as ChatGPT which is a valuable resource for students, helping them find relevant sources, generate research topics, and summarize articles (Qadir, 2022).

Additionally, AI has found its way into hypothesis testing and data analysis. Machine learning algorithms can assist in data interpretation, identify patterns, and even suggest new research hypotheses based on existing datasets, contributing to more data-driven and insightful research (Messner, 2023).

Furthermore, AI has become an integral part of idea generation and creative brainstorming in the academic realm (Stokel-Walker & Van Noorden, 2023). Collaborative AI platforms help researchers brainstorm new concepts, providing fresh perspectives and innovative ideas, thereby fostering a culture of creativity and exploration in academia. As AI technologies continue to advance, their role in academia is expected to expand, further enhancing research and ideation processes.

7. Student Support and Services:

Within the domain of student support and services, artificial intelligence (AI) has become a valuable asset. As example Generative AI such as ChatGPT is improving efficiency of processes (UNESCO, 2023). ChatGPT can improve efficiency by reducing the time spent on tasks such as responding to queries from applicants, helping students to sign up for courses, complete course requirements, check administrative information, finding news, resources, and other information, sending reminders or notifications, translation of information for international students/staff. In addition, administrative services can be available 24/7, supported across different platforms or integrated into websites, used with social media, messaging services and learning management systems/virtual campuses (UNESCO, 2023).

According to a recent Deloitte report, integrating AI into core operations such as teaching, learning, and community service enables faculty and staff to enhance human interventions by providing highly personalized and accessible support at scale, while also delivering lessons and materials tailored to individual learning styles and preferences (Clark, Cluver, Fishman, & Kunkel, 2024).

California State University has its own AI chatbot called CSUNny (pronounced sea-sunny) as part of its Admissions & Records division, which is available to answer questions 24/7, and it's knowledge base has thousands of answers, gradually students can chat via text message but also it has a web-mode (California State University, 2024). Georgia State University has an AI-enhanced chatbot called Pounce to help new students get started, as example with finances and registration, which also sends texts related to assignments, academic support and course content (O'Brien, 2023). Furthermore, research done by the same university shows that students who used Pounce were more likely to obtain a grade B or higher.

At the University of Murcia in Spain, an AI chatbot named Lola was introduced to help students with all sorts of campus and academic questions. Lola did a great job, chatting with 4,609 students and answering over 38,708 questions correctly more than 91% of the time. This was all before AI tools like ChatGPT became widely used, so things are only going to get better from here. Lola not only provided students with help outside of regular office hours but also gave the university staff more time to focus on other tasks instead of constantly answering queries (O'Brien, 2023). Nova Southeastern University in Florida had valuable data but didn't know how to use it to reduce student dropouts. The university collaborated with Aible, an AI technology, to identify students at risk of leaving and by focusing on these students, they reduced dropouts by 17% in just 15 days. The collaboration helped them turn raw data into useful predictive models, making their efforts more effective (O'Brien, 2023).

Most impressively, Georgia Tech has developed Jill Watson, an AI teaching assistant, back in 2016 initially to act as a 24/7 support for Computer Science majors which later in 2019 was deployed in four courses including a non-computer science one namely Biology and by 2020 it became available as a teaching assistant for any course. It is also customized to train users on the VERA AI, an ecology modeling system (Georgia Institute of Technology, 2020). The Virtual Ecological Research Assistant (VERA) at Georgia Tech, developed with the Smithsonian's Encyclopedia of Life, uses AI and Natural Language Processing to facilitate inquiry-based and model-based learning, enabling users to create and simulate ecological models, explore ecological questions, and predict environmental changes (Georgia Institute of Technology, 2024).

Throughout this section, it was discussed how AI significantly enhances interactive learning experiences for students by providing personalized educational content that caters to individual learning needs. For teachers, AI offers valuable insights that enable the fine-tuning of teaching strategies, thus fostering more effective and adaptive pedagogical approaches. Additionally, AI aids administrative staff by automating routine tasks, thereby allowing them to focus on strategic activities that directly benefit students and the institution. This comprehensive integration of AI not only streamlines operations but also fosters a collaborative and efficient educational ecosystem. In essence, AI serves as a catalyst for innovation, enhancing the roles of all educational stakeholders—students, teachers, and administrative staff—while promoting a more dynamic and responsive educational environment. By leveraging AI, institutions can achieve greater personalization, efficiency, and overall improvement in the quality of education, thus preparing for the future of learning.

CONCLUSION AND WAY FORWARD

In conclusion, this chapter provided an exploratory analysis of AI integration in higher education, aiming to inspire further research and collaboration while fostering sustainable educational practices. The integration of AI has the potential to revolutionize various facets of teaching, learning, and administration. AI has demonstrated so far significant benefits such as enhanced adaptability, personalized learning, improved accessibility and administrative efficiency. By leveraging AI, educators can create dynamic and inclusive learning environments that cater to diverse student needs. Also, AI is a strategic investment that can drive innovation, improve educational outcomes, and promote sustainability within educational institutions. In contrast, there are many challenges such as explainability, reliability, data privacy, and the potential for misinformation which must be carefully addressed.

As institutions navigate the evolving landscape of AI, it is crucial to design and implement strategies to maximize its full potential while aligning with higher education goals and institutional goals. While also addressing ethical concerns, ensuring AI's positive contribution to the evolution of higher education, acting from a knowledge-based approach not fear based.

Nobel laureate Marie Curie said famously "Nothing in life is to be feared, it is only to be understood. Now is the time to understand more so we can fear less." The successful implementation of AI requires a thorough understanding of AI technologies and their applications. Similar to how proper training is essential before driving a car to avoid accidents, all higher education stakeholders especially leaders, must be well-versed in AI to leverage its full potential safely and effectively. This understanding should inform comprehensive guidelines, ethical considerations, and the direction of ongoing research. Thus, understanding the complexities and potential of Artificial Intelligence in higher education specifically is important for the future of higher education in the context of this rapid technological advancement era we are living in because AI is not just a fade, it is here to stay.

In a nutshell, we could say that AI literacy is the next big thing of this decade, becoming the new alphabetic literacy. This was also a common theme among some of the various sections of this chapter . Thus, to facilitate this, the following AI literacy model is proposed for higher education stakeholders which categorizes AI literacy into three levels: Foundation, Application, and Innovation. Each level specifies the competencies required and identifies the target audience to ensure comprehensive AI education and implementation across the institution:

Table 3. Proposed AI Literacy Integration Model for Higher Education

Level	Competencies	Target audience
AI Foundation Level	• Basic understanding of AI concepts and terminology • Awareness of common AI applications in daily life • Understanding of ethical considerations and data privacy	• All students • All faculty • All administrative staff
AI Application Level	• Practical knowledge of AI tools and techniques • Ability to use AI applications in educational settings • Understanding of AI ethics in practice	• Faculty staff • Administrators
AI Innovation Level	• Deep understanding of AI algorithms (e.g. fuzzy logic, Bayesian networks, decision trees) • Leadership in AI ethics and governance	• Institutional leaders • AI specialists • Researchers

This framework insures that all stakeholders, from students to institutional leaders, are equipped with the necessary knowledge and skills to leverage AI effectively and ethically. By fostering a deep understanding of AI, institutions can move from a place of fear to one of empowerment and innovation, securing a sustainable and inclusive future for higher education in the new step of the digital age: the AI age.

REFERENCES

Al-Radaideh, Q. A., Al-Shawakfa, E., & Al-Najjar, M. I. (2006). Mining Student Data Using Decision Trees. Retrieved from https://www.researchgate.net/publication/242076740_Mining_Student_Data_Using_Decision_Trees

Alam, A. (2022). Intelligent Tutoring Robots. *International Conference on Advanced Computing and Intelligent Technologies (ICACIT) 2022* (pp. 400-401). Springer. DOI: 10.1007/978-981-19-2980-9_32

Apoki, U. C., Hussein, A. A., Al-Chalabi, H. M., Badica, C., & Mocanu, M. L. (2022). The Role of Pedagogical Agents in Personalised Adaptive Learning: A Review. *E-learning Personalization Systems and Sustainable Education.* DOI: 10.3390/su14116442

AutoTutor. (2023). https://adulted.autotutor.org/course/view.php?id=2&section=1. Retrieved from https://csal.autotutor.org

Babu, G., & Ontario, W. (2023). *Managing the Strategic Transformation of Higher Education through Artificial Intelligence.* Change Management and Innovation Strategies for Digital–Quantum Business Transformation., DOI: 10.3390/admsci13090196

Bozkurt, A. (2023, March 21). Generative artificial intelligence (AI) powered conversational educational agents: The inevitable paradigm shift. *Asian Journal of Distance Education*, 18(1), 4. Retrieved September 9, 2023, from http://www.asianjde.com/ojs/index.php/AsianJDE/article/view/718

Bozkurt, A., & Sharma, R. C. (2023). Challenging the Status Quo and Exploring the New Boundaries in the Age of Algorithms: Reimagining the Role of Generative AI in Distance Education and Online Learning. *Asian Journal of Distance Education*, 18(1). http://asianjde.com/ojs/index.php/AsianJDE/article/view/714/397

Bozkurt, A., Xiao, J., Lambert, S., Pazurek, A., Crompton, H., Koseoglu, S., . . . Honeychurch, S. (2023). Speculative Futures on ChatGPT and Generative Artificial Intelligence (AI): A Collective Reflection from the Educational Landscape. *ODU Digital Commons.* Retrieved from https://digitalcommons.odu.edu/cgi/viewcontent.cgi?article=1212&context=teachinglearning_fac_pubs

California State University. (2024). *Admissions and Records csunny Chat Support.* Retrieved from Division of Academic Affairs: https://www.csun.edu/admissions-records/csunny

Cardona, M. A., Rodriguez, R. J., & Ishmael, K. (2023). Artificial Intelligence and the Future of Teaching and Learning. Office of Educational Technology, U.S Department of Education. Retrieved from file:///C:/Users/ldospinescu/Desktop/ai-report.pdf

Chassignol, M., Khoroshavin, A., Klimova, A., & Bilyatdinova, A. (2018). Artificial Intelligence trends in education: a narrative overview. *Procedia Computer Science*. Retrieved from https://www.sciencedirect.com/science/article/pii/S1877050918315382

Chaudhry, M. A., & Kazim, E. (2021). Artificial Intelligence in Education (AIEd): A high-level academic and industry note 2021. *AI and Ethics*. Advance online publication. DOI: 10.1007/s43681-021-00074-z PMID: 34790953

Clark, C., Cluver, M., Fishman, T., & Kunkel, D. (2024, April 19). *Government and Public Services*. Retrieved from Latest Trends in Higher Education: https://www2.deloitte.com/us/en/insights/industry/public-sector/latest-trends-in-higher-education.html

Collins, B. (2023, October 10). Harnessing the Potential of Learning Analytics Across the University. *Wiley University Services*. Retrieved from https://universityservices.wiley.com/potential-for-higher-education-learning-analytics/

Coursera. (2023). Retrieved from In the context of higher education, the mentioned tools have diverse applications across several categories. Coursera, Top Hat, and Smart Sparrow contribute to curriculum design, content development, pedagogical strategies, and learning environments. Grad

Culbertson, M. J. (2015, June 19). Bayesian Networks in Educational Assessment. DOI: 10.1177/0146621615590401

Davé, P. (2023, November 28). *Emerging Technologies*. Retrieved from Columbia University: https://etc.cuit.columbia.edu/news/revolutionary-intersection-ai-and-immersive-learning-higher-education

Desai, A. N. (2020, June 3). Artificial Intelligence: Promise, Pitfalls, and Perspective. [JAMA]. *Journal of the American Medical Association*, 323(24), 2448. Advance online publication. DOI: 10.1001/jama.2020.8737 PMID: 32492093

Echo360. (2023). Retrieved from https://echo360.com/

Fazlollahi, A. M., Yilmaz, R., Winkler-Schwartz, A., Mirchi, N., Ledwos, N., Bakhaidar, M., . . . Del Maestro, R. F. (2023, September 19). AI in Surgical Curriculum Design and Unintended Outcomes for Technical Competencies in Simulation Training. *JAMA Network*. Retrieved September 26, 2023, from https://jamanetwork.com/journals/jamanetworkopen/article-abstract/2809667

Firat, M. (2023). What ChatGPT means for universities: Perceptions of scholars and students. *Journal of Applied Learning*, 6(1). https://journals.sfu.ca/jalt/index.php/jalt/index

Georgia Institute of Technology. (2020, January 24). *Research.* Retrieved from Georgia Tech Research: https://research.gatech.edu/jill-watson-ai-pioneer-education-turns-4

Georgia Institute of Technology. (2024). *GVU Center.* Retrieved from Georgia Tech: https://gvu.gatech.edu/research/projects/virtual-ecological-research-assistant-vera

Georgieva, M. (2023, June 2023). Retrieved from https://www.scalahosting.com/blog/moodle-vs-blackboard-head-to-head-comparison/

Gradescope. (2023). Retrieved from https://www.gradescope.com/

Griffin, A. (2015, October 8). *Independent UK.* Retrieved from https://www.independent.co.uk/tech/stephen-hawking-artificial-intelligence-

Grimalt-Álvaro, C., & Usart, M. (2023, April 18). Sentiment analysis for formative assessment in higher education: a systematic literature review. *Journal of Computing in Higher Education.* Retrieved from https://link.springer.com/article/10.1007/s12528-023-09370-5

Gu, H., & Deng, X. (2022, May 23). Markov Model-Based Curriculum Association Classification Model and Student Achievement Prediction. DOI: 10.1155/2022/3047208

Hammershaimb, L. (2020, March 10). Retrieved from https://www.aace.org/review/artificial-intelligence-and-animated-pedagogical-agents/

Hat, T. (2023). Retrieved from https://tophat.com/

Hendrycks, D., Mazeika, M., & Woodside, T. (2023). An Overview of Catastrophic AI Risks. *Cornell University.* doi:/arXiv.2306.12001DOI: 10.48550

Hsu, S., Wenting Li, T., Zhang, Z., Fowler, M., Zilles, C., & Karahalios, K. (2021). Attitudes Surrounding an Imperfect AI Autograder. DOI: 10.1145/3411764.3445424

Igbokwe, C. I. (2023, March 24). Application of Artificial Intelligence (AI) in Educational Management. [IJSRP]. *International Journal of Scientific and Research Publications*, 13(3). Advance online publication. DOI: 10.29322/IJSRP.13.03.2023.p13536

Ippakayala, V. K., & El-Ocla, H. (2017). OLMS: Online learning management system for e-learning. *World Journal on Educational Technology*. iTalk2Learn. (2023). Retrieved from https://www.italk2learn.com/

Johnson, L., Becker, S. A., Cummins, M., Estrada, V., Freeman, A., & Hall, C. (2016). *NMC Horizon Report: 2016 Higher Education Edition.* The New Meda Consortium. Retrieved from https://www.learntechlib.org/p/171478/

Ka Yuk Chan, C., & Tsi, L. H. (2023). The AI Revolution in Education: Will AI Replace or Assist Teachers in. Retrieved October 1, 2023, from https://arxiv.org/ftp/arxiv/papers/2305/2305.01185.pdf

Kahoot! (2023, October 31). Retrieved from The Kahoot! AI question generator can save you time by automatically creating high-quality, varied questions based on your chosen topic, adjusting difficulty based on student performance, providing instant feedback with correct answers, and scaling to acc

Kahoot! (2023). Retrieved from https://kahoot.com/

Kastrati, Z., Dalipi, F., Imran, A. S., Nuci, K. P., & Wani, M. A. (2021, April 28). Sentiment Analysis of Students' Feedback with NLP and Deep Learning: A Systematic Mapping Study. *MDPI*. DOI: 10.3390/app11093986

Khan Academy. (2023). Retrieved from https://blog.khanacademy.org/teacher-khanmigo/

Khomeiny, A. T., Kusuma, T. R., Handayani, A. N., Wibawa, A. P., & Supadmi Irianti, A. H. (2020). Grading System Recommendations for Students using Fuzzy Mamdani Logic. DOI: 10.1109/ICOVET50258.2020.9230299

Klutka, J., Ackerly, N., & Magda, A. J. (2020). *AI in Higher Education.* Learning House: Wiley. Retrieved from https://universityservices.wiley.com/wp-content/uploads/2020/12/201811-AI-in-Higher-Education-TLH-with-new-bage.pdf

Labadze, L., Grigolia, M., & Machaidze, L. (2023, October 31). Role of AI chatbots in education: Systematic literature review. *International Journal of Educational Technology in Higher Education*, 20(1), 56. Advance online publication. DOI: 10.1186/s41239-023-00426-1

Lin, C.-C., Huang, A. Y., & Lu, O. H. (2023, August 28). Artificial intelligence in intelligent tutoring systems toward sustainable education: a systematic review. *Springer Open*. DOI: 10.1186/s40561-023-00260-y

Lin, Y., & Yu, Z. (2023). Extending Technology Acceptance Model to higher-education students' use of digital academic reading tools on computers. *International Journal of Educational Technology in Higher Education*, 20(1), 34. Advance online publication. DOI: 10.1186/s41239-023-00403-8

McKenzie, L. (2019, November 17). *Digital Learning*. Retrieved from Insidehighered.com: https://www.insidehighered.com/digital-learning/article/2019/11/18/pearson-introduces-aida-ai-powered-calculus-tutor

Messner, W. (2023, February 2). Hypothesis Testing and Machine Learning: Interpreting Variable Effects in Deep Artificial Neural Networks using Cohen's f 2. Retrieved from https://arxiv.org/ftp/arxiv/papers/2302/2302.01407.pdf

Mustapha, S. (2023, September 29). Predictive Analysis of Students' Learning Performance Using Data Mining Techniques: A Comparative Study of Feature Selection Methods. *Applied System Innovation*, 6(5), 86. Advance online publication. DOI: 10.3390/asi6050086

O'Brien, C. (2023, September 14). *8-universities-leveraging-ai-to-drive-student-success*. Retrieved from Digital Marketing Institute: https://digitalmarketinginstitute.com/blog/8-universities-leveraging-ai-to-drive-student-success

Open, A. I. GPT 4 model. (2024). Retrieved from www.chatgpt.com

Pane, J. F., Steiner, E. D., Baird, M. D., Hamilton, L. S., & Pane, J. D. (2017). *Insights on Personalized Learning Implementation and Effects.* RAND Corporation. Retrieved from https://www.rand.org/pubs/research_reports/RR2042.html

Pearson. (2024, January). *Explore Pearson.* Retrieved from Pearson: https://www.pearson.com/explore-pearson.html

Peng, H., Ma, S., & Spector, J. M. (2019). *Personalized adaptive learning: an emerging pedagogical approach enabled by a smart learning environment.* Smart Learning Environments., DOI: 10.1186/s40561-019-0089-y

Phobun, P., & Vicheanpanya, J. (2010). Adaptive intelligent tutoring systems for e-learning systems. *Procedia: Social and Behavioral Sciences*, 4065-4068(2), 4064–4069. Advance online publication. DOI: 10.1016/j.sbspro.2010.03.641

Powerschool. (2024). *Clasroom.* Retrieved from www.powerschool.com: https://www.powerschool.com/classroom/schoology-learning/

Qadir, J. (2022, December 30). Engineering Education in the Era of ChatGPT:Promise and Pitfalls of Generative AI for Education. *Techrxiv.* DOI: 10.36227/techrxiv.21789434.v1

Roberts, R., Flin, R., Millar, D., & Corradi, L. (2021). *Psychological factors influencing technology adoption: A case study from the oil and gas industry.* Elsevier., DOI: 10.1016/j.technovation.2020.102219

Rus, V. (2023). *The University of Memphis*. Retrieved from https://www.memphis.edu/iis/projects/deeptutor.php

Rus, V., Niraula, N., & Banjade, R. (2015). DeepTutor: An Effective, Online Intelligent Tutoring System That Promotes Deep Learning., *37*. Retrieved from https://ojs.aaai.org/index.php/AAAI/article/view/9269

SciSpace. (2023). Retrieved from https://typeset.io/resources/scispace-literature-review-workspace/

Seo, K., Tang, J., Roll, I., Fels, S., & Yoon, D. (2021, October 26). The impact of artificial intelligence on learner–instructor interaction in online learning. *International Journal of Educational Technology in Higher Education*. Retrieved from https://educationaltechnologyjournal.springeropen.com/articles/10.1186/s41239-021-00292-9

Siminto, Akib, Hasmirati, & Widianto, D. S. (2023). Educational Management Innovation by Utilizing Artificial Intelligence in Higher Education. *Jurnal Manajemen Pendidikan*, 284-296. DOI: 10.31958/jaf.v10i1.6007

Society for Learning Analytics Research. (2023). Retrieved from https://www.solaresearch.org/about/what-is-learning-analytics/

Sparrow, S. (2023). Retrieved from https://www.smartsparrow.com/

Stokel-Walker, C., & Van Noorden, R. (2023, February 9). THE PROMISE AND PERIL OF GENERATIVE AI. *Springer*. Retrieved from https://www.nature.com/articles/d41586-023-00340-6

Treceño-Fernández, D., Calabia-Del-Campo, J., Bote-Lorenzo, M. L., Gómez-Sánchez, E., Luis-García, R., & Alberola-López, C. (2020, October). Integration of an intelligent tutoring system in a magnetic resonance simulator for education: Technical feasibility and user experience. *Computer Methods and Programs in Biomedicine*, 195, 105634. Advance online publication. DOI: 10.1016/j.cmpb.2020.105634 PMID: 32645627

UNESCO. (2023, April). *ChatGPT and Artifical Intelligence in Higher Education*. United Nations Educational, Scientific and Cultural Organization. Retrieved from https://www.iesalc.unesco.org/wp-content/uploads/2023/04/ChatGPT-and-Artificial-Intelligence-in-higher-education-Quick-Start-guide_EN_FINAL.pdf

UNESCO. (2023). *Guidance for Generative AI in education and research*. Retrieved from https://unesdoc.unesco.org/ark:/48223/pf0000386693

U.S Department of Education. (2016, June). Retrieved from https://ies.ed.gov/ncee/wwc/Docs/InterventionReports/wwc_cognitivetutor_062116.pdf

Valenzuela-Fernández, L., Garro-Aburto, L., & Ocaña-Fernández, Y. (2019). Artificial Intelligence and Its Implications in Higher Education. Retrieved from https://eric.ed.gov/?id=EJ1220536

Veerappa, M., & Rinzivillo, S. (2023, July). Explainable AI. *Ercim News*. Retrieved from https://ercim-news.ercim.eu/images/stories/EN134/EN134-web.pdf

Walton, D. (2023, April 5). Personalized Cognitive Tutoring using Davinci-003 API for Adaptive Question Generation and Assessment. doi:https://doi.org//arXiv.2304.02772DOI: 10.48550

Wang, H., Tlili, A., Huang, R., Cai, Z., Li, M., Cheng, Z., & Zhu, X. (2023). Examining the applications of intelligent tutoring systems in real educational contexts: A systematic literature review from the social experiment perspective. *Education and Information Technologies*, 28(7), 9113–9148. DOI: 10.1007/s10639-022-11555-x PMID: 36643383

Woodall, T. (2023, December 15). Retrieved from https://news.osu.edu/using-ai-to-pinpoint-hidden-sources-of-clean-energy-underground/

KEY TERMS AND DEFINITIONS

Adaptive Learning: The use of AI systems to adjust learning experiences according to the individual needs of the student.

AI Literacy: The knowledge and skills required to understand and use AI technologies effectively.

Artificial Intelligence (AI): The simulation of human intelligence by machines, particularly computer systems.

Automated Essay Scoring (AES): The use of AI to evaluate and score written student essays.

Bayesian Networks: Graphical models that represent probabilistic relationships among variables. Used in adaptive testing.

Chatbot: An AI-driven application designed to simulate conversation with users, widely used for student support.

Cognitive Tutor: AI-based software designed to mimic human tutoring by adapting instructional materials to the needs of individual learners.

Decision Trees: A predictive model used for decision-making based on a tree-like graph of decisions and their consequences.

Deep Learning: A subset of machine learning involving neural networks with many layers, used in complex AI applications like image and speech recognition.

Explainable AI (XAI): AI systems designed to provide clear, human-understandable explanations for their decisions and actions.

Fuzzy Logic: A form of logic that handles approximate reasoning rather than fixed and exact logic, used in AI for adaptive learning and grading systems.

Gamification: The application of game-design elements in non-game contexts, such as education, to enhance student engagement.

Generative AI: A type of AI that can create new content, such as text, images, or music, often used in educational content creation.

Intelligent Tutoring Systems (ITS): AI-based systems that offer personalized learning experiences and feedback to students.

Knowledge Graphs: A database that uses a graph-structured data model to represent and store information, often used for semantic learning.

Learning Analytics: The measurement, collection, and analysis of data on learners to optimize their learning outcomes.

Learning Management System (LMS): Software applications used to administer, document, track, and deliver educational courses.

Learning Pathways: Customized routes designed to guide students through learning objectives and content at their own pace.

Machine Learning (ML): A subset of AI where systems automatically improve their performance through experience.

Markov Models: Statistical models used to predict future states in a system based on its current state, often applied in predicting student behavior.

Natural Language Processing (NLP): A field of AI that enables machines to understand and process human language.

Neural Networks: A set of algorithms modeled after the human brain, used in AI to recognize patterns and make decisions.

Open Learner Model (OLM): A representation of a learner's knowledge that is shared between the system and the learner to foster self-regulation.

Personalized Learning: Tailoring learning experiences to meet the specific preferences, needs, and progress of individual students.

Predictive Analytics: The use of data, AI, and statistical algorithms to predict future outcomes based on historical data.

Reinforcement Learning: A type of machine learning where agents learn to make decisions through trial and error, optimizing for long-term rewards.

Sentiment Analysis: The process of using NLP to identify and extract subjective information from text, often used to assess student feedback.

Chapter 2
Playing to Win: A Case Study of Gamified Digital Transformation in the News Industry

Omar M. Badreldin
https://orcid.org/0009-0004-7658-8768
Nile University, Egypt

Alaa-Eldin M. Adris
https://orcid.org/0009-0007-1405-6047
Nile University, Egypt

Heba Labib
Nile University, Egypt

ABSTRACT

Digital transformations often struggle due to employee resistance and low engagement. Gamification, incorporating game elements into non-game settings, tackles these issues. This study explored gamification's impact on a news organization's digital transformation. A gamification technique implemented with 164 journalists for five months. Results showed a significant increase in digital transformation metrics, indicating successful behavior change. Points, rankings, and badges were the most effective elements. Surveys confirmed this, with 80% of participants finding it appealing and 83.7% more interested in learning new digital methods. This research shows gamification can be a powerful tool for successful digital transformation by driving positive behavior changes, increasing engagement, and reducing resistance to change. These findings offer valuable insights for organizations undergoing digital transformations.

DOI: 10.4018/979-8-3373-1132-6.ch002

INTRODUCTION

1.1 Background and Motivation

There is an increasing business and social need for companies to offer their services in digital formats, especially after the COVID-19 pandemic, and in order to transform the business operations and revenue model to be based on digital platforms, this requires going through a successful digital transformation process.

In their 2020 study, (Wade & Shan, 2020) concluded that the aggregated failure rate for such projects is 87.5%. These rates vary depending on organizational size and industry. Larger, traditional companies that have not undergone prior digitization and digitalization processes experience the highest failure rates, ranging from 91% to 96%. In contrast, smaller technology-based firms report lower failure rates, generally below 70%, indicating that these companies are more adept at successfully implementing digital transformation initiatives.

The factors contributing to the failure of digital transformation initiatives typically fall into three broad categories:

1- Technology related factors
2- Process/approach used for implementation.
3- The human and cultural factors

In a recent study (Ramesh & Delen, 2021), a 1.3 trillion dollar spent on digital transformation projects worldwide, and with this high failure rates, nearly over a trillion dollar is lost in the process in addition to the buildup resistance for the similar future digital transformation projects, which usually increases the failure rates for next digital transformation projects.

The most overlooked factors are those related to human behavior and organizational culture, especially the ability to alter or change human behavior to accept new digital-based workflow and operations. (Forbes Technology Council, 2021)

Human resistance to change is a common factor that contributes to the failure of digital transformation projects. This factor is not unique to digital transformation but can also be found in the scientific literature of other disciplines and is often overlooked during the implementation of digital transformation projects.

Gamification techniques have been widely employed as catalysts across various sectors, including formal education, marketing, healthcare, and tourism. These concepts have proven effective in enhancing engagement and accelerating processes within these industries.

However, the integration of gamification into digital transformation efforts has been relatively limited. This study explores the potential role of gamification as a catalyst in the digital transformation process, aiming to assess its effectiveness as an accelerator in a digital transformation process.

1.2 Digital Transformation in The News Industry

The news industry is undergoing a massive digital transformation, bringing both opportunities and challenges. Traditional business models, workflows, and cultures are being disrupted by new digital technologies and changing audience behaviors (Lewis & Westlund, 2015). Major challenges include declining print circulation, loss of advertising revenue, increasing competition from digital media, and difficulties monetizing online content (Newman et al., 2019). These economic pressures have led to newsroom budget and staff cuts, even as demand for 24/7 digital news grows.

Adapting to digital also requires retraining journalists in new multimedia skills and reducing legacy biases favoring print (Newman et al., 2019). Many news workers lack digital expertise and resist changing entrenched practices. Outdated workflows prevent integration of emerging digital media. Legacy attitudes that value print journalism over digital persist.

Cultural resistance and lack of digitally experienced resources make digital transformation difficult. News executives must balance print legacy preservation with digital transition despite economic instability and discomfort with change (Newman et al., 2019). Strategic management of organizational change is crucial for sustainable success. Approaches like incentives, training, transparency, and participative leadership can help overcome barriers (Lewis & Westlund, 2015). But fundamental shifts in mindset, skills, and culture are unavoidable as news organizations evolve for the digital age.

This research is based on an experiment in the news industry, where more than three previous digital transformation projects failed to achieve their goals or sustain the change momentum for enough duration to consider it a successful and stable transformation.

The shift in advertising spending from print to digital in the news industry put direct pressure on news organizations to diversify their sources of revenue (Nel, 2010). The loss of traditional/print revenue requires news publishers to revise their current content production cycle and business model to sustain their business.

The digital transformation of traditional news organizations requires journalists to produce new media/content specifically for digital platforms (i.e., interactive maps, charts, voting polls, timeline for storytelling, etc...) and enrich editorial text content with relevant keywords, links, and tags.

The success of digital transformation process for a news organization could be measured by the organization's ability to transform its daily editorial operations to create more digital content in different formats and change its business model to rely more on digital revenues.

The research captured key data points for measuring digital content five months before and after applying gamification.

The news industry's digital transformation faces challenges like declining print revenue and resistance to new digital workflows. Gamification can help by using elements like points and rewards to motivate staff to adopt digital skills and overcome resistance, enhancing engagement and participation.

1.3 Gamification Concept

Gamification leverages people's natural motivations for gaming, achievement, competition, status, self-expression, and altruism to drive engagement with productive activities (Nah et al., 2019). While related to games, gamification is distinct in its focus on "real-world" applications over just entertainment.

Key game elements applied in gamification include points, levels, leaderboards, badges. These game features are strategically implemented within non-game contexts using gamification frameworks like the Mechanics-Dynamics-Emotions (MDE) model to tap intrinsic and extrinsic motivations and achieve target behaviors or outcomes (Hartmann & Gommer, 2021).

Gamification has demonstrated success across education, marketing, crowdsourcing, fitness, sustainability, and workplace training (Friedrich et al., 2020; Torlaković et al., 2022). Gamification shows promise as a motivational tool, but requires thoughtful design based on established frameworks to have true impact.

1.4 Research Gap

Researchers have widely acknowledged that behavioral and cultural factors, particularly resistance to change, play a significant role in the failure of digital transformation projects. Gamification, which has proven effective in overcoming similar barriers in various sectors, offers a promising approach to addressing these challenges within digital transformation initiatives.

However, there is limited research on the application of gamification specifically within digital transformation processes. This study seeks to fill this gap by documenting a real-life experiment where gamification techniques were applied to a digital transformation project in the news and media industry. By exploring the potential of gamification to mitigate resistance and drive engagement, this research aims to provide insights into its role as a facilitator in digital transformation efforts.

1.5 Research Problem

Digital transformation is a critical imperative for businesses and organizations in today's rapidly changing world. However, the failure rate for Digital transformation implementations is alarmingly high, at nearly 87.4%. One of the primary reasons for these failures is related to behavioral and organizational culture issues, such as resistance to change, lack of employee engagement.

Gamification offers a promising solution to behavioral challenges and can significantly improve the success of digital transformation (DT) initiatives. It effectively motivates individuals and teams to achieve key goals, such as boosting productivity, enhancing learning outcomes, and fostering behavioral change. But despite the potential benefits of gamification for digital transformation, there is a dearth of research on this topic.

This research is conducted to explore the outcomes of using gamification as an agent to increase the digital transformation success rate and sustain this transformation for enough duration to consider it a permanent change.

1.6 Research Question(s)/Hypothesis

This research is examining the impact of applying gamification concepts, design elements and techniques to alter the human behavior and lower resistance to change, which should result in a positive contribute to the digital transformation process.

The research aims to assess the initial hypothesis of the following research question.

RQ1. To what extent there are differences, between the key data points of the digital platforms' engagement using gamification for five months and those before using gamification?

$H1_{01}$. There is no significant difference in the digital transformation key data points after applying gamification.

$H1_{a1}$. There is a significant difference in the digital transformation key data points after applying gamification.

RQ2: How did the gamification experience enhanced engagement in the digital transformation process?

To answer the RQ1, a quantitative method was used by defining the key digital transformation data points in the news industry and baselined those data points collected for five months before gamification implementation. Followed by a post-implementation collection of the same data points tracked for five months, then a statistical analysis using paired t-tests was employed to compare the results before and after the gamification.

Results was interpreted in relation to the null and alternative hypotheses, quantifying any significant differences and exploring alternative explanations if no significance is found.

RQ2 was addressed through a survey as a qualitative method, analyzing user perceptions of gamification elements and their impact on engagement, with thematic analysis providing insights into how gamification enhances the digital transformation process.

Both quantitative and qualitative findings was triangulated to draw comprehensive conclusions of the research findings and identify key gamification elements and potential resistance points influencing the success of a digital transformation process.

1.7 Significance

The initial literature review indicates a notable gap in research concerning the application of gamification within digital transformation processes. A search on Google Scholar revealed that only 1 to 3 papers linking these two topics were published annually between 2018 and 2023, highlighting the scarcity of academic exploration in this area.

This study aims to address this gap by evaluating the impact of gamification on digital transformation in the news and media sector, an area that remains largely unexplored. Through a real-world experiment, the research seeks to bridge the concepts of gamification and digital transformation, offering valuable insights for both practitioners and scholars, and contributing to the body of knowledge.

LITERATURE REVIEW

2.1 Introduction

Digital transformation (DT) is often glorified for its potential to transform business operations, create new revenue streams, significantly reduce costs, and improve people's lives.

With all those promises, businesses and organizations approach digital transformation with widely different perceptions, expectations, and sometimes misunderstandings of what is considered a digital transformation and what is not.

Across various business verticals, the failure rate for implementing digital transformation (DT) is nearly 87.4%, resulting in significant losses of time and money. One of the primary reasons for these high failure rates is related to behavioral and organizational culture issues that are challenging to overcome.

Gamification has been widely adopted in other domains such as education and marketing to address similar behavioral issues. Therefore, there is a high probability that gamification could reduce the failure rates of DT.

2.2 Digital Transformation

2.2.1 Basics and Definitions

Digital transformation (DT) has been conceptualized in various ways, with no universally accepted definition due to its complexity and multifaceted nature. (Vial, 2021) reviewed 282 sources and identified 23 distinct definitions, underscoring the ongoing debate. Despite the differences, key components such as organizational change, technological innovation, and the scope of transformation are central to most definitions. Researchers like (Vial, 2021)and (Gong & Ribiere, 2021) emphasize that DT is not merely about adopting new technologies but involves a fundamental change process that drives systemic improvements across organizations, business networks, and societies.

2.2.2 Digitization, Digitalization, and Digital Transformation

Contributing to the confusion, the terms digitization, digitalization, and digital transformation are often used interchangeably, despite each referring to distinct scopes and outcomes. However, these terms represent different processes.

Digitization is basically taking analog information, such as a printed photo, and encoding it into zeroes and ones so that computers can store, process, and transmit. The critical takeaway from this term is that only elements such as printed photos and papers are digitized, not the process or business cycle itself.

Digitalization is often linked with digitization, as the former depends on digitization efforts for implementation. It has multiple definitions, encompassing business processes, social life, and occasionally business models. However, the most commonly accepted definition is "the process of utilizing digitized information to streamline specific operations." A typical example of digitalization is the automation of certain aspects of a business process. (Bloomberg, 2018).

Therefore, digitization and digitalization alone do not result in a comprehensive change, new business models, or disruptive innovation, which falls primarily under the scope of digital transformation.

2.2.3 Digital Transformation

Digital transformation is more than just digitizing resources and workflow; it must create additional value and revenues from digital assets. (Gong & Ribiere, 2021)

The shift toward the adoption of digital platforms has become not only a strategic necessity but also an inevitable response, particularly following the global disruption caused by the COVID-19 pandemic. Organizations have found that embracing digital solutions is the only viable way to sustain their operations and maintain satisfactory service delivery to their customers or citizens. Without such a transformation, many businesses and public entities would have struggled to meet the evolving demands of a digitally dependent society, risking operational inefficiencies, customer dissatisfaction, and even potential closure. (Guandalini, 2022)

Organizations use digital technologies to alter the value creation paths they have previously relied upon to remain competitive. To that end, they must implement structural changes and overcome barriers hindering their transformation. These changes lead to positive impacts for organizations and, in some instances, for individuals and society (Vial, 2021).

2.2.4 New Trends in Digital Transformation

The recent advancements in digital transformation are primarily driven by the integration of emerging technologies such as artificial intelligence (AI), blockchain, and the Internet of Things (IoT), which are significantly altering business processes across industries. One key trend is the increasing role of AI in automating complex tasks, enhancing decision-making, and providing predictive insights, thus enabling organizations to optimize operations and drive innovation. AI technologies, particularly machine learning and natural language processing, are being used in financial institutions to digitize front, middle, and back office processes, improving efficiency and customer service (Butler, 2020). Additionally, AI enhances data analysis, allowing businesses to better understand customer behavior, streamline operations, and make data-driven decisions (Wang et al., 2022).

Another notable trend is the adoption of blockchain and distributed ledger technologies (DLT), which are transforming business models by offering greater transparency, security, and trust in digital transactions. Blockchain's decentralization, immutability, and auditability make it particularly valuable in sectors such as finance, supply chain management, and digital identity verification. The integration of blockchain with AI and IoT is creating more secure and efficient ecosystems, enabling automated processes such as smart contracts and enhancing regulatory compliance through technologies like RegTech (Wang et al., 2022). This convergence of technologies is reshaping industries by fostering new business models, reducing operational costs, and creating innovative customer engagement strategies.

2.2.5 Digital Transformation and Organizational Culture

Digital transformation necessitates the collaboration of various teams within an organization, requiring a concurrent shift in the internal organizational culture to support this process. Organizational culture, comprised of shared behavior patterns, values, and beliefs, serves as the foundation for how employees make decisions and respond to challenges. It is seen as the set of norms that shape the collective work environment, guiding the behavior of organizational members in achieving institutional goals. These cultural norms are crucial in determining how individuals within the organization adapt and respond to changes, particularly those driven by the goals of digital transformation. (Firican, 2023)

2.2.6 Sources of Organization Culture

The primary sources of organizational culture are the beliefs and assumptions of the organization founders, which later is altered by the learning experience of the organization members. (Pathiranage et al., 2020)

The organization's founders can make a significant impact in the early stages of the organization's culture by setting the business, operational goals, and founders' personal beliefs. Furthermore, as the organization continues to operate, the day-to-day learning experiences alter the organization's culture to match or resist modified business and operational goals. (Nguyen & Aoyama, 2014)

Similarly, the news organization's internal culture impacts how the journalists perceive changes to digital platforms. The segregation between editorial and other business activities of news organizations has long been a fundamental norm of journalism and is so ingrained into the fabric of the organization that they become part of the organization's values (Cornia et al., 2020).

The development of organizational culture is often rooted in the company's approach to innovation and its ability to adapt to digital disruptions. The culture reflects the collective response of employees to challenges, emphasizing behaviors that support flexibility, creativity, and continuous innovation. Entrepreneurial agility, in particular, plays a critical role in shaping a dynamic culture that is capable of responding effectively to market changes and technological advancements. Organizational culture, therefore, emerges as a combination of formal structures, informal interactions, and leadership directives, all of which contribute to the norms and values that guide decision-making processes and employee behaviors within the organization (Karimi & Walter, 2021) .

A common theme across multiple studies points to the need for organizations to cultivate a willingness to take risks and to experiment with digital technologies on a small scale before scaling these successful experiments to the rest of the organization (Li, 2020).

Organizations can encourage learning through small, incremental, and iterative changes while sustaining their ability to change long-term plans based on the outcomes of those incremental experiments as well as ongoing changes in their environment (Barann et al., 2019).

2.2.7 Organizational Culture as a Barrier to Digital Transformation

Internal organizational culture significantly impacts its ability to implement/adopt digital transformation; one of the internal culture issues is inertia which could constrain the force for innovation through digital technologies. (Svahn et al., 2017). In the Kodak case (Lucas & Goh, 2009) illustrated how the core capabilities of well-established organizations could become core rigidities that prevent the radical transformation of digital technologies. Another barrier to DT is the resistance that employees can demonstrate when disruptive technologies are introduced in the organization. The resistance is usually related to the ways and pace at which technologies are introduced into an organization, which could result in what practitioners highlight as "innovation fatigue" as one of the causes of resistance. (Chung et al., 2017).

2.2.8 Other Barriers to Successful Digital Transformation

Several non-cultural factors significantly impede digital transformation efforts. For example, legacy technology systems are a technology barrier, particularly in industries like finance, where outdated infrastructure limits the adoption of advanced technologies such as AI and cloud computing. These legacy systems increase operational costs and hinder innovation, making it difficult for organizations to adapt to modern requirements (Diener & Špaček, 2021).

Additionally, data security and regulatory concerns pose major challenges. As organizations handle increasing amounts of sensitive data, the risk of cyberattacks rises, and compliance with regulations such as GDPR becomes more complex, especially in sectors like healthcare and finance (Diener & Špaček, 2021)

Financial constraints also restrict digital transformation, particularly for smaller organizations that struggle to invest in new technologies and train their workforce while maintaining outdated systems. Lastly, a lack of strategic clarity and leadership further delays transformation efforts, as fragmented approaches without a clear vision lead to inefficient implementation (Jones et al., 2021). Addressing these barriers requires integrated strategies and investment in both technology and talent.

2.3 Gamification

2.3.1 Basics and Definitions

Gamification is generally defined as the application of characteristics and design techniques of games into non-gaming contexts (Perryer et al., 2016). Nick Pelling coined the term gamification in 2002, but it has not shown a wide adoption until late 2010.

The term "gamification" relates to games, not play (or playfulness), where "play" can be conceived as the broader, looser category, containing but different from "games". The "gaming" term captures playing structured by rules and competitive strife toward goals (Deterding et al., 2011)

2.3.2 Gamification and Motivational Theory

One of the most significant challenges in the digital transformation of traditional journalism is motivating journalists to adopt new tools and adjust their daily routines. Motivation, broadly defined as the underlying reasons that drive individuals to engage in activities with varying levels of interest and commitment, plays a critical role in this process. When an individual is motivated, they become "energized" or "activated" toward a specific goal, leading to meaningful participation in the desired transformation. In the context of journalism, fostering this motivation is crucial for successful adaptation to digital workflows and technological integration. (Hartt et al., 2020)

Motivation can be broadly categorized into two types: intrinsic and extrinsic. Intrinsic motivation refers to engaging in an activity for the inherent satisfaction derived from the task itself, without the expectation of external rewards. It is driven by internal factors such as personal growth, passion, or enjoyment of the work. In contrast, extrinsic motivation stems from external sources, including rewards, recognition, financial incentives, or social approval. Extrinsic motivation is characterized by performing an activity to achieve a separable outcome, such as a promotion or a financial reward. These two types of motivation are pivotal in shaping employee behavior and contentment, as both play a significant role in influencing how employees engage with their tasks and how satisfied they feel within their organizational context (Mardanov, 2020).

Gamification could be linked to intrinsic and extrinsic motivation types, as gamification concepts are directly linked with extrinsic motivation by providing external sources of reward such as game elements and rewards such as points, badges, and ranking. While the gameplay itself is linked to intrinsic motivation since playing games are often assumed to have inherent motivational power, which rises when participants become immersed in the gamified experience, and the gameplay is deemed enjoyable. (Hartmann & Gommer, 2021)

In Gamification, the gamified experience refers to the setup where game design principles are used in non-game situations to change people's behavior to participate actively and strive toward achieving target goals.

2.3.3 The MDE Framework

In order to probably design a gamified experience, researchers developed several frameworks to support the building of such experiences. One of the most commonly used gamification frameworks is the Mechanics, Dynamics, and Emotions (MDE) framework. (Robson et al., 2015)

The first principle of the MDE framework is Mechanics, where the game rules, goals, and settings are specified and announced to gamified experience participants.

The second principle is Dynamics, which captures the player's behavior within the game mechanics. For example, how the player decides and moves his pieces in a chess game.

The third principle is Emotions, which is the psychological effect and player's state of mind (i.e., Joy, Anger) resulting from the player's behavior (Dynamics) and limited by the game mechanics.

2.3.4 Design Elements of Gamification

Gamification design elements include but are not limited to points, social interaction, leaderboards, progress bar, levels, narrative and storyline, and badges/ medals. However, the most essential and commonly used design elements are points, badges, and leaderboards. (Nah et al., 2019)

Points are used to give a reward for completing a specified activity. Points also provide a standard measurement for activities and an easy way to compare players' performance.

Badges graphically represent proof that a player has exceptional performance and are usually earned based on collecting several points or making an exceptional effort in achieving a goal and are rewarded in a visual badge that appears on his profile or next to his name.

Leaderboards are where player names are publicly displayed and sorted according to their performance and collected points.

The primary purpose of leaderboards is to trigger competition among players and show which players perform best. The following table (Nah et al., 2019) shows the commonly used gamification design elements to implement principles of gamification.

Table 1. Principles of Gamification and their Design Elements (Nah et al., 2019)

Principles of Gamification (CIG-SCARF)	Design Elements
Challenge (C)-opportunities for growth, learning, and development	Points, Levels, Badges, Leaderboards, Quest, Feedback/ Progress Bars, Performance Graphs, Prizes/Rewards/Bonuses, Rules, Marketplace
Interactivity (I)-potential for immediate feedback	Points, Quest, Feedback/Progress Bars, Performance Graphs, Avatars,Roleplay
Goal Orientation (G)-setting clear and systematic goals	Points, Levels, Badges, Leaderboards, Onboarding,Prizes/ Rewards/Bonuses,Customization/Personalization
Social Connectivity (S)-opportunities to interact with others	Leaderboards, Social Engagement Loops, [use of] Teams [for Collaboration],Avatars, Roleplay
Competition (C)-opportunities for social comparisons or winners to emerge	Points, Levels, Badges, Leaderboards, Quest, Feedback/ Progress Bars, Prizes/Rewards/Bonuses, Rules, Marketplace
Achievement (A)-recognition of effort and/ or accomplishment	Points, Levels, Badges, Leaderboards, Feedback/Progress Bars, Performance Graphs, Prizes/Rewards/Bonuses
Reinforcement (R)-structure of rewards based on performance	Levels, Badges, Leaderboards, Onboarding, Feedback/ Progress Bars, Prizes/Rewards/Bonuses
Fun Orientation (F)-creating interest, curiosity, and enjoyment	Quest, Prizes/Rewards/Bonuses, Rules, Marketplace, 3D Space, Avatars, Storyline (Narrative Content), Roleplay, Customization/Personalization

2.3.5 Applications of Gamification

Gamification concepts are applied to a wide range of industries to drive or promote the desired behavior of both employees and customers. For example, the motor giant Nissan introduced a gamified experience called "Nissan Carwings" for their electric car owners. (Torlaković et al., 2022).

The mobile application gamifies eco-driving in the form of global competition for the most [REMOVED INCLUDEPICTURE FIELD] environmentally friendly driving. Each driver receives a particular Eco score based on their driving stats. This ranking is compared to all drivers in the driver's region or the whole world.

There are daily, weekly, and monthly rankings, and drivers also can earn medals from bronze to platinum according to their driving performance. [REMOVED INCLUDEPICTURE FIELD]

Gamification was successfully used to overcome behavioral and cultural issues and make participants more engaged and motivated to participate in the learn/actively change process. (Friedrich et al., 2020)

Therefore, gamification is an excellent opportunity to overcome behavioral and cultural challenges while applying the digital transformation.

2.3.6 Gamification in Journalism

In the context of journalism, gamification has been increasingly explored as a method to boost user interaction and overcome resistance to change within media organizations. According to (Torrijos, 2020) gamification used to make news consumption more interactive and engaging, allowing audiences to participate actively in storytelling. This trend has been particularly prominent in sports media coverage of mega-events like the Olympics and Football World Cups, where news outlets create "newsgames" and gamified pieces that blend information with entertainment. These interactive elements not only attract larger audiences but also serve an educational purpose by providing explanatory, data-driven content, particularly for lesser-known sports. The author assert that designing online user engagement through gamified elements has the potential to transform the traditionally passive consumption of news into an active process, thereby contributing to the sustainability of digital media platforms.

Based on (Vos & Perreault, 2020) analysis of gamification in journalism, gamification is presented as a complex and evolving strategy that aims to enhance audience engagement within digital news platforms. Their study highlights how journalists have debated the role of gamification, with some viewing it as a shift towards infotainment, while others see it as a legitimate extension of journalism's traditional goals of informing the public. The authors explain that gamification is often framed within a normative discourse that seeks to balance journalistic integrity with audience participation.

2.3.7 The Broad use of Gamification

The application of gamification is not limited to journalism; it has seen successful implementations across various sectors, particularly in education and healthcare, which face similar challenges of engagement and motivation. In the education sector, (Manzano-León et al., 2021)explored the impact of gamification learning systems on student motivation and achievement. Their study revealed that gamified learning environments, which included points, medals, rankings, and narrative significantly improved students' motivation and academic performance. The findings demonstrate the effectiveness of gamification in sustaining long-term engagement, a key challenge also faced by media organizations in digital transformation efforts.

Gamification in healthcare also emerged as an effective strategy to encourage patient engagement and promote healthier lifestyles through interactive digital platforms. The healthcare gamification industry has seen significant growth, driven by the increasing popularity of health and fitness apps, which saw a tenfold rise in usage between 2015 and 2021. These apps, incorporating features such as exercise tracking, diet management, and even rehabilitation programs, aim to modify behavior and improve patient outcomes. (Ansari et al., 2022)

2.3.8 Overcoming Change Resistance through Gamification

Gamification can help organizations overcome resistance to change by making the adoption of new tools and processes more engaging and less intimidating. Change resistance often stems from fear of the unknown, disruption of established routines, or lack of engagement. By introducing game-like elements such as rewards, challenges, and recognition, gamification can create a more engaging and motivating environment, encouraging employees to adopt new tools and processes more willingly.

One way gamification helps overcome resistance is by increasing motivation and reducing the perceived difficulty of change. Through intrinsic motivators like competition, achievement, and personal growth, gamification can make new systems and practices appear more approachable and rewarding. For example, offering badges or progress points for mastering new technologies can instill a sense of accomplishment and provide employees with short-term goals, helping them to navigate the learning curve of new tools more effectively. This approach fosters a positive attitude toward change and promotes continuous engagement with digital platforms.

Moreover, gamification addresses another core aspect of change resistance, which is the lack of immediate feedback. Employees often resist change because they do not see the immediate benefits of adopting new behaviors. By implementing real-time feedback mechanisms, such as leaderboards or personal dashboards, gamification provides instant gratification, reinforcing the value of new practices. This immediate feedback helps to demonstrate the positive outcomes of the transformation, which in turn encourages further participation and reduces skepticism about the change.

Incorporating team-based challenges or collaborative elements can also help align employees with organizational goals. When team performance is tied to shared rewards, individuals are encouraged to work together to overcome challenges, fostering a sense of community and reducing the isolation that often accompanies organizational change. In this way, gamification can transform resistance into active participation by creating a positive, engaging environment where change is not seen as a threat, but as an opportunity for collective success.

By leveraging these principles, gamification can effectively mitigate resistance to change, particularly in large-scale transformations where employee buy-in is crucial to success. Through enhanced motivation, real-time feedback, and collaborative efforts, gamification transforms the adoption of new systems and processes into a rewarding and engaging experience, easing the transition and reducing resistance.

2.3.9 Gamification Might Fail

Gamification does not always guarantee success and may even fail in certain contexts. One primary reason for this is that gamification often focuses heavily on the "fun" aspect, potentially at the expense of more substantial outcomes like academic performance in educational sector or skill acquisition in organizations. For example, it was found that while gamification improved student motivation and classroom dynamics in educational sector, it did not always translate into better academic results. In fact, in some cases, students who engaged less with the gamified activities performed better academically, raising questions about whether the motivational benefits of gamification are at odds with actual performance gains (Hernández-Fernández et al., 2020).

Moreover, gamification may also fail when it is poorly aligned with organizational or educational objectives, especially if the design of gamified systems does not align with the broader goals of the organization, such as knowledge retention or skill development, it can lead to superficial engagement rather than meaningful transformation. Additionally, there is the risk of fostering unhealthy competition or disengagement when gamified elements such as leaderboards are overemphasized. This can demotivate participants, especially those who fall behind in rankings, and diminish the inclusive and collaborative environment necessary for digital transformation (Hammedi et al., 2021). Therefore, the success of gamification largely depends on how well it is designed and aligned with broader goals, as well as how it considers the diverse needs and motivations of participants.

2.3.10 The Long-Term Effects of Gamification

The long-term impact of gamification has been a subject of growing interest, particularly regarding its sustainability and effectiveness over time. Research shows that while gamification can lead to immediate improvements in engagement and motivation, its long-term success is influenced by various factors, including novelty effects, personalization, and the alignment of gamification elements with user preferences.

One key challenge in maintaining the effectiveness of gamification is the novelty effect, where initial enthusiasm for gamified systems tends to wear off over time. (Altmeyer et al., 2021) found that gamification increased participation in fitness courses in the short term, but the novelty wore off as users became accustomed to the game elements. This suggests that without continuous updates or adaptations, gamification may lose its motivational pull, reducing long-term effectiveness (Altmeyer et al., 2021).

However, personalized gamification can mitigate these effects by catering to individual user preferences. Studies show that when gamification elements are tailored to specific user types demonstrate sustained engagement. For example, users who received a suitable set of gamification elements based on their personality types increased their participation by over 30% in long-term studies, indicating that personalization is key to sustaining the positive impact of gamification over time (Altmeyer et al., 2021).

Additionally, gamification's long-term effects on knowledge retention and skill transfer are significant. Research by (Khasawneh et al., 2024) highlights that students who engage with gamified learning environments tend to retain their skills and knowledge over extended periods, with retention rates remaining high even a year after initial learning. This suggests that gamification not only boosts short-term engagement but can also have lasting benefits in educational settings.

While gamification has the potential for long-term success, its effectiveness depends on mitigating the novelty effect through personalization and ensuring that the gamified elements align with the users' intrinsic motivations and personal objectives.

RESEARCH METHODOLOGY

3.1 Introduction

The research utilized a mixed methods approach, combining an experimental design with a qualitative survey component. The experiment examined the impact of applying gamification techniques on the digital transformation process at a news organization over a five-month period. Quantitative performance metrics were tracked daily across the sample of 164 journalists for five months before and after the introduction of gamification elements. These metrics provided indicators of usage of digital platforms and adoption of digital workflows.

Additionally, a structured survey was administered to collect qualitative feedback from a subset of 45 participants regarding the influence of gamification on their motivation and participation. The combination of experimental performance data and survey feedback enabled analysis of both the measurable impacts of gamification on digital transformation as well as insights into participants' experiences.

This mixed methods approach integrating an experimental study with a qualitative survey provides both numerical and narrative evidence regarding the use of gamification for digital transformation. The results shed light on the influence of gamification techniques on user participation, productivity, and motivation in a digital integration context.

3.2 Research Design

This study consists of two parts. The first part is an experiment conducted by the researcher, where gamification concepts were applied to a digital transformation process over a period of five months.

The second part of the study is a structured survey that aims to capture feedback from experiment participants regarding how gamification impacted their interest in participating in a digital transformation process.

The experiment used the quantitative data points illustrated in "3.4 Data collection" section to compare five months before and after applying gamification to a digital transformation process.

The survey collected qualitative feedback from experiment participants through structured questions to capture participants behavioral and emotional feedback after running the experiment.

3.3 Participants/Sample

The experiment total population consists of 200 journalists. The experiment had a total sample size of 164 participants who voluntarily joined, the sample represented different demographics such as gender, age groups and background studies.

The number of complete surveys received from participants is 45, which represents 27% of the experiment sample size.

3.4 Data Collection

The data collection was conducted daily for five months, capturing key data points that indicate the transformation from traditional operations to digital operations. These data points included the number of published articles, the volume of daily readership, and other relevant key data points.

The main measurement tools were the daily and monthly reports generated from the news publishing system five months before and after the run of the experiment. Additional surveys were conducted to capture insights from journalists, editorial managers participating in the experiment.

Table 2. Captured key data points.

S	Key data points	Importance
1	Number of accounts on digital platforms	Indicate the willingness of journalist to contribute in digital publishing by using their accounts
2	Number of monthly published articles	Measure volume of monthly articles by journalists
3	Monthly pageviews	Measure the number of articles viewed by readers, although this is subject to news breaks, but generally used to assess the total increase or decrease of readership for digital platforms.
4	Monthly pageviews from search engines	The number of articles viewed by readers searching on search engines (i.e.: Google). This measure the quality and adherence of digitally wrote articles to Search Engine Optimization (SEO) rules, which is a crucial aspect of writing for digital platforms.
5	Percentage of pageviews from search results to total pageviews	Measure the increase or decrease in the percentage of pageviews coming from search engine relative to total pageviews
6	Number of live videos	The number of live videos by journalists which is an important measure for the use of new journalism formats to tell a story using live videos.

These data points provided a quantitative measure for 164 journalists, demonstrating their progress before and after the application of gamification and measuring the overall impact of gamification on the digital transformation process.

Google Forms were utilized to conduct structured surveys that consisted of three sections: participant demographics, the impact of gamification on interest in participating in digital transformation, and open-ended questions that provided narrative for participants' experiences. The survey questions are available in the study appendix.

3.5 Data Analysis

The impact of gamification on the digital transformation process was evaluated through a two-stage analysis of the collected data.

3.5.1 Quantitative Analysis

For better understanding of the data under study, statistics were generated using the Minitab statistical package to characterize the data under study. Subsequently, a paired comparisons of the key data points before and after the gamification intervention were conducted to quantify and evaluate the numerical differences between the two stages.

Also, paired one tail t-tests were conducted to assess the statistical significance of the changes in key data points before and after gamification intervention to draw conclusive inferences regarding the effectiveness of the gamification intervention.

3.5.2 Qualitative Analysis

The conducted survey offered a valuable platform for discovering and assessing -from the eyes of the participants- on how gamification affected their performance during the digital transformation process and which gamification design elements and tools were preferred.

The closed-ended survey results were analyzed using Microsoft Excel for efficient data summarization and visualization. Subsequently, in the final thesis results section, a thematic analysis of the open-ended questions was undertaken to extract qualitative insights and uncover any valuable feedback from participant responses.

3.6 Ethical Considerations

A strict ethical code of conduct for interviews and surveys were followed, where clear details were communicated to the respondents about the purpose of the study, how the collected data is used and who have access to it.

A consent letter approving the use of the collected data in the scientific research is signed by the organization where the experiment takes place.

RESULTS

4.1 Gamification Experience Design

The digital transformation initiative started with a 2-month phase (1st interval), followed by a 3-month phase (2nd interval) and in both phases we settled on the Mechanics, dynamics, emotions (MDE) framework as the foundation for designing the Gamification experience.

Before conducting the experiment, the gamification experience was designed using the following design elements and game mechanics.

4.1.1 Game Design Elements

The essential design elements points, rank, avatars, dashboards, and leaderboards design elements were used as follows:

Table 3. Gamification design elements used in experiment.

S	Design element	Purpose
1	Points	They were used to show the number of collected points based on the number of pageviews generated from journalist articles.
2	Rank	The journalist rank among his collogues, and the rank is based on the total journalists' points.
3	Avatars	Each user is given a special avatar/title based on his points/rank
4	dashboards	Personal dashboard for each journalist to review his points, number of daily and publisher articles, rank, and top 10 articles that generated traffic.
5	Leaderboard	Show the Journalists names and profile photos, the size of the profile photo and name is based on each journalist's number of page views and rank.
6	Impact factor	The impact factor is used to assess the quality -in terms of pageviews- of the number of published articles for each journalist. (i.e., if a journalist has an impact factor of 5, this means the journalist has 5 times the impact of the average journalist)

4.1.2 Game Mechanics:

The main driver in designing gamification mechanics is aspiring to achieve a high journalist rank. It creates a competition between journalists to increase their number of published articles, points, and, therefore, their rank.

The journalist's ranks are assessed over three different timeframes; the first is the live rank which is evaluated every minute throughout the day, and the second is the daily rank which is evaluated for the past 24 hours. And the monthly rank is the accumulated rank at the end of the calendar month.

Additional mechanics were added by creating virtual clubs that could only be joined if journalists achieved high rank, like the "Club 50" for the highest 50 ranked journalists and "Club Millionaires" for journalists that generated more than 1 million pageviews from their articles.

The rewards, a crucial part of the gamification mechanics, are present in two types; the first one is the recognition among peers by having a high rank or being a member of one of the virtual clubs. The second reward is converting those collected points into financial rewards.

4.1.3 Rules Setup

The first rule in the game mechanics provides the limits and boundaries for the gamification experience; the first constraint is the minimum number of articles that should be published by each journalist, which is set according to the type and size of the editorial team in each section and assessed daily and monthly.

The second constraint is the duration for each game round, which is set to one calendar month. All journalists' statistics are reset to a fresh start, giving an equal opportunity for each journalist in the new game round.

4.2 Experiment Participants Demographics

The mean number of journalists interested in using digital platforms for the five months prior to implementing gamification was 57 journalists. And after applying gamification concepts, the total number of journalists participating in digital platforms increased to 170 journalists by the end of the two intervals.

The below table shows the breakdown by month for new journalists applying to join and participate in digital platform.

Table 4. breakdown of new journalists joining digital platforms

Month	**Number of new journalists joining the digital platforms**	**Interval**
July 2020	35	1st Interval
August 2020	36	
September 2020	27	2nd Interval
October 2020	6	
November 2020	2	

Participant gender analysis by the end of the two intervals

Table 5. Participant gender analysis by the end of the two intervals

	Males	**Females**
Percentage	74%	26%

4.3 First Interval and The Introduction of Gamification

Before launching the 1st interval phase, the organization kept a record of the critical Key Data Points (KDPs), which is considered the baseline for measuring the positive or negative change after applying Gamification.

The following table summarize the captured KDP's, their significance and interpretation:

Table 6. Digital transformation captured KDPs

S	Key Data Points (KDPs)	Interpretation	Significance
1	Number of monthly used digital platforms accounts	Indicates the number of journalists using the digital platforms	If the number increases, this means more interest from journalists to participate in the digital platforms.
2	Number of monthly articles published on the digital platforms	Indicates the total volume of articles published by journalists on the digital platforms	If the number increases, this means more interest in publishing content on digital platforms.
3	Monthly pageviews on the digital platforms	Indicates the number of pageviews generated from publishing articles on digital platforms.	As the number increases, this means an increase in viewability and reaching more readers. It directly results in more digital revenues, better brand awareness, and higher digital news market share.
4	Monthly pageviews generated from search engines results	The number of pageviews resulted from readers searching on search engines (i.e.: Google) and visiting our digital platforms.	This indicates the relevancy of published articles to reader's interests and measures the journalist's adherence to the technical requirements while writing content to be search engine friendly.
5	Percentage of pageviews from search results to total pageviews	Measure the increase or decrease in the percentage of pageviews coming from search engine to total pageviews	As the percentage increases, this indicates more relevant content to readers is published with adherence to the technical search engines friendly requirements.
6	Number of videos broadcasted on social networks	Measure the number of videos created and published by journalists	The indicate journalists' interest in producing content in new formats.

4.4.1 Outcomes of the First Interval

At the end of the two months pilot, the number of journalists using the digital platforms increased by 142%, considered the most significant increase ever recorded in all previous failed digital transformation projects. The number of published articles per month also increased by 37%.

Table 7. Impact of using gamification during first interval on KDP's compared to baseline

S	Key Data Points (KDPs)	Impact after pilot run
1	Number of monthly used digital platforms accounts	Increase by 142%
2	Number of monthly articles published on the digital platforms.	Increase by 37%
3	Monthly pageviews on the digital platforms	Increase by 12%
4	Monthly pageviews generated from search engines results	Increase by 17%
5	Percentage of pageviews from search results to total pageviews	4% increase
6	Number of videos broadcasted on social networks	86% increase

The organization also observed bluffing from the journalists by adding multiple contributing authors to the same article to give them free points/pageviews without contributing to the article writing, this required a change in the game dynamics rules by splitting the article pageviews/points across all the contributing authors instead of crediting the same pageviews for each contributing author.

Another bluffing was also noticed, where journalists tend to split the news content over multiple articles to increase their daily and monthly productivity numbers, which results in low pageviews for each article and offers a poor reader experience by having to read multiple articles to get the full details that could have been written in one article.

To overcome this issue, a new criterion named "impact factor" was developed to correlate the number of articles written by the same author to the generated pageviews from those articles. To achieve a high impact factor, journalists must write fewer articles that contain more meaningful content that attract more readers.

4.5 Second Interval (Rerun of The Gamification Experience)

With the apparent success of the pilot project, the newspaper management decided to rerun the gamification experience for three months, taking into consideration the notes from the pilot phase.

Moreover, by the end of the three months, there was a noticeable and sustained positive change in the journalists' interest to continue using digital platforms.

4.5.1 Outcomes of the Second Interval

The number of journalists using the digital platforms increased by 21% over the pilot phase and 247% over the period before running the pilot gamification experience.

The number of published articles increased by 58% over the pilot phase and 116% over the period before running the pilot gamification experience.

Table 8. Impact of gamification on KDP's compared to pilot and baseline.

S	Key Data Points (KDPs)	Impact compared to pilot run	Impact compared to baseline
1	Number of monthly used digital platforms accounts	Increase by 21%	Increase by 247%
2	Number of monthly articles published on the digital platforms	Increase by 58%	Increase by 116%
3	Monthly pageviews on the digital platforms	Increase by 9%	Increase by 14%
4	Monthly pageviews generated from search engines results	Increase by 13%	Increase by 30%
5	Percentage of pageviews from search results to total pageviews	Increase by 2%	Increase by 4%
6	Number of videos broadcasted on social networks	Increase by 18%	Increase by 85%

The study relied on both quantitative and qualitative data sets. The quantitative data was collected daily for each individual journalist's KDPs and aggregated monthly. Additionally, qualitative data was gathered through interviews to gain insights into the application of gamification to the digital transformation process.

4.6 Key Data Points (KDPs) Statistics

Key Data Points (KDPs) statistics five months before applying gamification, and after first and second intervals where gamification were applied.

To evaluate the statistical significance of KDPs before and after gamification, a one-tailed paired T-test was conducted comparing each KDP five months prior to and following gamification implementation.

The table below presents the statistical significance of changes in key data points (KDPs). The KDPs "number of active accounts", "number of published articles", and "pageviews from search" indicate a statistically significant difference, while the KDPs "total number of pageviews" and "number of videos" did not show statistically significant changes.

Table 9. Paired one tail T-Test for KDPs

Key Data Points (KDPs)	P-value
Number of active accounts	0.001166275
Number of published articles	0.008660202
Total pageviews	0.179078731
Pageviews from search	0.000791716

continued on following page

Table 9. Continued

Key Data Points (KDPs)	P-value
% of search pageview	0.009681918
Number of videos and live broadcasting	0.1099871

Table 10. Percentage of mean increase before and after gamification

Percentage of positive mean increase comparing 5 months before and after gamification	
Number of active accounts	**203%**
Number of published articles	80%
Total pageviews	16%
Pageviews from search	53%
% of search pageview	33%
Number of videos and live broadcasting	46%

This suggests that there is strong evidence that gamification led to a noticeable increase in the number of participating journalists and the quality of editorial content measured by "Number of active accounts", "number of published articles", "pageviews from search" which are the core KDPs for digital transformation process; but the mean of total pageviews increased by 16% and the number of produced videos by 46% are not considered statistically significant.

4.7 Survey Results

After finalizing the two intervals, a survey was distributed to participants to capture their feedback on the experiment. The survey is composed of three sections: basic information, gamification-related questions, and an open-ended section for feedback.

The survey respondents are 45 respondents who answered 8 multiple choice and 2 open ended question. The results of the survey analysis are presented below.

The gender distribution of the survey participants mirrors that of the news organization population, with approximately 70% male and 30% female respondents. This alignment with the target population enhances the generalizability and representativeness of the survey findings.

The majority of respondents (80%) confirmed that adding a dashboard screen that showed their points, progress, and rank motivated them to create and publish more digital content. Nearly 75% agreed that the presence of a daily leaderboard that showed the names of the top-ranked journalists made them more engaged in creating digital content. The same percentage agreed that the monthly leaderboards were beneficial in pushing them to continue producing more digital content.

Additionally, 83% of respondents indicated that the competition made them more interested in learning new ways to create digital content. Nearly 81% reported that they noticed an improvement in their digital content production efficiency compared to the period before the experiment.

The survey also queried participants, what they thought were the best elements used in the new experiment. The results showed that the most popular elements were:

- Showing the number of views (37 votes, 94.9%)
- Showing the ranking (31 votes, 79.5%)
- Publishing leaderboard (29 votes, 74.4%)

While the least popular elements were:

- Using symbols to represent the best (8 votes, 20.5%)
- Awarding badges (9 votes, 23.1%)

The survey included additional open-ended questions which are:

- What is your evaluation of the new experience and to what extent did you benefit from it?

The following is a sample of the responses:

- o "The experience was excellent. It showed the colleagues' progress and created a challenge among them."
- o "It was very good and made a big difference in understanding the readings."
- o "It was excellent. I benefited greatly from knowing the ranking, the number of readings, and the spirit of competition."
- o "It was excellent. There was a new type of competition that we were not used to."
    - What are your recommendations for improving the experience in the next period?

The following is a sample of the responses:

- o "More details to clarify the growth details of the numbers, for example showing the visitor source"
- o "Adding new ideas."

- o “Adding a ranking by editorial section.”
- o “It lacks more comparisons and data from previous months.”

The respondents appreciated the way dashboards and leaderboards showed the progress and created a challenge among them. They also found that it made a big difference in understanding the readings and they benefited greatly from knowing the ranking, the number of readings, which increased the spirit of competition.

However, some respondents suggested that the experience could be improved by providing more details to clarify the growth details of the numbers, such as showing the visitor source. They also suggested adding new ideas and adding a ranking by editorial section. They also mentioned that it would be helpful to have more comparisons and data from previous months.

Overall, the respondents were positive about the experiment and found it to be beneficial. However, they also provided some suggestions for how it could be improved. These suggestions could be helpful in making the experience even more effective.

The findings of the survey suggest that the new gamification experiment was effective in motivating journalists to create and publish more digital content. The elements that were most popular with the journalists were those that gave them visibility into their performance and allowed them to compete with their peers.

DISCUSSION

The research employed an experiment which was conducted in a news/media organization that had previously experienced several unsuccessful digital transformation projects. The study involved the implementation of gamification techniques in the digital transformation process. Data was collected through daily key data points (KDPs) and monthly aggregations, as well as interviews with journalists who were involved in the digital transformation process. The data was analyzed using both quantitative and qualitative methods to identify the impact of gamification on the success rate of the digital transformation process.

The experiment was conducted in two intervals. The first interval was conducted over a period of two months and involved a pilot implementation of gamification techniques in the digital transformation process. The second interval was conducted over a period of three months and involved changes in the gamification techniques implemented in the first interval.

This study contributes to scientific knowledge by providing empirical evidence on how gamification techniques can be leveraged to drive successful digital transformations in news/media organizations. It also highlights how KDPs can be used to evaluate success rates and provides insights and understanding of how gamification techniques can be used to drive successful digital transformations.

5.1 Summary of The Key Findings

The results section presents data that indicates a significant positive increase in the KDPs during the first interval. This suggests that the gamification techniques used were effective in introducing and changing the journalists' behavior to accept the digital transformation process. In the second interval, with the modified gamification techniques, the KDPs suggests that the modified gamification techniques were able to sustain and build upon the positive changes captured in the first interval.

The significant positive increase in the KDPs during in first interval also suggests that the gamification techniques were successful in increasing engagement and motivation among journalists.

In the second interval, with the modified gamification techniques, the KDPs continued to increase and provided a solid foundation that the changes captured in the first interval were sustained and built upon in the second interval.

The data presented in the results section supports the initial hypothesis that gamification techniques can be an effective tool for accelerating a successful digital transformation process by introducing new behaviors and change existing ones, leading to increased engagement and motivation among journalists. The results also confirm that gamification can be used to sustain these changes over time. These findings are noteworthy for organizations seeking to go through a digital transformation process. They suggest that gamification can be used to effectively increase engagement and motivation among employees and decrease resistance to change during the digital transformation process.

The results of this study are in line with those of other researchers in various fields. For example, (Manzano-León et al., 2021) conducted a systematic literature review to investigate the impact of gamification in education on student motivation and academic performance. The review found that gamification has a positive impact on academic performance, commitment, and motivation of students.

In the fitness field, (Tu et al., 2019) conducted research on the impact of gamification on physical activity and concluded that gamification can have a positive impact on consumers' physical activity levels. By incorporating game elements such as reward points, badges, and virtual goods, fitness apps can make staying physically active more enjoyable and increase consumers' engagement in physical activity.

The survey respondents' answers confirmed that the gamification concepts applied appealed to nearly 80% of the experiment participants and changed the behavior of 83.7% of participants to become more interested in learning new techniques and methods for producing digital content. Additionally, 81.9% reported an increase in their productivity after the experiment compared to the previous period.

The most effective gamification design elements used were Points followed by Ranking, Leaderboards, badges, and prizes.

5.2 Limitations

The study has some limitations that should be considered when interpreting the results. Firstly, the study was conducted in a single organization in the news/media sector and may not be directly generalizable to organizations in other business verticals.

Secondly, the study did not investigate the impact of gamification techniques on other aspects of digital transformation such as technology adoption or organizational culture. Thirdly, the study used a limited set of KDPs and did not explore other potential outcomes of gamification techniques.

Thirdly, while efforts were made to control all aspects of the experiment, external factors in real-life settings may positively or negatively impacted the results. For example, a breaking news or a change in Google's search engine article ranking factors, Facebook reach could have impacted some KDPs for dew days, and those factors are outside the control of the researcher.

Finally, the occurrence of bluffing to increase points during the first interval posed a threat to the fairness of the gamification experience. However, this was addressed by modifying the game mechanics in the second interval. It is likely that other forms of cheating or bluffing may still exist, and these need to be addressed to ensure that the gamification experience is fair and flawless.

There is also a limitation in research scope as it did not investigate the impact of gamification techniques on other stakeholders such as readers or advertisers. also, the study did not investigate the impact of gamification techniques on journalists' well-being or job satisfaction.

CONCLUSIONS

This study aimed to investigate the potential of gamification as a mechanism to enhance the success rate of digital transformation (DT) initiatives within the news and media sector, an industry facing substantial challenges in adopting digital processes. The research involved the application of gamification techniques over a five-month period, assessing their impact on behavioral changes and engagement among journalists.

The results indicate that gamification can significantly support digital transformation by promoting increased engagement, motivation, and behavioral adaptation. Game elements such as points, rankings, and leaderboards led to statistically significant improvements in Key Data Points (KDPs) during both experimental intervals, particularly in the initial phase. Additionally, survey responses highlighted that a majority of participants found gamification appealing and useful in fostering the adoption of new digital practices. This suggests that gamification can mitigate resistance to change and help sustain the momentum necessary for successful digital transformation.

Despite the positive outcomes, the study acknowledges several limitations. Conducted within a single organization and focused on specific industry-related KDPs, the findings may not be fully generalizable across different sectors. Furthermore, the study did not examine the broader impacts of gamification on elements such as technology adoption, organizational culture, or employee well-being. These limitations present opportunities for future research, particularly in diverse business contexts, to explore the comprehensive effects of gamification on organizational performance and cultural transformation.

In conclusion, this research contributes to the body of knowledge on gamification by providing empirical evidence of its efficacy in facilitating digital transformation within the news/media sector. It offers practical insights for practitioners seeking to leverage gamification to drive successful organizational change, while also emphasizing the need for further investigation into its broader applications across different industries and contexts.

REFERENCES

Altmeyer, M., Schubhan, M., Krüger, A., & Lessel, P. (2021). *A long-term investigation on the effects of (personalized) gamification on course participation in a gym*. http://ceur-ws.org

Ansari, S. Khot1, R., Shaikh, Z., Yadav, P., & Pirani, Z. (2022). A Comprehensive Review of Gamification in Healthcare: Incentives in Mobile Healthcare App. *International Research Journal of Engineering and Technology*. www.irjet.net

Barann, B., Hermann, A., Cordes, A. K., Chasin, F., & Becker, J. (2019). Supporting Digital Transformation in Small and Medium-sized Enterprises: A procedure model involving publicly funded support units. *Proceedings of the Annual Hawaii International Conference on System Sciences, January*, 4977–4986. DOI: 10.24251/HICSS.2019.598

Bloomberg. J. (2018). *Digitization, Digitalization, And Digital Transformation: Confuse Them At Your Peril*. Forbes.Com. https://www.forbes.com/sites/jasonbloomberg/2018/04/29/digitization-digitalization-and-digital-transformation-confuse-them-at-your-peril/

Butler, T. (2020). What's Next in the Digital Transformation of Financial Industry? *IT Professional*, 22(1), 29–33. DOI: 10.1109/MITP.2019.2963490

Chung, G. H., Choi, J. N., & Du, J. (2017). Tired of innovations? Learned helplessness and fatigue in the context of continuous streams of innovation implementation. *Journal of Organizational Behavior*, 38(7), 1130–1148. DOI: 10.1002/job.2191

Cornia, A., Sehl, A., & Nielsen, R. K. (2020). 'We no longer live in a time of separation': A comparative analysis of how editorial and commercial integration became a norm. *Journalism*, 21(2), 172–190. DOI: 10.1177/1464884918779919

Deterding, S., Dixon, D., Khaled, R., & Nacke, L. (2011). From game design elements to gamefulness: Defining "gamification." *Proceedings of the 15th International Academic MindTrek Conference: Envisioning Future Media Environments, MindTrek 2011, September*, 9–15. DOI: 10.1145/2181037.2181040

Diener, F., & Špaček, M. (2021). Digital transformation in banking: A managerial perspective on barriers to change. *Sustainability (Basel)*, 13(4), 1–26. DOI: 10.3390/su13042032

Firican, D. A. (2023). Digital Transformation and Digital Culture: A Literature Review of the Digital Cultural Attributes to Enable Digital Transformation. *Proceedings of the International Conference on Business Excellence, 17*(1), 791–799. DOI: 10.2478/picbe-2023-0073

Forbes Technology Council, & Forbes Expert Panel. (2021, June 15). *13 Industry Experts Share Reasons Companies Fail At Digital Transformation.* https://www.forbes.com/sites/forbestechcouncil/2021/06/15/13-industry-experts-share-reasons-companies-fail-at-digital-transformation/

Friedrich, J., Becker, M., Kramer, F., Wirth, M., & Schneider, M. (2020). Incentive design and gamification for knowledge management. *Journal of Business Research, 106*(November 2017), 341–352. DOI: 10.1016/j.jbusres.2019.02.009

Gong, C., & Ribiere, V. (2021). Developing a unified definition of digital transformation. *Technovation*, 102, 102217. Advance online publication. DOI: 10.1016/j.technovation.2020.102217

Guandalini, I. (2022). Sustainability through digital transformation: A systematic literature review for research guidance. *Journal of Business Research*, 148, 456–471. DOI: 10.1016/j.jbusres.2022.05.003

Hammedi, W., Leclercq, T., & Poncin, I., & Alkire (Née Nasr), L. (. (2021). Uncovering the dark side of gamification at work: Impacts on engagement and well-being. *Journal of Business Research*, 122, 256–269. DOI: 10.1016/j.jbusres.2020.08.032

Hartmann, A., & Gommer, L. (2021). To play or not to play: On the motivational effects of games in engineering education. *European Journal of Engineering Education*, 46(3), 319–343. DOI: 10.1080/03043797.2019.1690430

Hartt, M., Hosseini, H., & Mostafapour, M. (2020). Game On: Exploring the Effectiveness of Game-based Learning. *Planning Practice and Research*, 35(5), 589–604. DOI: 10.1080/02697459.2020.1778859

Hernández-Fernández, A., Olmedo-Torre, N., & Peña, M. (2020). Is classroom gamification opposed to performance? *Sustainability (Basel)*, 12(23), 1–20. DOI: 10.3390/su12239958

Jones, M. D., Hutcheson, S., & Camba, J. D. (2021). Past, present, and future barriers to digital transformation in manufacturing: A review. *Journal of Manufacturing Systems*, 60, 936–948. DOI: 10.1016/j.jmsy.2021.03.006

Karimi, J., & Walter, Z. (2021). The role of entrepreneurial agility in digital entrepreneurship and creating value in response to digital disruption in the newspaper industry. *Sustainability (Basel)*, 13(5), 1–26. DOI: 10.3390/su13052741

Khasawneh, Y. J. A., Khasawneh, N., & Khasawneh, M. A. S. (2024). Exploring the long-term effects: Retention and transfer of skills in gamified learning environment. *International Journal of Data and Network Science*, 8(1), 195–200. DOI: 10.5267/j.ijdns.2023.10.004

Lewis, S. C., & Westlund, O. (2015). Actors, actants, audiences, and activities in cross-media news work: A matrix and a research agenda. *Digital Journalism (Abingdon, England)*, 3(1), 19–37. DOI: 10.1080/21670811.2014.927986

Li, F. (2020). Leading digital transformation: Three emerging approaches for managing the transition. *International Journal of Operations & Production Management*, 40(6), 809–817. DOI: 10.1108/IJOPM-04-2020-0202

Lucas, H. C.Jr, & Goh, J. M. (2009). Disruptive technology: How Kodak missed the digital photography revolution. *The Journal of Strategic Information Systems*, 18(1), 46–55. DOI: 10.1016/j.jsis.2009.01.002

Manzano-León, A., Camacho-Lazarraga, P., Guerrero, M. A., Guerrero-Puerta, L., Aguilar-Parra, J. M., Trigueros, R., & Alias, A. (2021). Between level up and game over: A systematic literature review of gamification in education. *Sustainability (Basel)*, 13(4), 1–14. DOI: 10.3390/su13042247

Mardanov, I. (2020). Intrinsic and extrinsic motivation, organizational context, employee contentment, job satisfaction, performance and intention to stay. *Evidence-Based HRM*, 9(3), 223–240. DOI: 10.1108/EBHRM-02-2020-0018

Nah, F. F. H., Eschenbrenner, B., Claybaugh, C. C., & Koob, P. B. (2019). Gamification of enterprise systems. *Systems*, 7(1), 1–21. DOI: 10.3390/systems7010013

Nel, F. (2010). Where else is the money?: A study of innovation in online business models at newspapers in britain's 66 cities. *Journalism Practice*, 4(3), 360–372. DOI: 10.1080/17512781003642964

Newman, N., Richard Fletcher, with, Kalogeropoulos, A., & Kleis Nielsen, R. (2019). *Reuters Institute Digital News Report 2019.*

Nguyen, N. T. D., & Aoyama, A. (2014). Achieving efficient technology transfer through a specific corporate culture facilitated by management practices. *The Journal of High Technology Management Research*, 25(2), 108–122. DOI: 10.1016/j.hitech.2014.07.001

Pathiranage, Y. L., Jayatilake, L. V. K., & Abeysekera, R. (2020). A literature review on organizational culture towards corporate performance. *International Journal of Management. Accounting & Economics*, 7(9), 522–544.

Perryer, C., Celestine, N. A., Scott-Ladd, B., & Leighton, C. (2016). Enhancing workplace motivation through gamification: Transferrable lessons from pedagogy. *International Journal of Management Education*, 14(3), 327–335. DOI: 10.1016/j.ijme.2016.07.001

Ramesh, N., & Delen, D. (2021). Digital Transformation: How to Beat the 90% Failure Rate? *IEEE Engineering Management Review*, 49(3), 22–25. DOI: 10.1109/EMR.2021.3070139

Robson, K., Plangger, K., Kietzmann, J. H., McCarthy, I., & Pitt, L. (2015). Is it all a game? Understanding the principles of gamification. *Business Horizons*, 58(4), 411–420. DOI: 10.1016/j.bushor.2015.03.006

Svahn, F., Mathiassen, L., & Lindgren, R. (2017). Embracing digital innovation in incumbent firms: How Volvo Cars managed competing concerns. *Management Information Systems Quarterly*, 41(1), 239–253. DOI: 10.25300/MISQ/2017/41.1.12

Torlaković, A., Veselinović, L., & Bičo Ćar, M. (2022). The Application of Gamification in a Business Environment: The Most Common Themes and Elements. *Lecture Notes in Networks and Systems, 472 LNNS*, 996–1002. DOI: 10.1007/978-3-031-05230-9_117

Torrijos, J. L. R. (2020). Gamification of sports media coverage: An infotainment approach to olympics and football world cups. *Communicatio Socialis*, 33(1), 29–44. DOI: 10.15581/003.33.1.29-44

Tu, R., Hsieh, P., & Feng, W. (2019). Walking for fun or for "likes"? The impacts of different gamification orientations of fitness apps on consumers' physical activities. *Sport Management Review*, 22(5), 682–693. DOI: 10.1016/j.smr.2018.10.005

Vial, G. (2021). Understanding digital transformation. *Managing Digital Transformation*, 13–66. DOI: 10.4324/9781003008637-4

Vos, T. P., & Perreault, G. P. (2020). The discursive construction of the gamification of journalism. *Convergence (London)*, 26(3), 470–485. DOI: 10.1177/1354856520909542

Wade, M., & Shan, J. (2020). Covid-19 has accelerated digital transformation, but may have made it harder not easier. *MIS Quarterly Executive*, 19(3), 213–220. DOI: 10.17705/2msqe.00034

Wang, Z., Li, M., Lu, J., & Cheng, X. (2022). Business Innovation based on artificial intelligence and Blockchain technology. *Information Processing & Management*, 59(1), 102759. Advance online publication. DOI: 10.1016/j.ipm.2021.102759

Chapter 3
AI-Driven Innovations for Sustainable Tourism

Rina Daline
https://orcid.org/0000-0002-1979-0781
Christ University, India

Joby Thomas
https://orcid.org/0000-0002-4493-1048
Christ University, India

Mallika Sankar
Christ University, India

ABSTRACT

Technology and Tourism blend and bestow a plethora of productive outcomes to the Tourism industry. Sustainable tourism refers to sustainable practices associated with the environment, economy, and social aspects and should strike a balance between them to achieve long-term sustainability. The United Nations rolled out 17 sustainable development goals to protect the planet by 2030. On the other hand, the deterioration of tourism resources arises as a significant concern as they enter into the decline stage of the life cycle. Technological applications helped to tackle this deteriorating situation. One of the immersive technologies Virtual Reality aids in improving sustainable tourism and proved to be a suitable one for environmentally sustainable travel. The virtual tours of the vulnerable tourist destinations mitigate the threatening situation by allowing the destination to rejuvenate and thus contribute towards its sustainability.

DOI: 10.4018/979-8-3373-1132-6.ch003

INTRODUCTION:

According to the United Nations Environment Program (UNEP) and United Nations World Tourism Organization (UNWTO), Sustainable tourism can be defined as tourism that is responsible for the present and future impacts on the environment, economy, and social. It strives toward fulfilling the requirements of tourists, host communities, the environment, and the tourism industry (UNEP & UNWTO, 2005). Sustainable tourism is all about the sustainable practices adopted by the tourism industry to increase the positive impacts and decrease the negative impacts ("What is Sustainable Tourism?", n.d.). The negative impacts consist of economic leakage, mass tourism beyond the carrying capacity, environmental degradation, etc. The positive impacts created by sustainable practices contribute to natural and cultural heritage conservation, employment creation, income generation, wildlife conservation, etc. (Frent, 2016).

Virtual Reality emerges as a potential instrument in promoting the Sustainability of tourism destinations. It promotes Sustainability in terms of both environmental and socio-values perspectives. It also promotes the continuation of tourist destination areas and prevents them from entering the decline stage (Gegung, 2021). The Covid-19 Pandemic helps in rejuvenating the tourism destinations in the decline stage. Due to the shutdown of tourism destinations, the negative impacts contributed by various tourism activities have been mitigated to a greater extent. Virtual Reality tourism enables environment-friendly travel at a lower price. The Sustainability of the tourist destination is made intact with virtual travel by enhancing the lifecycle of the destination (Wiltschier & Clarke, 2016). In addition to that, Virtual Tourism bolsters sustainable tourism by mitigating greenhouse gas emissions and transportation pollution (Junyu Lu et al., 2022) and rejuvenates the destination into a green, resilient and inclusive one. Virtual tourism upscales a feeble destination into a more resilient one, and the negative impacts are brought down more splendidly than the traditional way of visiting.

In Amusement Parks, Virtual Reality has been incorporated with roller coaster rides. The prime purpose is to enhance the immersive experience for the thrill-seekers and to increase the Sustainability of the product. The life cycle of the roller coaster product may be renewed with the incorporation of technology applications (Malcolm, 2019). The heritage monuments stand out as an epitome of our ancestor's intelligence and help us to know about their culture. Many of the ancient traces have been studied with the help of historical sites or monuments. But many heritage sites are in the decline stage from a global perspective. Natural catastrophe or human conflicts could be one of the main reasons for the deterioration of heritage sites. This problem was addressed by technology expert CyAark, a California-based Three Dimensional (3D) laser company. Their primary objective is to safeguard the world's heritage through 3D and Virtual Reality technologies. The archaeological wonders who are found to be in peril are scanned with the assistance of 3D laser scanning technologies and then re-created in the Virtual Platform. The endangered world's heritage sites could be sustained by shifting them to Virtual space ("A virtual reality of endangered", 2019).

Tourism is finding better approaches to capture tourists' attention in all reasonable ways. It has come up with an imaginative thought to win the certainty of the clients. Smart tourism offers intellectual access to large destinations, museums, UNESCO World heritage sites, ancient monuments, etc. (Bill Zhu, 2020). Novel technologies have been incorporated into the tourism and hospitality industry. Hospitality offers real-time promotions, conferences, automatic parking, smart check-in, and check-out practices to enhance service quality. As most of the operations are turning digital, the hospitality industry leverages the network to 5G and wifi-6 for quick and seamless operations. Artificial Intelligence is the core of technology and plays an important role in automating the operation of the hospitality industry (Lau, 2020). The Hospitality Industry is going through a digital transformation to win back the trust and confidence of the customers. Robots are launched for humanless services and maintain social distancing protocols (Subai et al., 2020).

Virtual Reality

Virtual Reality is one of the prominent immersive technologies (Fritz et al., 2005) that allows users to immerse and access information in the virtual interactive world (Kounavis et al., 2012). Virtual Reality offers a fully immersed experience to the user in a computer-simulated environment (Fritz et al., n.d.) where it alters the interfaces between the user and technology through expressive information (Erenay & Hashemipour, 2003). The system-generated environment stimulates a comprehensive three-dimensional environment (Mitchell, 2020).

Virtual Reality is a computer-generated stimulus that transports people from the physical to the virtual environment (Hobson & Williams, 1995), and they can experience the imaginary environment through their virtual Presence. The virtual environment acts as an enabler to escape people from the physical environment (Desai, 2014). However, escapism from day-to-day activities is a part of Virtual Reality. Instead, it helps stimulate the senses and enables an opportunity for virtual interaction (Guttentag, 2010). Immersion and Presence are the two major components of Virtual Reality (Slater and Sanchez-Vives, 2016). Immersion is objective, while Presence is subjective.

Panorama of Virtual Tourism

Virtual Tourism has been used predominantly to create and enhance the demand for a tourism product. Promotions and Marketing activities via virtual tours enhance the destination images and influence people to visit the real world (Jarratt, n.d.). Global tourist attractions, Museums, Ancient Monuments, UNESCO World heritage sites, National Parks, Zoos, etc., are upgraded to a Virtual environment to make them available for all (Kemska, 2021). The Vulnerable heritage sites and Cultural assets can be safeguarded with the assistance of Virtual Reality. The replacement of typical tourism with virtual tours to such places will protect the resources (Voronkova, 2018). Virtual tours to Forbidden Heritage places have got a tremendous demand among travelers (Kemska, 2021).

Sustainability and Virtual Reality Tourism:

Virtual Reality has great potential in various aspects of the tourism industry like education, marketing, cultural heritage, and sustainability. It stands out from other immersive technologies as it provides immersive and interactive experiences along with dissemination of information which is lacking in other technologies (Yung &Khoo-Lattimore, 2019). The 3D Visualization features the possible sustainable development practices associated with environmental, economic, and social aspects of a destination (Bryan et al. 2011)

Virtual Reality (VR) has brought in tremendous applications in the field of Tourism. VR technology has been used in the phases of planning, management, entertainment, preserving heritage resources, marketing, promotions, etc. (Wiltschier & Clarke, 2016). Tourism resources are the prime attractions that connect the tourist and the tourism. Tourism resources are perishable and non-renewable in nature. Any tourism product will go through different stages in the tourism life cycle. A tourism destination, for instance, would go through 6 stages commencing from Exploration, Involvement, Development, Consolidation, Stagnation, Decline, or Rejuvenation (Butler, 1980). A tourism product will go into a decline stage when it is overused or explored beyond the limit. Overcrowding is another major problem that constitutes the decline of a destination. A destination's life cycle could be extended with the aid of technological applications.

Sustainable development concerns fulfilling the needs of the present scenario without depleting the resources of the future generations (Brundtland 1985; Ryan 2002). Sustainable tourism measures will be devised based on the sustainable development concepts. Achieving sustainable tourism through technological applications should add value to the sustainable measures. Virtual Reality, when incorporated into tourism, bestows the following benefits related to its sustainability. The VR technology should mitigate the negative impacts of tourism-related social, economic, and environmental factors. This immersive technology can be used as a powerful tool for the preservation and conservation of natural resources, and it should be ethical in nature, which enables ethical behavior towards the natural environment and resources. Virtual Reality not only helps in conservation but also addresses the overtourism challenges by enabling an online virtual tourism experience (Bal & Czalczynska-Podolska, 2020).

Tourism Product

A tourism product is a mix of various combinations of elements and components that aim to satisfy the customers' needs. It is an amalgamation of services and experiences that were bound together to cater the needs of the customers (Ashworth & Voogt, 1990). The tourism industry comprises both tangible and intangible products. For example, accommodations, hotels, air tickets, transfers, are considered tangible products or components of tourism products that can be easily seen and touched, and experienced. On the other hand, attractions, and amenities, are a few examples of intangible tourism products. The tangible and intangible products are complementary to each other and work together to provide a complete travel experience to their customers. They are considered an integral part of the tourism experience. These products differ in their characteristics for example in terms of physical presence but aim to enhance customer satisfaction as a whole (Albayrak, 2010).

Tourism product is designed to provide complete satisfaction to the customers from the origin to the destination and back to the origin. It focuses majorly on customer satisfaction by providing experiences at the pre-trip, in-trip, and post-trip periods. As higher customer satisfaction is the key element for the successful operation of the tourism industry, tourism products are designed to achieve that. The tourism products can be defined as the collective sum of physical and psychological satisfaction associated with the customers (Tourists) during all stages of travel. For example, a tourist destination can be considered as an aggregation of individual products and experiences that result in the overall experience of that destination (Murphy et al., 2000). Tourism products can be tangible, intangible, or a combination of both forms. In the tourism industry, the majority of the products coexist together (Vassiliadis, 2008).

Tourism Product Life Cycle

Tourism Area Life Cycle (TACL) is the prominent and most influential model for evaluating tourism products, explaining product growth, and tourist and tourist area life cycle as a whole. It is considered as one of the most successful and robust frameworks used in the field of tourism. A tourism product, for example, a tourist destination can be evaluated with the help of the Tourism Area Life Cycle model. This conceptual framework assesses the tourism product starting from the origin till the last stage (Baum, 1998). The tourism product life cycle framework evolved from the product life cycle model used primarily in the areas of business and management. This product life cycle model has undergone a makeover to evaluate and explain the development, and changes that take place in tourism destinations. The Tourism Product Life Cycle was designed to study the various stages of the tourist destinations. The stages proposed were exploration, involvement, development, consolidation, stagnation, rejuvenation or decline (Butler, 2024).

Exploration

Exploration is the first and foremost stage of the tourism product life cycle. This stage provides opportunities for new tourism products to flourish and get discovered by tourists. A tourist destination at this stage will be explored by a few tourists who are willing to visit new or unexplored destinations. Tourists looking for unique natural landscapes, historic places, pristine beaches, ethnic groups, and cultural importance will be visiting those destinations. Since those destinations are in the discovery stage there will be limited accessible options to visit. Those destinations are yet to be developed with infrastructure, amenities, and accessibility features. Even local communities are unaware of the business opportunities and the benefits tourism provides. This stage provides opportunities for the destination and local community to transform. Environmental, economic, and social impacts associated with the destination are less at this exploration stage. A serene and optimistic relationship exists between the tourists and the local community (Truyols, 2023).

Role of Virtual Reality at the Exploration Stage

Since the tourism product has limited discoverability to tourists at this stage, Virtual Reality technology can be used as a marketing tool to promote the aspiring destination. A tourist destination can be recreated in a virtual space and showcased to many people. This will create an interest among people to visit the new destination in the real world. VR will act as an influencing tool to explore new tourism products.

Involvement

Involvement is the second stage of the tourism area life cycle. At this stage, the new tourist destination will be visited by a good number of visitors. The number of tourists will be gradually increasing in visiting the destination compared to the exploration stage. The local community will be aware of the benefits bestowed by tourism activities and will start engaging in them. Accommodations, restaurants, and guest houses started to emerge in the destinations to cater to the needs of the tourists. The government will focus on tourism-related infrastructure development and similarly investors start to show interest in future development projects.

Role of Virtual Reality at the Involvement Stage

Virtual reality can be greatly used in tourism promotional activities. As the stakeholders show great interest in establishing tourism activities it is important to draw the attention of the tourists. The government can promote the tourist destination through Virtual Reality on the Ministry of Tourism website and the state tourism website. Virtual video tours can be posted on the government tourism portals to grab the attention of visitors. This will motivate travellers or tourists to visit the tourist destination in the real world. Virtual reality can be used to motivate and influence the travel decisions of people. Virtual reality attracts potential visitors to visit the destination and indirectly pays the way for the economy's income generation.

Development

The third stage of the tourism product life cycle leads to an increase in visitors with the increase in promotional activities by tourism companies, Destination Management Organizations, and tourism promoters. Infrastructures will be given full importance at this stage where accessibility such as transportation will be established to ease the difficulties of the visitors in reaching out to the destination. Many tourism enterprises and companies started to take over the tourism businesses from the local community as the destination entered into the development stage. This phase grabs the attention of many foreign investments which leads to the launch of new international hotel chains, food courts, etc. Their ultimate aim is to provide a high quality of service in return for high profit from the destination.

Role of Virtual Reality at the Development Stage

This is the phase where the potential of the destination is overlooked by global companies and start establishing their premises at the destination. Virtual Reality has proved to be one of the best immersive technologies and this helps in high-end marketing and promotional strategies. The large chain enterprises can implement Virtual Reality to upscale their businesses. For example, a hotel can use VR to promote its rooms and premises through VR tours. This helps in increasing the booking rate and visibility of the property.

Consolidation

In, the fourth stage of the tourism product life cycle tourism becomes the main feature for the economic growth of the local community. The host community will be majorly dependent on tourism activities as they become the main source of income. Since the destination becomes tourism-dependent increasing tourist visits and increasing per capita expenditure of the tourist becomes the prime thing to maintain the income flow. Due to increased tourism flow, infrastructure may be vandalized, and the relationship between locals and tourists may worsen at this stage of the cycle.

Role of Virtual Reality at the Consolidation Stage

Virtual Reality can be used to sustain the income flow during the off-season as well by promoting products, handicrafts, and souvenirs on the virtual platform. Virtual stores can be set up to sell tourism products during the off-season. This could retain the sales and tourist expenditure even during the off-season. Virtual Reality technology can generate income for the economy even if the tourists don't visit the destination physically.

Stagnation

The fifth stage of the tourism product life cycle will hold the maximum carrying capacity of the people. When the tourism product reaches this stage, demand begins to reduce and stagnates. At this point, the tourist destination might utilize its maximum potential, and tourists will no longer be attracted to the product. As the tourists have drained the destination's maximum potential, economic, social, and environmental problems will rise at this stage. The destination will start to lose its charm and fascination due to over-tourism and over-exploitation of resources. Mass tourism will pressurize the environment and negatively impact the destination, such as habitat destruction, soil erosion, pollution, overconsumption of resources, water contamination, displacement of locals, etc. Sustainability issues will reach the park at this stage. In addition to that, retaining old tourists and attracting new tourists might be the biggest challenge for tourism enterprises.

Role of Virtual Reality at the Stagnation Stage

Virtual reality can safeguard the natural resources and other negative impacts created by tourists to a great extent. As the land carrying capacity is exceeded at this stage, Virtual Reality can save the destination by encouraging virtual tourism. Virtual Tourism is defined as the use of technology to experience and explore places in a virtual space without being physically present at the destination. This virtual tourism will create an illusion closer to the real experience so the tourists can feel they are physically transformed there. Virtual tourism offers many benefits to the destination such as cost-friendliness, ease of accessibility, and safety and most importantly it ensures the sustainability of the destination. Since sustainability is the major issue of the destination at the stagnation stage, Virtual Reality may be the best solution to tackle this issue.

Rejuvenation /Decline

The last stage of the tourism product life cycle follows one of the two scenarios. It will go either into the rejuvenation stage or the decline stage. Rejuvenation allows the destination to heal for some time until it comes back to its original form so that it can welcome tourists after some time. The main objective of the rejuvenation is to make the destination green, sustainable, and resilient for the future. In the other case, a tourist destination may left abandoned due to destruction which leads them to enter into the decline stage. Declination also accounts for unexpected events such as the pandemic which ultimately restricts the tourist's visit.

Role of Virtual Reality at the Rejuvenation/Decline Stage

Virtual Reality technology helps to stop the tourist destination enter into the decline stage. When the destination's resources are completely exploited it can encourage tourists to visit the destination online. Virtual tourism can be a better alternative to traditional travel when the destination undergoes rejuvenation. On the other hand, if the destination already entered into abandoned or declined stage, the destination can be recreated online for future generations. 3D scans and pictures of the destination can be collected and imported into virtual technology to recreate the destroyed destination in the virtual space for the future.

Figure 1. Role of Virtual Reality across various stages of Tourism Product Life Cycle

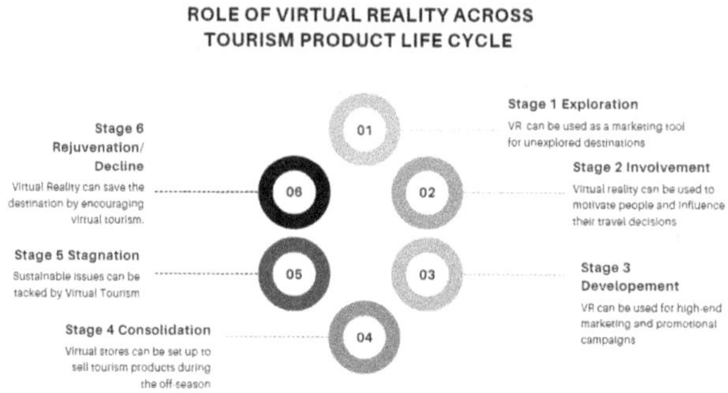

Significance of the Study:

Tourism resources are the enticing features or attractions that build up a constructive tourism destination. A tourist product will survive as long as the resources are available to the people. The fading of tourism resources is the inception of decline. Tourism activities and natural disasters exacerbate the tourism products and hence push them towards the decline stage of the life cycle. Overexploitation is the major concern of tourism resources. The perishable nature of tourism resources also substantiates it. In addition to that, mass tourism or overcrowding at a destination can exceed its carrying capacity leading to deprivation of its resources. All the aforementioned factors contribute to the decline of tourism products. This could be controlled potentially with the help of technological applications. *The potential benefits of Virtual Reality are humongous, and this study helps to understand the various benefits the technology bestows through Sustainability. Several studies have talked about sustainable issues at the tourism industry (Budeanu, 2016;Jovicic, 2014; Streimikiene, 2021) while the research related to Virtual Reality sustainable solutions was not studied. Therefore, this book chapter will address this research gap by providing AI-driven sustainable solutions to the tourism industry.*

Objectives:

- To analyze the role of Virtual Reality technology in sustainable tourism
- To identify how Virtual Reality helps in extending the life cycle of a tourism product.
- To promote sustainable tourism through Virtual Reality Marketing.

Theoretical Background

The theory of Sustainability emphasizes enabling a long-lasting economy and society on a global level. Sustainability should strike a balance between environmental, social, and economic policies to maintain a long-lasting global economy.

Theories on Sustainability aim to incorporate social responses to the issues associated with culture and environment. Social responsibility is integrated with environmental problems and cultural issues. The economic model and Ecological models are designed to sustain both financial capital, natural resource and biological resources, and ecological integrity (Jenkins, 2009).One of the main theories of Sustainability is Green Economics. This theory focuses on the interconnected relationship between the environment and the people. The green economy theory is associated with ecological economics as it states that natural resources have economic value, which is measurable, and they strive towards sustainable development (Halton, 2021). According to the United Nations Environment Programme (UNEP), the Green economy is considered to be "low carbon, resource-efficient, and socially inclusive" ("Green economy", n.d.)

Need for Sustainability in Tourism

Tourism is one of the important factors contributing to the global economy, wealth, employment, and infrastructure development. The tourism industry has been continuously showing positive export growth for the last seven years and has contributed 29% of global export services (UNWTO, 2019). Despite this progress, the industry faces hurdles in achieving the long-term sustainable development goals since it contributes to various environmentally sustainable issues. The industry contributes to pollution, over-development, over-tourism, climate change, etc.

The tourism industry contributes a significant percentage of the global carbon footprint. It is accounted for emitting around 8% of the global greenhouse gases (Lenzen et al., 2018). This calls for the situation to mitigate the global greenhouse gas emissions by adopting environment-friendly practices that lead to the reduction of harmful emissions. Sustainable tourism practices should align with long-term environmental sustainable goals. Sustainability is highly emphasized in tourism as it creates an alarming situation with respect to climate change and associated environmental issues. These issues will be mitigated by possible lifestyle changes and technology that help in reducing global emissions and carbon footprint through novel sustainable practices (Kantar Sifo, 2019).

Attaining Economic Sustainability through Virtual Reality

Virtual Tourism is mainly used for marketing a tourism product, say a tourist destination. They highlight the hidden beauty of a destination and influence tourists to visit the destination in the physical world. An unexplored destination can be promoted through Virtual Reality, which in turn increases the tourist flow (Luhar, 2021). This indirectly contributes to the economy by bringing in novel job opportunities to the local community and new accommodation and transport services to cater to the needs of the tourists. Tourist spending on the destination will create a multiplier effect and contribute to the growth of the economy. Thus Virtual tours act indirectly as a catalyst to boost the multiplier effect on the economy.

High Infrastructure cost is listed to be one of the negative economic effects of tourism (Frent, 2016). This could be omitted in the case of Virtual tours, which require only operational costs. The tourism development fund supported by the government related to infrastructure can be spent wisely for education and health (UNEP, 2016b).

Virtual tourism promotions can help in selling the local crafts or souvenirs of the destination. Geotourism festival has launched virtual booths to promote local businesses. They created an opportunity to sell the local souvenirs on the virtual platform. One of the local businesses is able to sell 200 products through a virtual booth (Adetunji, 2020). Creating a space for local businesses on a virtual platform helps them to scale their business globally by attracting a humongous number of customers and increasing sales.

Negative Socio-Cultural impacts generated by tourism

The social and cultural differences between the tourists and the host community or within the communities lead to negative socio-cultural impacts on the local community (Goeldner and Ritchie, 2006). A meticulous search has been carried out to identify the type of negative social impacts of tourism imposed on the local community. The negative impacts have been classified as a proportion of people in public places, public spaces, Noise, Traffic, crime, Alcohol consumption, Drugs, Inadequate behavior, prostitution, rise in property values, rents, the ratio of permanent residents at the destination, Indifferences in social and moral values, Friction between locals and tourists and reduced quality of life of locals (Deery, Jago, et al., 2012).

Some of the local community are reluctant to invite tourists and share the public spaces with them due to a range of inconveniences caused by them. The troubles may include late-night socializing, parties, and late-night boating, which all contribute to exasperating noise pollution for the local residents. Crime rate increases with the insurgent number of tourists who indulge in drugs, prostitution, and alcohol consumption. Inadequate behavior of tourists troubles the local residents and creates hatred between them (Frent, 2016). In certain situations, there is a possibility of locals shifting to different regions due to tourist arrivals and tourist activities that trouble their lifestyle (Archer, Cooper, et al., 2005). The normal lifestyle of the local residents is questioned with the tourism and the activities associated with it. Differences in language, culture, ethnicity, religion, and lifestyle give rise to cultural clashes between the local community and the tourists. These indifferences act as a hindrance and create discomfort between them (UNEP, 2016b). Cultural shock is another major challenge caused by the cultural indifferences between the locals and tourists.

Tourism is also blamed for modifying or losing the local identity. Loss of authenticity becomes a critical issue when a destination goes through a major changeover to satisfy the diverse needs and demands of the tourists. A high risk of commodification arises out of this since it will be difficult to manage all the demands of the tourists (Frent, 2016).

Sustaining the socio-cultural values in a tourist destination faces a major challenge where there is an influence of western culture over the host culture. It is considered to be one of the detrimental effects left on the host culture. The adoption of western culture by the host community through demonstration and interaction with the tourists leads to the demonstration effect (Winter, 2008).

Ameliorating Socio-Cultural Sustainability through Virtual Reality

All the inconveniences and troubles such as Inadequate tourist behavior, disturbing local lifestyle, cultural shock, cultural indifferences, and the associated issues and crime rate caused by tourism will be eliminated through Virtual Reality. A tourist destination that is not welcomed by the host community can be shifted to a virtual platform by enabling virtual tours of that destination. The destination will be recreated on a virtual platform with the help of Virtual Reality immersive technology and enables tourists to experience the attractions in their own place rather than disturbing the local community. Virtual destinations will bring a lot of benefits to the community in terms of safeguarding their culture and social life from disruption. Virtual tours can be replaced for destinations that experience severe social and cultural impacts from tourism activities.

The impact of the demonstration effect could be controlled with the help of Virtual Reality as virtual travel keeps the host culture intact. There is less probability of the mix of host cultures losing its authenticity. Virtual tours could be preferred for places where the influence of the demonstration effect is huge and arduous to control (Gegung, 2021). Overall, virtual tours will be a win-win situation for both the local community and the tourists. The destination will be left intact free from all illegal activities, and the crime rates will eventually drop down, assuring the normal healthy life of the local residents. The demonstration effect can be mitigated effectively with the help of virtual reality technology.

Negative Environmental Impacts Caused by Tourism

Tourism activities are directly proportional to environmental sustainability. Poor and deficient tourism development plans will contribute to the destruction of a destinations' environment (Archer, Cooper, et al., 2005). The environment is adversely affected by littering, congestion of humans and transportation, soil erosion, floods, water scarcity, pollution, damage to wildlife habitats, threats to the ecosystem, etc. (Mason, 2003). Bike or Automobile tourism plays a significant part in contributing to the enormous amount of pollution (Snak, Baron, et al., 2001). While Accessible tourism is flourishing on one side, on the other hand, it deteriorates the natural environment. The pollution caused by automobiles greatly destroyed the grasslands, vegetation, and flora. Automobiles not only destroy the flora but also contaminate the atmosphere as well (Denszi, Ciangă, et al., 2001). In addition to that, certain vehicles, such as motorboats, snowmobiles, etc., cause noise pollution.

Bran, Marin et al. (1998) classifies the different types of negative environmental impacts caused by the tourism industry. They are pollution, composition changes in flora and fauna, erosion, natural resources deprivation, and visual impact.

ENVIRONMENTAL SUSTAINABILITY BY VIRTUAL REALITY

Creating Awareness Through Virtual Experience

The critical environmental issues like climate change, pollution, and forest fire are difficult to relate to or imbibe deep impact due to urban lifestyle and inappropriate measurables. The immersive technologies offer an amazing opportunity to bring the environmental issues closer to people with the intention of educating them and creating awareness to mitigate the crisis. The places affected by human activities will be offered in a virtual platform to have a closer experience of the negative impacts. Virtual Reality is able to establish an emotional connection with environmental issues. It offers the environment in a super visual way that creates an emotional connection and ultimately ends up creating awareness. VR videos about the most critical environmental issues narrated in a fascinating and memorable way will enhance the awareness about it successfully.

Virtual Reality helps in understanding environmental issues more vividly than any other traditional method. Enabling a spatial presence with the environment (for example, a vulnerable tourist destination) helps to remember the destination information longer than any other way. Memory is an important element in keeping up the awareness of critical environmental issues. The memory element will be improved through Virtual Reality Interactions. It has been found that interactions in Virtual Reality can induce the memory factor by 15% more than any other normal videos. In addition, the Virtual Reality experiences will create emotional engagement 27% more than the traditional videos ("Virtual reality experiences", n.d.). The multimedia content in virtual reality tourism creates two types of memories. One memory is related to the complete satisfaction of the tour, while the other memory is associated with the tourism experience.

Virtual Tours to Vulnerable Destinations

Virtual Tours offers an alternative solution to visit fragile destinations without damaging the environment and ecosystem. Tourism is one of the major contributors to the deterioration of destinations. Tourism activities, tourists visiting more than the carrying capacity, and over-exploitation of resources all contribute to the destruction of the destination. If appropriate measures are not taken, they will enter into a decline stage and eventually end up in a situation where they are no longer a tourism product. This catastrophe can be avoided by the influence of technology. Virtual tours can stimulate the replica of the destinations and allow people to visit the destination virtually. Over some time, the destination will go for rejuvenation, heal itself by balancing the ecosystem and get ready to welcome visitors. Tourism Managers have agreed that Virtual tourism is the best solution for fragile destinations and places with sensitive ecosystems. Henceforth, virtual travel will protect the endangered destinations by switching from traditional travel (Kask, 2018).

Virtual Reality in Optimizing Ecological Landscapes

Virtual Reality is used in managing ecological landscapes. It analyses wetland landscapes in the past and shows light on the dynamic changes it has gone through over a period of time. It basically recreates the wetland landscape and provides the dynamic changes related to the ecological landscape pattern. The immersive technology helps in landscape planning, design, and guiding practices. Thus the VR technology systematically assists in representing, planning, designing, and managing the whole ecological landscape and improves the appropriate utilization of ecological resources (Guanmei & Zhen Wang, 2022).

Table 1. Description of various global sustainable practices by Virtual Reality

Title of the Paper/ Article	Geography/ Scope of the study	Tourism Product/ Services	Issues	Sustainable Practices through Virtual Reality
Virtual reality enhanced roller coasters and the future of entertainment – audience expectations (Burt & Louw, 2019)	Western Countries and Europe	Amusement Park (Roller coasters)	● High financial investment ● Decline in visitors due to aging of rides ● Dismantle older roller coasters	● Incorporation of Virtual Reality technology in the existing roller coaster rides ● Renewing the product life cycle by launching VR overlay experience ● Enhancing customer experience ● Revamp the existing roller coster at less cost without tearing down or rebuilding which requires huge investment.
The Use of Virtual Reality to Promote Sustainable Tourism: A Case Study of Wooden Churches Historical Monuments from Romania (Caciora, 2021)	Romania, Europe	Cultural Heritage (Wooden Churches)	● Need for the preservation, conservation and awareness of tangible heritage culture by sustainable practices among the tourism stakeholders ● Lack of awareness about the importance of wooden churches to tourism and the benefits it bestow to the local economy	● Launching a website and upload virtual tours of the wooden churches. ● Virtual tourism created with the amalgamation of multimedia elements such as 3D models, Photographs, audio, video, panoramic images, text. ● Enabling a prominent online way of visiting fragile and endangered tourist attractions. ● Creating awareness among the tourism stakeholders about the significance of tangible heritage culture (Eg: Wooden churches)
The Application of Virtual Reality Technology in the Coordination and Interaction of Regional Economy and Culture in the Sustainable Development of Ecotourism (Guanmei & Zhen Wang, 2022).	Henan Province, China, Asia	Ecological resources (Wetland Landscapes)	Relationship between VR and economic culture and ecotourism sustainability	● Adopting VR technology to examine the changing patterns of ecological landscape by reproducing the landscape, identifying the dynamic changes, planning the landscape and designing and guiding the plan respectively. ● Create a systematic management plan of ecological landscapes through VR and enhance the appropriate usage of ecological resources.

continued on following page

Table 1. Continued

Title of the Paper/ Article	Geography/ Scope of the study	Tourism Product/ Services	Issues	Sustainable Practices through Virtual Reality
The use of 3D visualization for sustainable tourism planning (Kask et al., 2018).	Estonia, Europe	Beach Destination (Võrtsjärv shore)	● Inefficient Sustainable Tourism planning at Võrtsjärv shore. ● Lack of participation in sustainable tourism planning among the local communities. ● Ineffective use of Human resources, Time and Financial needs.	● Use of 3D Visualization techniques to design the sustainable tourism planning of Võrtsjärv shore. ● It incorporates visuals, objects, user interface, details and design of the destination with the help of 3D Visualizations. This technology assists in proper optimization of resources (Human, Finance and Time). ● Modeling the sustainable development project of Võrtsjärv shore by projecting buildings, landscape development,infrastructure, and new constructions more precisely for future planning and execution. ● 3D Visualization highlights the future scenario indicators with respect to environmental, social, and economic.

Promoting Sustainable Tourism through Virtual Reality Marketing (VRM)

The Sustainability of a destination can be measured by three elements such as longevity of the stay at a destination, interest or desire to visit a destination and the experience associated with the destination. The longevity of the destination is directly related to the life cycle of the tourism product (Tourist destination). More longevity refers to the larger life cycle of a tourism product. The second element emphasizes the interest in the destination. This will be created through promotions, marketing, word of mouth, social media handles government portals, Destination Management Organizations, Tourism websites, etc. The interest can be enhanced through intriguing destination promotions, enticing marketing, offers, discounts, etc. As long as people have an interest and likeness towards the destination, its Sustainability will be maintained. The third important sustainable element focuses on experiences associated with the destination. Good and pleasant experiences lead to likeness and interest enhancement towards a tourism destination. These experiences can be attained at two levels (i.e.) physical experience at the destination and virtual experience away from the destination. Virtual Reality provides a uniquely immersive experience (Beck & Egger, 2018) which acts as an ignition tool to visit the destination in the physical world. The immersive experience attained through Virtual Reality increases the visit intention (Huang et al., 2013) and thereby contributes to the Sustainability of that destination. As a whole, longevity, interest, and experience are dependent on each other and emerge as the significant elements contributing to the Sustainability of a destination.

The evolving information and communication technologies revamp the way tourism products are promoted to the market. Virtual Reality has proved its capability in marketing and distinguished itself from the traditional ways in terms of providing product information along with experiences (Guttentag, 2010). The technology bestows cognitive information and immersive experience in a computer-simulated multisensory environment. VR offers a piece of unique experiential information compared to others (Beck & Egger, 2018)

Virtual Reality emerged as a powerful tool in marketing and attracted potential customers through effective communication. It helps to engage the customers in a novel way (Beck & Egger, 2018). VR Marketing of destinations is noted to be the most researched Virtual Reality application (Guttentag, 2010). With the help of immersive marketing experiences, the VR technology is able to reduce the perceived risk of the online tourism products and services, leading to improved decision making (Cho & Fesenmaier, 2001; Cho, Wang & Fesenmaier, 2002; Lui, Piccoli, Ives, 2007; Huang, Backman, Backman & Moore, 2013; Huang, Backman, Backman, & Chang, 2015; Beck & Egger, 2018).

The VR has immense potential in the field of destination promotion as it creates an awareness about the destination and instigates people to experience the destination (Guttentag, 2010; Marchiori, Niforatos & Preto, 2017), tries and purchase tourism products ((Tussyadiah et al., 2017)), creates a destination imaging on people's mind (Pantano & Servidio, 2009), hunt for information and indulge in decision making (Huang et al., 2015) and influences people's intention to visit the destination in the real world (Huang et al., 2013).

Virtual reality Marketing is identified to be a form of sustainable tourism and helps to increase the economic condition of a tourist destination. Bali, an island in Indonesia, has adopted Virtual Reality Marketing practices for the niche market. They targeted the tourists or consumers, senior citizens, and pregnant ladies who had visited Bali earlier and wished to visit again. Marketing approaches through Virtual Reality focus on the niche market who are willing to take up travel to Bali destinations. The tourism market enables a single marketing concept and links the various tourism product and service providers like restaurants, spas, transportation, hotels, airlines, car rentals, etc. (Nyoman et al., 2021).

DISCUSSION:

The immersive technologies Augmented reality and Virtual Reality used in the areas of conservation, entertainment, education, awareness, etc. Ecological conservation arose as a promising area where the AR and VR helps to protect the environment through various actions. These technologies help to intrigue climate awareness and increase prevention projects among people. Virtual Reality has been used predominantly at destination promotions and attractions. Destination promotion through Virtual Reality is the enticing and unique feature of this immersive technology. It provides an immersive experience by portraying the tourist destination in a different dimension rather than the physical one. The 360 degree videos, images and other multimedia elements blended with interactive experience creates an immersive experience for the people. Virtual Reality highlights the enticing and hidden beauty of the destinations through the virtual experience (Shaikh et al., 2018). These VR experiences influence the people to visit the destination in the real world and helps in the decision making process. It increases the tourist flow to a destination. On the other hand, Virtual tours mitigate the tourist flow when the destination is vulnerable or fragile. Virtual reality is argued to be one of the prominent immersive technologies to bolster the vulnerable destinations by allowing them to go through a rejuvenation stage rather than landing up in the decline stage of the tourism product life cycle. The VR technology helps in extending the longevity of the destination by transforming them into a resilient one and contributing to its sustainability as well (Guttentag, 2010; Shaikh et al., 2018).

Virtual Reality (VR) has bestowed enormous benefits in limiting the tourism resources to enter into the decline stage. It also plays a vital role in controlling the demonstration effect, protecting heritage sites, and extending the longevity of various tourism products.Virtual Reality achieves sustainability through the complete merge of three elements. They are likely to have longevity, experience and interest. The lifecycle of a tourism product, experience with the tourism product and interest grown towards the tourism product contributes to sustainable tourism. The three elements longevity. experience and interest influence each other and they are directly proportional to the sustainability of a destination or a tourism product.

Sustainable tourism development is responsible for the present and future impacts associated with the economic, environmental, and social aspects. Sustainable practices should be implemented in such a way it maintains an equilibrium between these three aspects (UNEP & UNWTO, 2005). Tourism creates negative impacts on the environment, economy, and social and cultural elements such as pollution, climate change, destination going into a decline stage, etc. These negative impacts can vanish with the incorporation of Virtual Reality into tourism practices. The flow chart describes how the negative impacts can be controlled through Virtual Reality Technology.

SWOT Analysis of Virtual Reality in Tourism

The SWOT analysis of VR in tourism in the post-pandemic has been analyzed as follows. Immersion is the key factor that helps to transport customers into the virtual world. Virtual Reality is found to be more immersive than any other immersive technologies and it creates an optimistic attitude towards the tourist destination. Sensory levels play an important role in the immersion experience. The distinct levels of sensory information creates differentiation in attitude, presence and visit intentions of a destination (Alyahya & McLean, 2021).

Authentic experience, cognitive response and affective response have a positive impact on the Virtual Reality tourists and influences them to visit the destination in the real world (Kim et.al, 2020). Immersion, Multi Sensory Fidelity (Mohit & Madhurima, 2021) and engagement constitutes to the higher Virtual reality presence and this feature helps in differentiating traditional media from Virtual Reality. This distinguishing feature of Virtual reality helps in positive emotions and influences higher revisit intention to a destination (Ryan et.al, 2021). Technology Embodiment is closely associated with immersion as the technology integrates into user experience and enhances the tourist experiences (Tussyadiah et.al, 2018).The acceptance level of Virtual tours depends on the range of factors. The prime attributes are Ease of accessibility, level of entertainment and the usefulness of tour (El-Said, 2021). Virtual Tourism has its own limitations or challenges to overcome. They are found to be technology aware, Use of technology, learning time, and desire to switch from the corporeal world (Ryan & Catheryn, 2019). Virtual reality promotions with higher telepresence create a positive revisit intention among travelers. Higher level of telepresence with a lower level of social presence proves to create a stronger connection with the destination and enhances the revisit intention (Ying et.al, 2021).

The concept of Virtual Tourism is blended with sustainable tourism as it helps reduce greenhouse gas emissions through transportation. The lack of physical travel in Virtual tourism cuts down vehicle usage and aids tremendously in reducing greenhouse gas emissions. Virtual tourism is popularised only among the tech-savvy and promotional activities have to be done to reach the wider audience. Cultural landscapes are preferred over natural landscapes in the Virtual tours and the Museum tops the list of cultural landscapes (Junyu et.al, 2022). Virtual Reality Marketing emerged as a sustainable form of tourism in Bali focussing on a niche markets to increase their economic growth (Nyoman et.al, 2021).

Table 2. SWOT Analysis of Virtual Reality in Tourism

Strengths	Weaknesses	Opportunities	Threats
Connects people with real world despite travel restrictions	Lack of real experiences	Virtual Tour	Technology vanishes some tourism jobs
Cost effective, Safety and environment friendly travel	Lack of intense memories associated with the tourist destinations	Leads to sustainable tourism practices	Multiplier effect could be curbed
Highly Immersive technology	Accessibility	Positive influence on customer purchase decision	Aversion towards physical travel
Higher Telepresence reduces psychological illness	Technology awareness	Mitigates greenhouse gas emissions by virtual travel	Regions depended on tourism income suffers with technology travel replacement
Authentic experience, cognitive response and affective response	Technology glitches	Marketing and Promotions	Livelihood of tourism dependent regions will be questioned
Immersion, Multi-Sensory Fidelity and engagement	Learning time	Mitigate demonstration effect, protecting heritage sites, and extending the longevity of various tourism products.	Leads to addiction and isolation
Positive emotions	Desire to switch from the corporeal world	Reduces negative economic, socio-cultural and environmental impacts caused by tourism	Threat to local tourism businesses
Disconnect from real world	Limited interaction with people	Revisit intention of a destination	Lack of socio-cultural interaction.
Distinct sensory information	High investment required	Replace dangerous sites of Dark Tourism	Non-competition among tourism providers
Try before purchase	Mental Disorders	Increase tourist flow	Ethical issues
Knowledge enhancement	Physiological and cognitive impacts	Contribute to economic growth	Data Piracy
Travel to past and forbidden places	Social Isolation	Conservation of vulnerable destinations and natural resources	Manipulation

CONCLUSION

Tourism resources are the appealing features of tourism. A tourism destination without its resources is lifeless. The major concern arises when the sustainability of the tourism product or resources is questioned. The socio-cultural and environmental sustainability can be achieved with the assistance of technology. Tourism offers the incorporation of Virtual Reality in its activities by leveraging its longevity and thus contributes to the sustainability feature. Virtual tours help in extending the lifespan of a tourist destination, mitigating the impact of the demonstration effect on the host culture, increasing the longevity of tourism products like roller coaster rides, and safeguarding the endangered world archaeological and heritage sites. The contribution of Virtual Reality toward sustainability is enormous and will continue proactively in the following years.

Virtua Reality can be effectively used in all the stages of the tourism product life cycle such as exploration, involvement, development, consolidation, stagnation, and rejuvenation or decline. This immersive technology is mainly used to market tourism products, to promote destinations, and to provide virtual tours. Virtual Tourism has become an alternative way of travel whenever a destination needs rejuvenation. The chapter has tried to prove that sustainability of the destination can be achieved through Virtual Reality and explained the same through various examples throughout the chapter. Virtual Reality helps to improve economic, environmental, and social benefits of the local community to a greater extent. It also helps in optimizing resources, reducing greenhouse emissions, channelizing products, enhancing immersive experiences etc.

Virtual Reality Marketing has revamped the traditional marketing strategies of the tourism industry. The VR marketing has bestowed many sustainable outcomes such as improved purchase decisions, revisit intention, positive travel decisions, etc. VR plays a significant role in the planning phase or pre-trip phase. Prior knowledge about a destination helps them to plan effectively. Experiential economy in tourism is achieved through Virtual Reality. It enables an opportunity to try the tourism product before purchasing it. Overall VR provides many benefits to the tourism industry and one of the most important benefits is increased revenue generation. Tourism companies who have incorporated VR in their operations stand ahead of their peers which makes them unique, attract more potential customers and simultaneously increasing their sales.

Virtual Reality being a highly immersive technology provides various sustainable benefits to the tourism industry such as launching virtual tours, influencing positive travel decisions of the customers, mitigating negative impacts on the economy, environment, and social aspects of a tourist destination, conserving natural resources, rejuvenating vulnerable destinations, improving economic benefits, launching virtual stores. This book chapter has provided AI-driven innovations and strategies for sustainable tourism and future studies can focus on evaluating these strategies and finding out the prominent solution among them. The future scope can also compare the conventional marketing strategies of the tourism industry against the Virtual Reality marketing strategies.

REFERENCES

A virtual reality of endangered. (2018, April 19). *The Hans India.*https://www.thehansindia.com/posts/index/Young-Hans/2018-04-19/A-virtual-reality-of-endangered-heritage-sites/375029?infinitescroll=1

Adetunji, J. (2020, December 2).Can virtual tourism save local businesses in tourist destinations? TheConversation.https://theconversation.com/can-virtual-tourism-save-local-businesses-in-tourist-destinations-149537

Albayrak, T., Caber, M., & Aksoy, S. (2010). Relationships of the Tangible and IntangibleElements of Tourism Products with OverallCustomer Satisfaction. *International Journal of Trade. Economics and Finance*, 1(2), 140.

Alyahya, M., & McLean, G. (2021). Examining Tourism Consumers' Attitudes and the Role of Sensory Information in Virtual Reality Experiences of a Tourist Destination. *Journal of Travel Research.* Advance online publication. DOI: 10.1177/00472875211037745

Ashworth, G. J., & Voogt, H. (1990). *Selling the city.*

Bal, W., & Czalczynska-Podolska, M. (2020). The Stages of the Cultural Landscape Transformation of Seaside Resorts in Poland against the Background of the Evolving Nature of Tourism. *Land (Basel)*, 9(2), 55. DOI: 10.3390/land9020055

Baum, T. (1998). Taking the exit route: Extending the tourism area life cycle model. *Current Issues in Tourism*, 1(2), 167–175. DOI: 10.1080/13683509808667837

Beck, J., & Egger, R. (2018). Emotionalise Me: Self-reporting and Arousal Measurements in Virtual Tourism Environments. In Stangle, B., & Pesonen, J. (Eds.), (pp. 3–15). Information and Communication Technologies in Tourism. [Electronic resource], DOI: 10.1007/978-3-319-72923-7_1

Brundtland, G. H. (1985). World Commission on Environment and Development. *Environmental Policy and Law*, 14(1), 26–30. DOI: 10.1016/S0378-777X(85)80040-8

Bryan, B. A., Crossman, N. D., King, D., & Meyer, W. S. (2011). Landscape Futures Analysis: Assessing the Impacts of Environmental Targets under Alternative Spatial Policy Options and Future Scenarios. *Environmental Modelling & Software*, 26(1), 83–91. DOI: 10.1016/j.envsoft.2010.03.034

Budeanu, A., Miller, G., Moscardo, G., & Ooi, C. S. (2016). Sustainable tourism, progress, challenges and opportunities: An introduction. *Journal of Cleaner Production*, 111, 285–294. DOI: 10.1016/j.jclepro.2015.10.027

Burt, M., & Louw, C. (2019). Virtual reality enhanced roller coasters and the future of entertainment – audience expectations. *World Leisure Journal*, 61(3), 183–199. DOI: 10.1080/16078055.2019.1639274

Burt, M., & Louw, C. (2019). Virtual reality enhanced roller coasters and the future of entertainment – audience expectations. *World Leisure Journal*, 61(3), 183–199. DOI: 10.1080/16078055.2019.1639274

Butler, R. (2024). Tourism destination development: The tourism area life cycle model. *Tourism Geographies*, •••, 1–9. DOI: 10.1080/14616688.2024.2325932

Butler, R. W. (1980). The concept of tourism area cycle of evolution: Implications for management of resources. *The Canadian Geographer. Geographe Canadien*, 24(1), 5–12. DOI: 10.1111/j.1541-0064.1980.tb00970.x

Caciora, T., Herman, G., Alexandru, I., Baias, S., Ilie , D., Josan, I., & Hodor, N. (2021). The Use of Virtual Reality to Promote Sustainable Tourism: A Case Study of Wooden Churches Historical Monuments from Romania. *Remote Sensing (Basel)*, 13(9), 1758. DOI: 10.3390/rs13091758

Cho, Y.-H., & Fesenmaier, D. (2001). A new paradigm for tourism and electronic commerce: experience marketing using the virtual tour. In Buhalis, D., & Laws, E. (Eds.), *Tourism Distribution Channels: Practices, Issues and Transformation* (pp. 351–370). Continuum.

Cho, Y.-H., Wang, Y., & Fesenmaier, D. (2002). Searching for experiences: The web-based virtual tour in tourism marketing. *Journal ofTravel & Tourism Marketing*, 12(4), 1–17. DOI: 10.1300/J073v12n04_01

Desai, P. R., Desai, P. N., Ajmera, K. D., & Mehta, K. (2014). A Review Paper onOculus Rift-A Virtual [IJETT]. *International Journal of Engineering Trends and Technology*, 13(4), 175–179. DOI: 10.14445/22315381/IJETT-V13P237

El-Said, O., & Aziz, H. (2021). Virtual Tours a Means to an End: An Analysis of Virtual Tours' Role in Tourism Recovery Post COVID-19. *Journal of Travel Research*. Advance online publication. DOI: 10.1177/0047287521997567

Erenay, O., & Hashemipour, M. (2003). Virtual Reality in Engineering Education: A CIM Case Study. *The Turkish Online Journal of Educational Technology*, 2(2), 51–56. Retrieved May 2014, from http://www.tojet.net/articles/v2i2/228.pdf

Frent, C. (2016). An overview on the negative impacts of tourism.Journal of tourism – studies and research in tourism, (22).

Fritz, F., Susperregui, A., & Linaza, M. T. (2005) Enhancing Cultural Tourism experiences with Augmented Reality Technologies, The 6th International Symposium on Virtual Reality, Archaeology and Cultural Heritage VAST 2005, http://public-repository.epochnet.org/publications/VAST2005/shortpapers/short2005.pdf, accessed: May 2014.

Fritz, F., Susperregui, A., & Linaza, M.T., op.cit, p.2

Gegung, Eunike. (2021). Jurnal Kepariwisataan Indonesia 15 (1) (2021) THE USE OF VIRTUAL REALITY TO INCREASE TOURISM DESTINATION SUSTAINABILITY AND HOW USERS PERCEIVE THE AUTHENTICITY OF VR EXPERIENCES Jurnal Kepariwisataan Indonesia 15 (1) (2021). 15. 9-15.

Green economy. (n.d.). United Nations Environment Programme. https://www.unep.org/regions/asia-and-pacific/regional-initiatives/supporting-resource-efficiency/green

Guttentag, D. A. (2010). Virtual reality: Applications and implications for tourism. *Tourism Management*, 31(5), 637–651. DOI: 10.1016/j.tourman.2009.07.003

Halton, C. (2021, August 26). Green economics. *Investopedia.*https://www.investopedia.com/terms/g/greeneconomics.asp

Hobson, J. S. P., & Williams, A. P. (1995). Virtual Reality: A new horizon for the tourism industry. *Journal of Vacation Marketing*, 1(2), 125–135.

Huang, C.-Y., Backman, K. F., Backman, S., & Chang, L. L. (2015). Exploring the implications of virtual reality technology in tourism marketing: An integrated research framework. *International Journal of Tourism Research*, 18(2), 116–128. DOI: 10.1002/jtr.2038

Huang, C.-Y., Backman, K. F., Backman, S., & Moore, D. (2013). Exploring user acceptance of 3D virtual worlds in travel and tourism marketing. *Tourism Management*, 36, 490–501. DOI: 10.1016/j.tourman.2012.09.009

Huang, G., & Wang, Z. (2022). The Application of Virtual Reality Technology in the Coordination and Interaction of Regional Economy and Culture in the Sustainable Development of Ecotourism. *Mathematical Problems in Engineering,* vol. 2022, Article ID 9847749, 13 pages. DOI: 10.1155/2022/9847749

Jarratt, D. (2021, April). An exploration of webcam-travel: Connecting to place and nature through webcams during the COVID-19 lockdown of 2020. *Tourism and Hospitality Research*, 21(2), 156–168. DOI: 10.1177/1467358420963370

Jenkins, W. (2009). Sustainability Theory. *Berkshire*.https://www.berkshirepublishing.com/assets_news/sustainability/Spirit_SustainabilityTheory.pdf

Jovicic, D. Z. (2014). Key issues in the implementation of sustainable tourism. *Current Issues in Tourism*, 17(4), 297–302. DOI: 10.1080/13683500.2013.797386

Kask, S. (2018). Virtual reality in support of sustainable tourism. Experiences from Eastern Europe. (Doctoral thesis, Estonian University of Life Sciences, Tartu, ISBN 978-9949-629-61-9). DOI: 10.15159/emu.35

Kask, S., Kullb, T., & Orruc, K. (2018). The use of 3D visualization for sustainable tourism planning. *Journal of Baltic Studies*, 2018(3), 371–385. Advance online publication. DOI: 10.1080/01629778.2018.1455719

Kemska, A. S. (April, 2021). VIRTUAL TOURISM IN UKRAINE - FEATURES AND PROSPECTS. POLIT. Challenges of science Today. http://fmv.nau.edu.ua/wp-content/uploads/2021/04/%D0%9F%D0%BE%D0%BB%D1%96%D1%82-2021.pdf#page=274

Kim, M. J., Lee, C.-K., & Jung, T. (2020). Exploring Consumer Behavior in Virtual Reality Tourism Using an Extended Stimulus-Organism-Response Model. *Journal of Travel Research*, 59(1), 69–89. DOI: 10.1177/0047287518818915

Kounavis, C. D., Kasimati, A. E., & Zamani, E. D. (2012). Enhancing the Tourism Experience through Mobile Augmented Reality: Challenges and Prospects. *International Journal of Engineering Business Management*, 4, 1–6. Retrieved May 2014, from http://cdn.intechopen.com/pdfs/38051/InTech Enhancing_the _tourism_experience_through_mobile_augmented_reality_challenges_and_prospects.pdf. DOI: 10.5772/51644

Lau, A. (2020). New technologies used in COVID-19 for business survival: Insights from the Hotel Sector in China. *Information Technology & Tourism*, 22(4), 497–504. DOI: 10.1007/s40558-020-00193-z

Lee, H. S. (2018). A study on the sustainability of the Ramsar wetland ecotourism sites in Korea: Making a list of evaluation indicators and measurement variables. *International Journal of Tourism and Hospitality Research*, 32(2), 99–117. DOI: 10.21298/IJTHR.2018.02.32.2.99

Lenzen, M., Sun, Y., Faturay, F., Ting, Y.-P., Geschke, A., & Malik, A. (2018). The carbon footprint of global tourism. *Nature Climate Change*, 8(6), 522–528. DOI: 10.1038/s41558-018-0141-x

Lu, J., Xiao, X., Xu, Z., Wang, C., Zhang, M., & Zhou, Y. (2022). The potential of virtual tourism in the recovery of tourism industry during the COVID-19 pandemic. *Current Issues in Tourism*, 25(3), 441–457. DOI: 10.1080/13683500.2021.1959526

Luhar.H., (2021), An Analytical Study on Economy Upliftment through Virtual Tourism with special reference to Vasai Virar City, -- Palarch's Journal Of Archaeology Of Egypt/Egyptology 18(7). ISSN 1567-214x

Lui, T.-W., Piccoli, G., & Ives, B. (2007). Marketing strategies in virtual worlds. *The Data Base for Advances in Information Systems*, 38(4), 77–80. DOI: 10.1145/1314234.1314248

Marchiori, E., Niforatos, E., & Preto, L. (2017). Measuring the media effects of a tourism-related virtual reality experience using biophysical data. In Schegg, R., & Stangl, B. (Eds.), *Information and communication technologies in tourism 2017* (pp. 203–215). Springer. DOI: 10.1007/978-3-319-51168-9_15

Mitchell, C. (2020). Virtual reality. Investopedia. Retrieved from https://www.investopedia.com/

Mohit, M., & Madhurima, D. (2021). Effects of multisensory virtual reality on virtual and physical tourism during the COVID-19 pandemic. *Current Issues in Tourism*. Advance online publication. DOI: 10.1080/13683500.2021.1978953

Murphy, P., Pritchard, M. P., & Smith, B. (2000). The destination product and its impact on traveller perceptions. *Tourism Management*, 21(1), 43–52. DOI: 10.1016/S0261-5177(99)00080-1

Pantano, E., & Servidio, R. (2009). An exploratory study of the role of pervasive environments for promotion of tourism destinations. *Journal of Hospitality and Tourism Technology*, 2(1), 50–65. DOI: 10.1108/17579881111112412

Ryan, C. (2002). Equity, Management, Power Sharing and Sustainability: Issues of the 'New Tourism.'. *Tourism Management*, 23(1), 17–26. DOI: 10.1016/S0261-5177(01)00064-4

Shaikh, S., & Bokde, K., INGALE, A., & Tekwani, B. (2018). Virtual Tourism. *International Research Journal of Engineering and Technology*, 5(4), 2044–2046.

Sifo, K. (2019). Mediebild, opinion och faktakoll i klimatfrågan. Retrieved from: https://www.kantarsifo.se/rapporter-undersokningar /mediebild-opinion-och-faktakoll-i-klimatfragan

Slater, M., & Sanchez-Vives, M. V. (2016). Enhancing Our Lives with Immersive Virtual Reality. *Frontiers in Robotics and AI*, 3, 1–47. DOI: 10.3389/frobt.2016.00074

Streimikiene, D., Svagzdiene, B., Jasinskas, E., & Simanavicius, A. (2021). Sustainable tourism development and competitiveness: The systematic literature review. *Sustainable Development (Bradford)*, 29(1), 259–271. DOI: 10.1002/sd.2133

Subai, Ada & Konya, K & Adim, Victor. (2020). Technological Innovation of Hotel Services: A COVID-19 Strategic Response. British Journal of Economics, Finance and Management Sciences 19, August 2020, Vol. 17 (2)

Subawa, N. S., Widhiasthini, N. W., Astawa, I. P., Dwiatmadja, C., & Ni, P. I. P. (2021). The practices of virtual reality marketing in the tourism sector, a case study of Bali, Indonesia. *Current Issues in Tourism*, 24(23), 3284–3295. Advance online publication. DOI: 10.1080/13683500.2020.1870940

Truyols, M. (2023, February, 24). The 6 phases of a tourist destination's life cycle. Mize. https://mize.tech/blog/the-6-phases-of-a-tourist-destinations-life-cycle

Tussyadiah, I. P., Wang, D., & Jia, C. H. (2017). Virtual reality and attitudes toward tourism destinations. In Schegg, R., & Stangl, B. (Eds.), *Information and Communication Technologies in Tourism 2017* [Electronic Resource]. (pp. 229–239)., DOI: 10.1007/978-3-319-51168-9_17

United Nations Environment Program. UNEP (2016b) Negative socio-cultural impacts from tourism,https://www.unep.org/resourceefficiency/Business/SectoralActivities/Tourism/FactsandFiguresaboutTourism/ImpactsofTourism/SocioCulturalImpacts/NegativeSocio-CulturalImpactsFromTourism/tabid/78781/Default.aspx, accessed July 29, 2016

Virtual reality experiences for sustainability & environmental awareness. (n.d.). *steamedegg*.https://steamedegg.co.uk/environmental-awareness-virtual-reality/

Voronkova, L. (2018). Virtual Tourism: On the Way To the Digital Economy. *IOP Conference Series. Materials Science and Engineering*, 463, 042096. DOI: 10.1088/1757-899X/463/4/042096

What is Sustainable Tourism? (n.d.).*GSTC*. https://www.gstcouncil.org/what-is-sustainable-tourism/

Wiltshier, P., & Clarke, A. (2016). Virtual cultural tourism: Six pillars of VCT using co-creation, value exchange and exchange value. In *Tourism and Hospitality Research* (pp. 1–12). SAGE Publications.

Winter. (2008). Demonstration effect. *UNBC Outdoor Recreation & Tourism Management*.https://www2.unbc.ca/sites/default/files/sections/outdoor-recreation-tourism-management/demonstration_effect.pdf

WTO. (2005).Making Tourism More Sustainable - A Guide for Policy Makers. *UNWTO*.https://www.e-unwto.org/doi/book/10.18111/9789284408214

Ying, T., Tang, J., Ye, S., Tan, X., & Wei, W. (2021). Virtual Reality in Destination Marketing: Telepresence, Social Presence, and Tourists' Visit Intentions. *Journal of Travel Research*. Advance online publication. DOI: 10.1177/00472875211047273

Yung, R., & Khoo-Lattimore, C. (2019). New realities: A systematic literature review on virtual reality and augmented reality in tourism research. *Current Issues in Tourism*, 22(17), 2056–2081. DOI: 10.1080/13683500.2017.1417359

Yung, R., & Khoo-Lattimore, C. (2019). New realities: A systematic literature review on virtual reality and augmented reality in tourism research. *Current Issues in Tourism*, 22(17), 2056–2081. DOI: 10.1080/13683500.2017.1417359

Yung, R., Khoo-Lattimore, C., & Potter, L. E. (2021). Virtual reality and tourism marketing: Conceptualizing a framework on presence, emotion, and intention. *Current Issues in Tourism*, 24(11), 1505–1525. DOI: 10.1080/13683500.2020.1820454

Zhu, B. (2020). How the tourism industry in China is finding new ways to engage customers.Capgemini.https://www.capgemini.com/2020/05/tourism-industry-engaging-customers/

ADDITIONAL READING

Daline, R., & Thomas, J. (2024). AI-Powered Marketing Strategies in the Tourism and Hospitality Sector. In *Utilizing Smart Technology and AI in Hybrid Tourism and Hospitality* (pp. 223–242). IGI Global. DOI: 10.4018/979-8-3693-1978-9.ch010

Daline, T. R., & Thomas, J. (2024, April). An Outlook on Sustainable Business Practices through Virtual Reality Marketing. In *2024 International Conference on Science, Engineering and Business for Driving Sustainable Development Goals (SEB4SDG)* (pp. 1-6). IEEE. DOI: 10.1109/SEB4SDG60871.2024.10630336

Yuce, A. (2022). Impact of Technology in Sustainable Tourism Development: Virtual Reality. In *Mobile Computing and Technology Applications in Tourism and Hospitality* (pp. 98–119). IGI Global. DOI: 10.4018/978-1-7998-6904-7.ch005

KEY TERMS AND DEFINITIONS

Immersion: The degree of one's sensory involvement in the virtual environment

Presence: The feeling of one being present in the virtual environment

Sustainable Tourism: Sustainable Tourism is concerned with the social, economic, and environmental aspects of tourism activities.

Tourism Product Life Cycle: It is a model that defines the various stages of a tourism product goes through in its life cycle.

Virtual Reality: Virtual Reality is a computer-simulated environment that allows people to escape from the real one.

Virtual Reality Marketing: A marketing strategy to promote products with the help of Virtual Reality technology.

Virtual Tourism: A destination is visited by technology-simulated experiences from one's comfort place.

Chapter 4
The Challenge of Innovation Measurement in Small and Medium Enterprises (SMEs):
Approaches and Challenges in Assessing Innovative Capabilities

Asmaa Ahmed Youssef
https://orcid.org/0009-0005-1291-0612
School of Management of Technology, Nile University, Egypt

ElHassan Anas ElSabry
https://orcid.org/0000-0001-6135-0280
School of Management of Technology, Nile University, Egypt

ABSTRACT

The measurement of innovation is critical for SME success, driving economic growth, competitiveness, and social progress. Accurately measuring innovation reveals insights into an SME's adaptability, product creation, and competitive edge. However, innovation's complexity challenges effective measurement frameworks. This publication explores diverse approaches to innovation measurement, highlighting methodologies' strengths and limitations for SME assessment. It defines innovation's impact on efficiency, satisfaction, competitive advantage, and resilience, exploring inputs, activities, outputs, and environmental factors. Challenges like valuing intangibles and global disparities are addressed, suggesting future research for enhanced frameworks and sector-specific indicators in innovation management. These advancements aim to improve the accuracy and relevance of innovation measurement,

DOI: 10.4018/979-8-3373-1132-6.ch004

supporting SMEs in continuous improvement and strategic decision-making.

BACKGROUND

The Challenge of Defining Innovation

The concept of innovation has captivated researchers and industry leaders due to its significant impact on organizational operations and strategy. The way innovation is defined within an organization shapes which activities are performed internally and which are outsourced, influencing the degree and nature of innovation. Different industries and scholars approach innovation from multiple perspectives, considering both radical and incremental changes in products, processes, and markets (Tidd & Bessant, 2018).

Defining Innovation: A Key to Effective Management

Defining innovation is crucial for effective innovation management. This definition affects the entire innovation process, determining how organizations foster and measure innovation. Innovation is inherently complex and multidimensional, influenced by various interacting factors. Organizations that possess the necessary resources, motivation to innovate, and a supportive climate for new ideas typically innovate more rapidly and successfully. Continuous transformation of knowledge and ideas into new products, processes, and systems is vital for benefiting the organization and its stakeholders (Drucker, 1985).

Understanding the "soft" aspects of innovation—such as defining and measuring it—is essential for building long-term, sustainable organizations. Innovation is often viewed as essential for organizational survival and competitiveness in dynamic environments (Schumpeter, 1934).

Complexity and Multifaceted Nature of Innovation

Innovation is a term that is widely used but often misunderstood. Its definition can be ambiguous and context-dependent, varying significantly across industries, academic disciplines, and cultural contexts. At its core, innovation involves transforming ideas or inventions into goods and services that create value or attract customers. However, this broad definition encompasses a spectrum of activities and outcomes, making it challenging to establish a singular, all-encompassing definition (OECD, 2018).

Innovation can be incremental or radical, product-oriented or process-focused, and it can occur in various domains such as technology, business models, organizational structures, or customer experiences. Incremental innovation refers to modest improvements to existing products or processes, while radical innovation involves breakthroughs that fundamentally change the market or create entirely new industries. Innovation also extends beyond technological advancements to include social innovations that address societal needs, such as new educational methods, healthcare models, and sustainability practices (Goffin & Mitchell, 2017).

Different Stakeholders, Different Perspectives

Stakeholders interpret innovation differently. For businesses, innovation often means developing new products or improving operational efficiency to gain a competitive edge. Policymakers view innovation as fostering an environment that encourages research and development to stimulate economic growth. Academics focus on the theoretical underpinnings and processes that drive innovative activities (Tidd & Bessant, 2018).

The rapid pace of technological change and the increasing interconnectedness of global markets further complicate the definition of innovation. What was once considered innovative can quickly become standard practice, highlighting the dynamic nature of innovation. This requires an adaptive and forward-thinking approach to understanding and defining innovation (OECD, 2018).

The Need for a Clear Definition

Despite these challenges, a clear definition of innovation is essential for effective measurement and management. Without a shared understanding of what constitutes innovation, organizations and policymakers may struggle to design strategies, allocate resources, or evaluate outcomes effectively. Therefore, it is crucial to approach the definition of innovation with a recognition of its complexity and a commitment to ongoing dialogue and refinement (Tidd & Bessant, 2018).

Practical Implications and Academic Insights

In practical terms, industries such as retail view innovation through the lens of strategy and change. Retailers often link innovation to achieving strategic goals and maintaining growth. They use concrete terms like "newness" and "reinvention" to describe their activities, focusing on continuous improvements rather than radical departures from existing practices (Goffin & Mitchell, 2017).

Academic literature also highlights diverse interpretations of innovation. Studies discuss how environmental, organizational, and managerial factors influence the adoption of innovation. Historical analyses show how the meaning of innovation has evolved over time and across different cultural and institutional settings (Tidd & Bessant, 2018).

In conclusion, defining innovation is a complex, multifaceted challenge with significant implications for organizational strategy and operations. Organizations must consider various perspectives and contexts in defining innovation, as this shapes their innovative activities and overall capability to adapt and thrive in competitive markets. A nuanced, flexible understanding of innovation is necessary to capture its various dimensions and ensure effective management and strategic alignment. Through continuous dialogue and refinement, stakeholders can develop a shared understanding of innovation, facilitating better measurement, resource allocation, and strategic planning. This, in turn, enhances an organization's ability to innovate, ensuring long-term success and competitiveness.

The Significant Impact of Innovative Activities

Driving Efficiency

Innovative activities play a pivotal role in enhancing organizational efficiency. By streamlining processes and reducing operational costs, innovation enables organizations to optimize workflows and achieve significant time and cost savings. Technological advancements and automation, for example, have revolutionized how businesses operate, leading to higher-quality outputs and improved productivity (Schumpeter, 1942).

Enhancing Customer Satisfaction

Innovation significantly impacts customer satisfaction by addressing emerging needs and preferences. Organizations that continuously innovate can introduce new and improved products or services that meet the evolving demands of their customers. This approach not only helps in retaining existing customers but also attracts new ones, thereby expanding the market base (Drucker, 1985).

Fostering Competitive Advantage

Innovative activities foster a competitive advantage by differentiating an organization from its competitors. Unique products, cutting-edge technologies, and novel business models create barriers to entry for other firms, securing a dominant market position. Companies that prioritize innovation can stay ahead of market trends, anticipating and responding to changes more effectively than their competitors (Porter, 1990).

Enhancing Organizational Resilience

Innovation contributes to organizational resilience and adaptability. In a rapidly changing global market, the ability to innovate allows organizations to pivot and adapt to new circumstances, mitigating risks and seizing opportunities. This agility is crucial for sustaining performance and achieving long-term success (Tidd & Bessant, 2020).

To conclude this, the impact of innovative activities is profound and multifaceted. By enhancing efficiency, customer satisfaction, competitive advantage, and adaptability, innovation is indispensable for organizations aiming to thrive in today's dynamic business environment.

Importance of Innovation Measurement

The measurement of innovation is indispensable for strategic decision-making, performance assessment, continuous improvement, transparency, and competitiveness. By implementing robust innovation metrics, organizations can better manage their innovation processes and achieve sustainable growth.

Enabling Strategic Decision-Making

Measuring innovation is critical for enabling strategic decision-making within organizations. By quantifying innovative activities, companies can identify which initiatives are yielding the most value and align their strategies accordingly. This data-driven approach helps in optimizing resource allocation, ensuring that investments are directed toward the most promising projects (Drucker, 2019).

Assessing Performance and Impact

Innovation measurement allows organizations to assess the performance and impact of their innovation efforts. Metrics such as the number of new products developed, time-to-market, and return on innovation investment provide insights into the effectiveness of innovation strategies. These metrics help in evaluating progress, identifying areas for improvement, and demonstrating the value of innovation to stakeholders (Pisano, 2015).

Facilitating Continuous Improvement

Innovation measurement fosters a culture of continuous improvement. By regularly monitoring and analyzing innovation metrics, organizations can identify trends, learn from failures, and replicate successes. This ongoing evaluation process encourages iterative learning and adaptation, which are essential for sustaining long-term innovation (Tidd & Bessant, 2020).

Enhancing Transparency and Accountability

Transparent innovation measurement enhances accountability within organizations. When innovation activities are tracked and reported, it creates a clear record of what has been achieved and what needs further attention. This transparency helps in setting realistic goals, managing expectations, and holding teams accountable for their contributions to innovation (OECD, 2018).

Supporting Benchmarking and Competitiveness

Measuring innovation also supports benchmarking against industry standards and competitors. Organizations can compare their innovation performance with peers, identify best practices, and adopt strategies that enhance their competitiveness. This comparative analysis helps in setting ambitious targets and driving continuous innovation (Keeley et al., 2013).

APPROACHES TO INNOVATION MEASUREMENT

Measuring innovation is a complex task that encompasses various inputs, activities, outputs, and environmental factors. By implementing comprehensive measurement frameworks, organizations can gain valuable insights into their innovation processes, make informed strategic decisions, and drive continuous improvement. This holistic approach to innovation measurement ensures that organizations remain competitive and capable of sustaining long-term growth in a constantly changing landscape (Björk, Frishammar & Sundström, 2023).

Measuring Innovation Inputs

Innovation inputs are the resources and conditions necessary for innovation to occur. These inputs include financial investments, human resources, infrastructure, and intellectual capital. Measuring these inputs provides insight into the resources allocated to innovation and their adequacy for supporting innovative activities.

Financial Investments

Financial investments are a primary input for innovation, encompassing budgets allocated to research and development (R&D), new product development, and other innovative initiatives. Metrics to measure financial investments include R&D expenditure as a percentage of sales, total R&D spending, and capital invested in innovation projects (OECD, 2018). These metrics help organizations track their commitment to innovation and compare it with industry benchmarks.

Human Resources

Human resources dedicated to innovation are critical for success. This includes the number of employees involved in R&D, the skills and expertise of innovation teams, and training programs aimed at enhancing innovation capabilities. Metrics such as the ratio of R&D personnel to total employees and investment in employee training for innovation provide insights into the human capital available for innovation (Gault, 2018).

Infrastructure

Infrastructure investments, including laboratories, technology platforms, and innovation hubs, are essential inputs for innovation. Measuring the adequacy and utilization of these facilities helps organizations ensure that their physical resources support innovation activities. Metrics include the number of innovation facilities, their usage rates, and the investment in upgrading infrastructure (Guzmán et al., 2018).

Intellectual Capital

Intellectual capital encompasses the knowledge, patents, and intellectual property that support innovation. Metrics such as the number of patents filed, the value of intellectual property, and the stock of technical know-how provide a measure of the intellectual assets that contribute to innovation (Hanel, 2018). These metrics help organizations understand the strength of their knowledge base and its potential to drive innovation.

Measuring Innovation Activities

Innovation activities are the processes and actions taken to develop new products, services, or processes. Measuring these activities helps organizations track their innovation efforts and identify areas for improvement.

R&D Activities

R&D activities are a core component of innovation. Metrics to measure R&D activities include the number of R&D projects, the stages of development they are in, and the success rate of these projects. Additionally, tracking the collaboration between R&D and other departments, as well as partnerships with external research institutions, provides a comprehensive view of R&D efforts (Archibugi & Filippetti, 2018).

Idea Generation and Management

The process of generating and managing ideas is crucial for innovation. Metrics such as the number of ideas submitted, the rate of idea acceptance, and the implementation rate of accepted ideas help measure the effectiveness of idea management systems. Tools like innovation management software can track these metrics and facilitate the idea-to-implementation process (Bessant et al., 2020).

Prototyping and Testing

Prototyping and testing are essential activities in the innovation process. Metrics to measure these activities include the number of prototypes developed, the duration of the prototyping phase, and the success rate of prototypes in testing. These metrics help organizations evaluate the efficiency and effectiveness of their prototyping and testing processes (Tidd & Bessant, 2020).

Collaboration and Networking

Collaboration and networking with external partners, such as universities, research institutions, and other companies, are vital for innovation. Metrics to measure collaboration include the number of collaborative projects, the diversity of partners, and the outcomes of these collaborations. These metrics provide insights into the extent and impact of external collaborations on innovation (Chesbrough, 2020).

Measuring Innovation Outputs

Innovation outputs are the tangible and intangible results of innovation activities. Measuring these outputs helps organizations evaluate the effectiveness of their innovation processes and the value generated from their investments.

New Products and Services

The development of new products and services is a primary output of innovation. Metrics to measure this output include the number of new products launched, the revenue generated from new products, and the market share of new products. These metrics help organizations assess the commercial success of their innovation efforts (Edison et al., 2018).

Process Improvements

Innovation often leads to improvements in existing processes. Metrics to measure process improvements include the number of processes improved, the cost savings achieved, and the efficiency gains realized. These metrics provide a measure of the impact of innovation on operational efficiency and effectiveness (Garcia & Calantone, 2018).

Intellectual Property

The creation of intellectual property, such as patents and trademarks, is another output of innovation. Metrics to measure intellectual property outputs include the number of patents filed and granted, the value of patents, and the income generated from licensing intellectual property. These metrics help organizations understand the value of their intellectual property portfolio (Hall & Harhoff, 2018).

Customer Impact

The impact of innovation on customers is a critical measure of success. Metrics to measure customer impact include customer satisfaction scores, net promoter scores (NPS), and customer retention rates. These metrics provide insights into how innovations are perceived by customers and their effect on customer loyalty and satisfaction (Kumar & Pansari, 2016).

Measuring Environmental/External Factors

Environmental and external factors play a significant role in shaping an organization's innovation activities and outcomes. Measuring these factors helps organizations understand the broader context in which they operate and how it influences their innovation efforts.

Market Conditions

Market conditions, including market size, growth rate, and competitive landscape, influence innovation activities. Metrics to measure market conditions include market growth rates, competitive intensity, and market demand for new products and services. These metrics help organizations identify opportunities and threats in the market environment (Porter, 1980).

Regulatory Environment

The regulatory environment, including laws, regulations, and standards, affects innovation. Metrics to measure the regulatory environment include the number of regulatory changes, the impact of regulations on innovation activities, and the cost of compliance. These metrics help organizations understand the constraints and opportunities created by the regulatory landscape (Blind, 2012).

Technological Trends

Technological trends, including advancements in technology and emerging technologies, influence innovation. Metrics to measure technological trends include the number of new technologies adopted, the rate of technological change, and the impact of new technologies on innovation activities. These metrics help organizations stay abreast of technological developments and leverage them for innovation (Christensen et al., 2018).

Societal and Cultural Factors

Societal and cultural factors, including consumer preferences, societal values, and cultural trends, shape innovation. Metrics to measure societal and cultural factors include consumer behavior trends, societal attitudes toward innovation, and cultural acceptance of new technologies. These metrics help organizations align their innovation efforts with societal and cultural expectations (Inglehart, 2018).

CHALLENGES IN INNOVATION MEASUREMENT

Innovation measurement is a complex task that presents various challenges due to its multifaceted nature. Measuring the value of intangible assets and assessing innovation in developing countries are significant challenges in innovation measurement. Intangible assets, such as intellectual property and organizational culture, play a critical role in driving innovation but are difficult to quantify (Lev, 2001; Teece, 2000). Various methods, including the balanced scorecard (Kaplan & Norton, 1992) and intellectual capital measurement (Edvinsson & Malone, 1997), can help address these challenges.

In developing countries, innovation measurement is complicated by factors such as limited data availability and the prevalence of the informal economy (OECD, 2010). Adapting innovation surveys, using qualitative methods, and developing frugal innovation metrics can provide valuable insights into innovation processes in these contexts (Guzmán et al., 2018).

By addressing these challenges, organizations and policymakers can better understand and foster innovation, driving economic growth and development.

The Value of Intangible Assets

Understanding Intangible Assets

Intangible assets are non-physical assets that contribute significantly to an organization's value and competitive advantage. These include intellectual property, brand reputation, employee skills, and organizational culture. Unlike tangible assets, intangible assets are not easily quantifiable, making their measurement a significant challenge in innovation assessment (Lev, 2001; Teece, 2000).

The Importance of Intangible Assets

Intangible assets are crucial in driving innovation. They often form the backbone of a company's innovation capabilities, fostering creativity, and enabling the development of new products and services. For instance, a company's culture of innovation and the skills of its employees are critical drivers of innovative activities (Lev, 2001; Teece, 2000). Furthermore, intellectual property such as patents and trademarks protect innovative ideas and contribute to a company's competitive edge.

Challenges in Measuring Intangible Assets

Measuring intangible assets involves several challenges:

Valuation Difficulties: Unlike tangible assets, the value of intangible assets is often subjective and context dependent. For instance, the value of a brand can vary significantly based on market perception and consumer behavior (Lev, 2001).

Lack of Standardized Metrics: There is no universally accepted method for measuring intangible assets. Different organizations and industries may use varying metrics, making comparisons difficult (OECD, 2010).

Dynamic Nature: Intangible assets are dynamic and can change rapidly. For example, employee skills and organizational culture evolve over time, complicating the measurement process (Teece, 2000).

Methods for Measuring Intangible Assets

Despite the challenges, several methods have been developed to measure intangible assets:

Balanced Scorecard: The balanced scorecard is a strategic planning and management system that includes intangible assets as part of its measurement framework. It assesses performance from multiple perspectives, including financial, customer, internal processes, and learning and growth (Kaplan & Norton, 1992).

Intellectual Capital Measurement: This approach focuses on assessing intellectual capital, which includes human capital (skills and expertise of employees), structural capital (processes and databases), and relational capital (relationships with customers and partners) (Edvinsson & Malone, 1997).

Brand Valuation Models: Various models, such as Interbrand's brand valuation model, estimate the value of a brand based on its financial performance, role in purchase decisions, and competitive strength (Interbrand, 2021).

Case Studies

Example 1: Google. Google is renowned for its intangible assets, particularly its innovative culture and intellectual property. The company invests heavily in employee development and fosters a culture of creativity and risk-taking. Google's numerous patents protect its innovative products and services, contributing to its competitive advantage (Edelman, 2011).

Example 2: Apple. Apple's brand is one of its most valuable intangible assets. The company's brand valuation is consistently high, reflecting its strong market presence and consumer loyalty. Apple's innovation-driven culture and intellectual property portfolio also play crucial roles in its success (Forbes, 2021).

Measuring Innovation in Developing Countries

Measuring innovation in developing countries requires a comprehensive approach that integrates quantitative surveys with qualitative insights and specialized metrics. Case studies and qualitative methods provide in-depth narratives of innovation processes, while innovation surveys offer systematic data collection on inputs, activities, and outcomes. Frugal innovation metrics and ecosystem assessments further enhance understanding by focusing on affordability, sustainability, and the enabling conditions for innovation. By employing these methods, policymakers, researchers, and development practitioners can effectively measure and support innovation initiatives that contribute to inclusive economic growth and social progress in developing countries.

The Importance of Innovation in Developing Countries

Innovation is crucial for economic growth and development, particularly in developing countries. It can drive productivity improvements, enhance competitiveness, and address social and environmental challenges (World Bank, 2010). However, measuring innovation in these contexts presents unique challenges.

While innovation holds significant promise for driving economic growth and addressing societal challenges in developing countries, effective measurement is essential to realize these benefits fully. Policymakers, researchers, and development practitioners must adopt tailored approaches to capture the multifaceted nature of innovation in these contexts. By overcoming measurement challenges and fostering an enabling environment for innovation, developing countries can leverage innovation as a powerful tool for sustainable development and inclusive growth (World Bank, 2010).

Economic Growth and Productivity Improvements

Innovation drives economic growth by spurring technological advancements, improving efficiency, and fostering new industries. In developing countries, where traditional economic sectors may dominate, innovation introduces new methods, products, and services that enhance productivity. For instance, innovations in agriculture can lead to higher crop yields, reduced wastage, and increased incomes for farmers. Similarly, advancements in manufacturing technologies can boost production efficiency and export competitiveness.

Enhancing Competitiveness

Innovation is a cornerstone of competitiveness in the global market. Developing countries can harness innovation to differentiate their products and services, penetrate new markets, and attract investment. By fostering an environment conducive to innovation, countries can build resilient industries capable of withstanding global economic fluctuations and technological disruptions. This, in turn, creates opportunities for job creation, skill development, and overall economic diversification.

Addressing Social and Environmental Challenges

Innovation in developing countries extends beyond economic benefits to address critical social and environmental challenges. For example, innovations in healthcare can improve access to affordable and effective medical treatments, thereby enhancing public health outcomes. Environmental innovations focus on sustainable practices that mitigate climate change, conserve natural resources, and promote ecological resilience.

Challenges in Measuring Innovation in Developing Countries

Despite its transformative potential, measuring innovation in developing countries presents unique challenges. These include:

Limited Data Availability: Developing countries often lack comprehensive data infrastructure and statistical systems to accurately capture innovation activities and outcomes. This can be due to limited resources for conducting surveys and collecting data, as well as inadequate infrastructure for data management (UNESCO, 2015).

Informal Economy: A significant portion of economic activity in developing countries occurs in the informal sector, which is not captured by traditional innovation metrics due to local socio-economic conditions, cultural factors, and institutional frameworks. This sector includes small-scale enterprises and informal workers, whose innovative activities are often overlooked (OECD, 2011).

Diverse Innovation Forms: Innovation in developing countries can take diverse forms, such as frugal innovation, which focuses on affordable solutions for low-income markets. These forms of innovation may not fit traditional measurement frameworks (Radjou & Prabhu, 2012).

Resource Constraints: Limited financial and technical capabilities can constrain innovation measurement efforts, and hinder the adoption of formal innovation measurement methods, such as surveys and technological indicators. Developing countries may lack the necessary expertise and infrastructure to implement comprehensive innovation measurement systems (World Bank, 2010).

Policy and Institutional Support: Effective measurement of innovation requires supportive policies, institutional frameworks, and regulatory environments. Developing countries may struggle with governance issues, corruption, and bureaucratic inefficiencies that impede innovation measurement and implementation (World Bank, 2010).

Methods for Measuring Innovation in Developing Countries

Despite the challenges, several approaches can be used to measure innovation in developing countries:

Innovation Surveys: Innovation surveys are fundamental tools for assessing innovation in developing countries. Models such as the Community Innovation Survey (CIS) and the Oslo Manual, developed by OECD and Eurostat, respectively, provide frameworks adaptable to diverse economic contexts (OECD/Eurostat, 2018). These surveys gather comprehensive data on various aspects of innovation, including inputs, activities, and outputs. By collecting information on innovation expenditures, sources of funding, and collaboration networks, they offer insights into the innovation landscape that quantitative data alone cannot fully capture.

Case Studies and Qualitative Methods: Qualitative methods, such as case studies and ethnographic research, complement quantitative surveys by delving deeper into the specific contexts and processes of innovation in developing countries (Yin, 2014). Case studies offer detailed narratives of innovation initiatives, highlighting the challenges, strategies, and outcomes that quantitative data may overlook. They provide a rich understanding of the socio-economic factors influencing innovation and the role of local knowledge and networks in shaping innovative practices.

Frugal Innovation Metrics: Metrics tailored to frugal innovation capture the unique characteristics of affordable and sustainable solutions developed in resource-constrained environments (Radjou & Prabhu, 2012). These metrics emphasize criteria such as cost-effectiveness, scalability, and environmental sustainability, reflecting the principles of frugal innovation. By focusing on accessibility and usability for low-income communities, frugal innovation metrics provide a nuanced understanding of innovation's impact on inclusive growth and development.

Innovation Ecosystem Assessments: Assessing the innovation ecosystem in developing countries involves evaluating the interconnected factors that influence innovation outcomes. This includes examining the role of government policies, educational institutions, infrastructure, and financing mechanisms in fostering an enabling environment for innovation (World Economic Forum, 2013). Innovation ecosystem assessments highlight the importance of supportive frameworks and collaborative networks in promoting innovation-driven growth and sustainable development.

Case Studies

Example 1: Kenya's Mobile Money Innovation

Kenya's M-Pesa represents a notable example of innovation in financial services tailored to the needs of developing countries. Launched in 2007 by Safaricom, M-Pesa enables users to perform financial transactions via mobile phones, including money transfers and bill payments (Jack & Suri, 2014). The innovation significantly enhances financial inclusion by providing access to banking services for millions of unbanked individuals. Measuring the impact of M-Pesa involves both quantitative metrics, such as transaction volumes and user adoption rates, and qualitative assessments of its socio-economic benefits. Studies have shown that M-Pesa has contributed to poverty reduction, improved household financial management, and spurred entrepreneurial activities in Kenya (Jack & Suri, 2014).

Example 2: India's Frugal Innovation

India has emerged as a global hub for frugal innovation, characterized by the development of affordable and sustainable solutions that address the needs of low-income populations. The Jaipur Foot is one such example of frugal innovation, providing a low-cost prosthetic limb that has restored mobility to thousands of amputees across India (Prahalad, 2010). Measuring the impact of frugal innovation involves assessing metrics beyond traditional economic indicators, such as affordability, accessibility, and user satisfaction. These innovations not only address critical societal challenges but also catalyze economic opportunities by creating new markets and employment prospects.

Policy Implications

Measuring innovation in developing countries has significant policy implications. Policymakers need reliable data on innovation activities to design effective policies and allocate resources efficiently. This includes supporting innovation in both the formal and informal sectors, fostering frugal innovation, and building the necessary infrastructure for data collection and analysis (UNESCO, 2015).

SUGGESTIONS FOR FUTURE RESEARCH

Developing a Comprehensive Framework that Could Be Used by Practitioners in Innovation Management Field

Innovation management is a dynamic and complex field that requires a structured approach to ensure sustainable competitive advantage. To aid practitioners in this field, future research should focus on developing a comprehensive framework that encompasses the entire innovation process. This framework should integrate key components such as strategy, process, culture, structure, and performance measurement while ensuring a strong linkage between academic research and practical application. Additionally, the framework should be tailored to be applicable in developing countries, taking into consideration their unique contexts and challenges.

Strategy

A clear innovation strategy aligns innovation efforts with the overall business strategy, ensuring that innovation initiatives support the company's long-term goals. Future research should investigate how organizations can effectively develop and communicate an innovation strategy that fosters alignment across all levels of the organization. This includes exploring the role of leadership in setting a vision for innovation and creating a supportive environment for innovative thinking (Tidd & Bessant, 2020). In developing countries, this strategy must also address local market needs, resource constraints, and socio-economic conditions (Dodgson, Gann, & Phillips, 2014).

Process

The innovation process involves a series of stages from idea generation to commercialization. Developing a standardized process model that can be tailored to different organizational contexts is crucial. Research should focus on identifying best practices for managing each stage of the innovation process, including ideation, evaluation, development, and market introduction. This involves studying methodologies such as design thinking, lean startup, and agile development, and determining how they can be integrated into a cohesive framework (Ries, 2011; Brown, 2008). For developing countries, the process model should emphasize frugality, scalability, and adaptability to local conditions (Radjou & Prabhu, 2012).

Culture

Organizational culture plays a significant role in fostering innovation. Future research should explore how to build and sustain a culture that encourages creativity, risk-taking, and collaboration. This includes examining the impact of organizational values, leadership styles, and employee engagement on innovation outcomes (Cameron & Quinn, 2011). Additionally, research should investigate how to overcome cultural barriers to innovation and promote a mindset that embraces change and continuous improvement. In the context of developing countries, fostering a culture of innovation may also involve addressing socio-cultural norms and encouraging grassroots innovation (Gupta, 2016).

Structure

The organizational structure must support innovation efforts. Future studies should examine different structural configurations, such as cross-functional teams, innovation hubs, and decentralized structures, to determine their effectiveness in facilitating innovation. Research should also explore the role of external networks and partnerships in enhancing the innovation capabilities of organizations (Chesbrough, 2003). This includes investigating how to manage open innovation and leverage external knowledge and resources effectively. In developing countries, structural configurations should account for resource limitations and the need for flexibility and resilience (Von Tunzelmann, 2009).

Performance Measurement

Measuring innovation performance is critical for assessing the effectiveness of innovation initiatives and making informed decisions. Future research should develop a comprehensive set of metrics that capture both the inputs and outputs of the innovation process. This includes quantitative measures such as R&D spending, the number of patents, and revenue from new products, as well as qualitative measures such as customer satisfaction and employee engagement (OECD/Eurostat, 2018). Research should also explore the use of advanced analytics and big data to gain deeper insights into innovation performance. In developing countries, performance metrics should reflect local priorities and capabilities (World Bank, 2010).

Integration of Digital Technologies

The integration of digital technologies into the innovation process is a critical area for future research. Technologies such as artificial intelligence, big data, and blockchain have the potential to transform how organizations innovate. Future studies should investigate how these technologies can be incorporated into the innovation framework to enhance efficiency, reduce costs, and improve decision-making (Brynjolfsson & McAfee, 2014). This includes exploring the use of digital tools for idea management, prototyping, and market testing. For developing countries, digital technologies should be accessible and scalable to ensure widespread adoption and impact (Graham & Mann, 2013).

Case Studies and Best Practices

To develop a practical framework for innovation management, future research should include extensive case studies of successful and unsuccessful innovation initiatives across different industries. This will provide valuable insights into the factors that contribute to innovation success and failure. Research should also compile best practices from leading organizations and identify common themes and strategies that can be applied in various contexts (Yin, 2014). In developing countries, case studies should focus on local innovations and highlight how constraints can be turned into opportunities (Prahalad, 2010).

Framework Validation

Once a comprehensive framework for innovation management is developed, it should be validated through empirical research. This involves testing the framework in real-world settings and evaluating its impact on innovation performance. Future research should include pilot studies in different organizational contexts to assess the feasibility and effectiveness of the framework. Feedback from practitioners should be incorporated to refine and improve the framework (Eisenhardt & Graebner, 2007). For developing countries, validation should consider diverse environments and resource conditions (OECD, 2011).

Developing Unique Sectoral-Relevant Indicators that could be Measured

Innovation is not a one-size-fits-all concept; different sectors have unique characteristics and requirements. Therefore, developing sector-specific indicators that accurately measure innovation performance is crucial for providing actionable insights. Future research should focus on identifying and validating these indicators across various industries, ensuring they are relevant to both developed and developing countries.

Manufacturing

In the manufacturing sector, innovation indicators should capture the efficiency and effectiveness of production processes. Future research should explore metrics such as production cycle time, yield rates, and energy consumption. Additionally, research should investigate the role of advanced manufacturing technologies such as 3D printing, robotics, and the Internet of Things (IoT) in driving innovation (Manyika et al., 2017). This includes developing indicators that measure the adoption and impact of these technologies on production efficiency and product quality. For developing countries, indicators should also consider resource availability and the potential for sustainable practices (World Economic Forum, 2016).

Healthcare

The healthcare sector is characterized by stringent regulatory requirements and a focus on patient outcomes. Innovation indicators in this sector should capture the impact of new treatments, medical devices, and healthcare delivery models on patient care. Future research should explore metrics such as patient satisfaction, treatment success rates, and cost-effectiveness (Porter & Lee, 2013). Additionally, research should investigate the role of digital health technologies, such as telemedicine and electronic health records, in driving innovation and improving healthcare outcomes (Raghupathi & Raghupathi, 2014). In developing countries, indicators should address accessibility and affordability of healthcare innovations (WHO, 2015).

Information Technology

The IT sector is driven by rapid technological advancements and short product life cycles. Innovation indicators in this sector should capture the speed and agility of innovation processes. Future research should explore metrics such as time-to-market, the number of software updates, and user adoption rates (Varian, 2010). Additionally, research should investigate the role of emerging technologies such as artificial intelligence, blockchain, and cybersecurity in driving innovation. This includes developing indicators that measure the adoption and impact of these technologies on IT products and services (Brynjolfsson & McAfee, 2014). For developing countries, indicators should reflect digital inclusion and technology adoption rates (Graham & Mann, 2013).

Energy

The energy sector is undergoing a transformation driven by the need for sustainable and renewable energy sources. Innovation indicators in this sector should capture the development and deployment of new energy technologies. Future research should explore metrics such as energy efficiency, carbon footprint, and the adoption of renewable energy sources (IEA, 2019). Additionally, research should investigate the role of smart grid technologies and energy storage solutions in driving innovation and enhancing the sustainability of energy systems (World Economic Forum, 2013). In developing countries, indicators should consider the availability of infrastructure and the impact on energy access (UN, 2015).

Retail

The retail sector is characterized by a focus on customer experience and supply chain efficiency. Innovation indicators in this sector should capture the impact of new retail technologies and business models on customer satisfaction and operational performance. Future research should explore metrics such as customer retention rates, inventory turnover, and sales per square foot (Hagberg et al., 2016). Additionally, research should investigate the role of e-commerce, omnichannel strategies, and data analytics in driving innovation. This includes developing indicators that measure the adoption and impact of these technologies on retail performance (Piotrowicz & Cuthbertson, 2014). In developing countries, indicators should address the integration of informal markets and local supply chains (UNCTAD, 2017).

Public Sector

Innovation in the public sector is essential for improving the efficiency and effectiveness of government services. Innovation indicators in this sector should capture the impact of new policies, programs, and technologies on public service delivery. Future research should explore metrics such as citizen satisfaction, cost savings, and service delivery times (Bloch & Bugge, 2013). Additionally, research should investigate the role of digital government initiatives, open data, and public-private partnerships in driving innovation. This includes developing indicators that measure the adoption and impact of these initiatives on public sector performance (Mergel, 2016). In developing countries, indicators should reflect efforts to improve governance and service delivery under resource constraints (World Bank, 2016).

Validation and Standardization

To ensure the reliability and comparability of sector-specific innovation indicators, future research should focus on validation and standardization efforts. This involves testing the indicators in real-world settings and refining them based on feedback from practitioners. Additionally, research should explore the development of standardized frameworks and guidelines for measuring innovation in different sectors. This will facilitate benchmarking and enable organizations to assess their innovation performance relative to industry standards (OECD/Eurostat, 2018). For developing countries, standardization should accommodate local contexts and encourage broad adoption (UNESCO, 2015).

Integration with Policy and Practice

For innovation indicators to be useful, they must be integrated into policy and practice. Future research should explore how these indicators can be incorporated into innovation management frameworks and used to inform decision-making. This includes developing tools and methodologies for collecting, analyzing, and reporting innovation data. Additionally, research should investigate the role of government policies and support programs in promoting the adoption of sector-specific innovation indicators. This includes exploring how policies can incentivize organizations to measure and report their innovation performance (OECD, 2011). In developing countries, integration should focus on building institutional capacity and fostering a supportive policy environment (World Bank, 2010).

Continuous Improvement

Innovation measurement is an ongoing process that requires continuous improvement. Future research should explore how sector-specific innovation indicators can be regularly updated and refined to reflect changes in technology, market conditions, and regulatory requirements. This includes developing mechanisms for ongoing monitoring and evaluation of innovation performance. Additionally, research should investigate the role of feedback loops and learning mechanisms in enhancing the effectiveness of innovation measurement (Tidd & Bessant, 2020). This will ensure that innovation indicators remain relevant and provide valuable insights for organizations. In developing countries, continuous improvement should emphasize adaptability and resilience (Radjou & Prabhu, 2012).

CONCLUSION FOR FUTURE RESEARCH

In conclusion, developing a comprehensive framework for innovation management and sector-specific innovation indicators are critical areas for future research. A well-structured framework can guide practitioners in managing the innovation process effectively, while sector-specific indicators can provide actionable insights for improving innovation performance. By linking academic research with practical application and tailoring frameworks to the unique contexts of developing countries, scholars and practitioners can enhance the understanding and practice of innovation management. This will ultimately lead to more effective and impactful innovation initiatives that drive economic growth and development worldwide.

REFERENCES

Archibugi, D., & Filippetti, A. (2018). *The handbook of global science, technology, and innovation*. Wiley.

Bessant, J., Tidd, J., & Pavitt, K. (2020). *Managing innovation: Integrating technological, market, and organizational change* (6th ed.). Wiley.

Björk, J., Frishammar, J., & Sundström, L. (2023). Measuring innovation effectively—Nine critical lessons. *Research Technology Management*, 66(2), 17–27. DOI: 10.1080/08956308.2022.2151232

Blind, K. (2012). The influence of regulations on innovation: A quantitative assessment for OECD countries. *Research Policy*, 41(2), 391–400. DOI: 10.1016/j.respol.2011.08.008

Bloch, C., & Bugge, M. M. (2013). Public sector innovation—From theory to measurement. *Structural Change and Economic Dynamics*, 27, 133–145. DOI: 10.1016/j.strueco.2013.06.008

Brown, T. (2008). Design thinking. *Harvard Business Review*, 86(6), 84. PMID: 18605031

Brynjolfsson, E., & McAfee, A. (2014). *The second machine age: Work, progress, and prosperity in a time of brilliant technologies*. WW Norton & Company.

Cameron, K. S., & Quinn, R. E. (2011). *Diagnosing and changing organizational culture: Based on the competing values framework*. John Wiley & Sons.

Chesbrough, H. W. (2003). *Open innovation: The new imperative for creating and profiting from technology*. Harvard Business Press.

Christensen, C. M., Raynor, M. E., & McDonald, R. (2018). What is disruptive innovation? *Harvard Business Review*, 93(12), 44–53. PMID: 17183796

Crossan, M. M., & Apaydin, M. (2010). A multi-dimensional framework of organizational innovation: A systematic review of the literature. *Journal of Management Studies*, 47(6), 1154–1191. DOI: 10.1111/j.1467-6486.2009.00880.x

Damanpour, F., & Schneider, M. (2006). Phases of the adoption of innovation in organizations: Effects of environment, organization and top managers. *British Journal of Management*, 17(3), 215–236. DOI: 10.1111/j.1467-8551.2006.00498.x

Dodgson, M., Gann, D. M., & Phillips, N. (2014). *The Oxford handbook of innovation management*. Oxford University Press. DOI: 10.1093/oxfordhb/9780199694945.001.0001

Drucker, P. F. (1985). *Innovation and entrepreneurship: Practice and principles*. Harper & Row.

Edelman, B. (2011). Google, Inc. Harvard Business School Case Study, 910-036.

Edison, H., Smørsgård, N. M., Wang, X., & Abrahamsson, P. (2018). Lean internal startups for software product innovation in large companies: Enablers and inhibitors. *Journal of Systems and Software*, 135, 69–87. DOI: 10.1016/j.jss.2017.09.034

Edvinsson, L., & Malone, M. S. (1997). *Intellectual capital: Realizing your company's true value by finding its hidden brainpower*. HarperBusiness.

Eisenhardt, K. M., & Graebner, M. E. (2007). Theory building from cases: Opportunities and challenges. *Academy of Management Journal*, 50(1), 25–32. DOI: 10.5465/amj.2007.24160888

Fagerberg, J., & Verspagen, B. (2009). Innovation studies—The emerging structure of a new scientific field. *Research Policy*, 38(2), 218–233. DOI: 10.1016/j.respol.2008.12.006

Forbes. (2021). The world's most valuable brands. Retrieved from https://www.forbes.com/powerful-brands/list/

Garcia, R., & Calantone, R. (2018). A critical look at technological innovation typology and innovativeness terminology: A literature review. *Journal of Product Innovation Management*, 19(2), 110–132. DOI: 10.1111/1540-5885.1920110

Gault, F. (2018). Defining and measuring innovation in all sectors of the economy. *Research Policy*, 47(3), 617–622. DOI: 10.1016/j.respol.2018.01.007

Gault, F. (2018). *Handbook of innovation indicators and measurement*. Edward Elgar Publishing.

Godin, B. (2015). *Innovation contested: The idea of innovation over the centuries*. Routledge., DOI: 10.4324/9781315855608

Goffin, K., & Mitchell, R. (2017). *Innovation management: Effective strategy and implementation* (3rd ed.). Macmillan International Higher Education. DOI: 10.1057/978-1-137-37344-1

Gopalakrishnan, S., & Damanpour, F. (1997). A review of innovation research in economics, sociology and technology management. *Omega*, 25(1), 15–28. DOI: 10.1016/S0305-0483(96)00043-6

Graham, M., & Mann, L. (2013). Imagining a silicon savannah? Technological and conceptual connectivity in Kenya's BPO and software development sectors. *The Electronic Journal on Information Systems in Developing Countries*, 56(1), 1–19. DOI: 10.1002/j.1681-4835.2013.tb00396.x

Gupta, A. K. (2016). *Grassroots innovation: Minds on the margin are not marginal minds*. Random House India.

Guzmán, G., Salazar, M., & Lucio, J. (2018). Measuring innovation in developing countries: Challenges and opportunities. In Khan, S., & Lutz, M. (Eds.), *Handbook of innovation in public services* (pp. 123–138). Edward Elgar Publishing.

Guzmán, G. M., Del Carmen Alzate, M., & Serna, L. (2018). Innovation and competitiveness in SMEs: The local experience in Medellin, Colombia. *Journal of Innovation & Knowledge*, 3(2), 60–67.

Hagberg, J., Sundstrom, M., & Egels-Zandén, N. (2016). The digitalization of retailing: An exploratory framework. *International Journal of Retail & Distribution Management*, 44(7), 694–712. DOI: 10.1108/IJRDM-09-2015-0140

Hall, B. H., & Harhoff, D. (2018). Recent research on the economics of patents. *Annual Review of Economics*, 4(1), 541–565. DOI: 10.1146/annurev-economics-080511-111008

Hanel, P. (2018). *Intellectual property rights and innovation in SMEs in OECD countries*. Edward Elgar Publishing.

Hanel, P. (2018). Intellectual property rights business management practices: A survey of the literature. *Technovation*, 28(9), 650–664. DOI: 10.1016/j.technovation.2008.02.010

Inglehart, R. (2018). *Cultural evolution: People's motivations are changing and reshaping the world*. Cambridge University Press. DOI: 10.1017/9781108613880

Interbrand. (2021). Best Global Brands 2021. Retrieved from https://interbrand.com/best-global-brands/

International Energy Agency. (2019). [International Energy Agency.]. *Energy Efficiency*, •••, 2019.

Jack, W., & Suri, T. (2014). Risk sharing and transactions costs: Evidence from Kenya's mobile money revolution. *The American Economic Review*, 104(1), 183–223. DOI: 10.1257/aer.104.1.183

Kaplan, R. S., & Norton, D. P. (1992). The balanced scorecard: Measures that drive performance. *Harvard Business Review*, 70(1), 71–79. PMID: 10119714

Keeley, L., Pikkel, R., Quinn, B., & Walters, H. (2013). *Ten types of innovation: The discipline of building breakthroughs*. Wiley.

Kumar, V., & Pansari, A. (2016). Competitive advantage through engagement. *JMR, Journal of Marketing Research*, 53(4), 497–514. DOI: 10.1509/jmr.15.0044

Lev, B. (2001). *Intangibles: Management, measurement, and reporting*. Brookings Institution Press.

Manyika, J., Chui, M., Bughin, J., Dobbs, R., Bisson, P., & Marrs, A. (2017). *Disruptive technologies: Advances that will transform life, business, and the global economy*. McKinsey Global Institute.

Mergel, I. (2016). Agile innovation management in government: A research agenda. *Government Information Quarterly*, 33(3), 516–523. DOI: 10.1016/j.giq.2016.07.004

OECD. (2010). *Handbook on deriving capital measures of intellectual property products*. OECD Publishing.

OECD. (2010). *Measuring innovation: A new perspective*. OECD Publishing., DOI: 10.1787/9789264059474-

Chapter 5
Understanding Barriers to Innovation in Non-Innovative SMEs in Egypt

Nahed Salem
https://orcid.org/0000-0003-3536-8502
The Graduate School of Management of Technology, Nile University, Egypt & Academy of Scientific Research and Technology, Egypt

Alaa-Eldin Adris
https://orcid.org/0009-0007-1405-6047
The Science Technology Innovation Policy Research Hub, Egypt & The Graduate School of Management of Technology, Nile University, Egypt

Hebatalla Kaoud
The Science Technology Innovation Policy Research Hub, Egypt & The Graduate School of Management of Technology, Nile University, Egypt

ABSTRACT

In today's globalized world, innovation is a key enabler that leads to economic growth, competitiveness, and sustainability. The gap between an innovative and any non-innovative SME in all regards is most probably widening, as much of small and medium business in Egypt are still to step into creative methodologies that many corporations have focused on to ensure their success in the market. By identifying the obstacles that prevent SMEs from being innovative, the research will attempt to determine what barriers hinder innovation in Egypt's non-innovative SMEs. This research seeks to identify the Sub-sectoral barriers in manufacturing sector specific to non-innovative SMEs to shed the light on these obstacles. This study is guided by this open ended question: What sub-sectoral barriers are hindering the innovative

DOI: 10.4018/979-8-3373-1132-6.ch005

capacity of non-innovative SMEs in the Egyptian manufacturing sector? This study utilizes an analysis of existing data from the Community Innovation Survey (CIS).

1.INTRODUCTION

In the actual context of a globalized world, innovation has been regarded as an important driver for economic growth, competitiveness, and sustainability (Menna et al., 2019). In all aspects, the innovative SMEs (above) gap with any non-innovative SME is most likely expanding as most small and medium-sized businesses in Egypt and many others have yet to take part in creative ways that a lot of corporations shifted their attention to guarantee competitive status.

SMEs (small-to-medium firms) are the backbone of established and emerging economies. This has long made them a world leader in innovation, wealth generation, economic activity, and job creation. They play a critical role in fostering innovation and in creating competition, generating nearly two-thirds of all jobs every year and are necessary for economic growth. Almost universally, SMEs are being hailed as the cutting edge of flexibility and agile response to market changes compared to lumbering large enterprises. Activities of innovation among SMEs range from the use of new technologies and business processes to new goods and services. Innovation is the main driver of competitiveness for SMEs. It enables these firms to promote their productivity, enhance product quality, reach new markets and respond to customer needs. In the context of SMEs, innovation is not limited to high-tech industries but involves different sectors, including manufacturing, services and agriculture.

1.1 Definition of Innovation and Measurement in SMEs

Innovation is defined as the introduction of a new or improved product or business process (or combination thereof), that is significantly different from the firm's previous products or business processes and that has been introduced on the market or brought into operation within the firm. Product innovation specifically entails a new or improved good or service that is substantially different from the firm's previous offerings and has begun being sold on the market, while a new or improved process involving one or more business functions signifies a departure from how the firm previously produced a product/service and which has been put into practice by the firm(Oslo Manual 2018).

The father of economics, Schumpeter (1934) has identified innovation as one of the main components of economic growth. Innovations make key contributions to economic growth, competitiveness, and sustainable development in a globalized world today. To remain competitive within the marketplace, innovation is being created by firms in all industries (Calantone et al., 2002; Lin, 2007). Small and Medium Enterprises (SMEs) are essential for any economy, also in Egypt, and their innovation performance is key to their survival over time (Alhaji et al., 2022; Perez & Radicic, 2022).

1.2 Importance of Innovation for SME Growth and Competitiveness

The studies by (Barbieri et al., 2019; Hervas-Oliver et al.,2021; Latifi et al., 2021; Apa et al., 2020; Supyuenyong and Swierczek, 2011) delineate essential insights into many issues that border on innovations and the organizational performance of firms, especially SMEs. In the study, (Barbieri et al., 2019) assess the impact of public funding on strategic choices of innovation at firms empirically with a sample dataset of Italian manufacturing firms. The results show that robust public funding is inclined toward composite make-and-buy since it improves absorptive capacity, bringing better innovative performance. (Hervas-Oliver et al., 2021) explored SME inbound open innovation, mainly covering process and product innovations. They found that internal resources and the type of technological innovation pursued shape inbound open innovation.

(Latifi et al.,2021) contributed to the stream by examining the role of Business Model Innovation as a driver of firm performance and noted that BMI is inherently associated with several risks and uncertainties per se. The authors chose data from a sample including 563 European SMEs to develop a conceptual model in which efficiency growth, organizational capabilities, and revenue growth mediate the relationship between BMI and firm performance in a fully mediated way. These findings provide excellent recommendations for company owners and managers if they wish to implement BMI concerning their firm's strategies. (Apa et al.,2020) examined the influence of university-industry collaborations on SMEs' innovation performance in Veneto. Their study emphasized that the different forms of cooperation play an essential role, together with the absorptive capacity of SMEs, in exploiting external knowledge toward better innovation outcomes.

While (Supyuenyong and Swierczek, 2011) assessed the impact of knowledge management capabilities on organizational performance in SMEs, from their sample of 30 Iranian SMEs, they established that knowledge acquisition, as well as sharing and application, all significantly enhance organizational performance. This echoes the previous literature on the requirement to manage human capital to refine KM capabilities. It suggests that strategic investments in such initiatives can result in substantive performance benefits. These contributions have gone jointly to give a much broader understanding of how public funding, innovation typologies, business model innovation, university collaborations, and knowledge management come together in explaining SMEs' innovation and performance.

Indicators should, therefore, be measured in bespoke ways that take into consideration the unique challenges and opportunities facing SMEs. Contrasting sharply with large companies, most SMEs rely on more flexible and agile approaches; therefore, innovation metrics would not apply as usual in such circumstances. This might further include a mix of both quantitative and qualitative measures in the analysis for an all-rounded view of innovation activities, which, among other things, contains surveys, case studies, and performance metrics. SMEs have also benchmarked industry standards and state-of-the-art practices about innovation measurement. This points out the gaps, strengths, and areas of improvement.

It provides valuable insights into the SMEs dealing with improving innovative capabilities, competitiveness, entry into new markets, and improving overall business performance. Innovation measurement in SMEs is therefore central to attaining long-term sustainability through a culture of continuous improvement. If understood and accurately measured, it can help the innovation activities of SMEs to overcome challenging issues of modern business landscapes and further explore new opportunities with sustainable growth.

1.3 Community innovation Survey (CIS)

In terms of policymaking, the Community Innovation Survey (CIS) of the European Union (Eurostat), which is conducted by national institutions in accordance with Commission Implementing Regulation (EU) No. 995/2012 of October 26, 2012 (Organisation for Economic Co-operation and Development OECD, 2005, Eurostat, 2015), is a well-known innovation survey that uses input- and output-oriented indicators. The technical characteristics and economic relevance of a company's new product are discussed using this questionnaire-based approach (Kleinknecht and Bain, 2016; Cricelli et al., 2016).

The issue of a scarcity of innovation data, however, affects a lot of institutions. Businesses are thought to be reluctant to respond to delicate inquiries concerning their innovation procedures. The OECD's "Oslo Manual 2005," which includes instructions for compiling and using data, is the most well-known manual of international innovation indicators.

In this regard, quantitative measures for process innovation were supplied with empirical evidence from the Community Innovation Survey German segment's data by (Rammer, 2023). At this stage, two major output dimensions were considered: the unit cost reductions due to process innovation and the percentage increase in sales share attributed to the enhanced quality of a firm's products because of new or improved process technologies. Indeed, these two dimensions show significantly how process innovation impacts the cost and demand functions of a firm. Using firm-level panel data, Rammer's methodology relates firm characteristics and performance to these output measures, demonstrating their relationships by theoretical expectations and that the proposed measures truthfully reflect the process innovation outcomes of firms.

Similarly, an empirical study was carried out by (Pangsy-Kania et al., 2023) in innovative manufacturing companies across several EU countries and Turkey using data from the CIS 2018, covering 2016-2018. The present study confirms hypotheses regarding the role of business strategies and heterogeneity of innovation outputs within clusters of companies; cluster analysis revealed that manufacturing firms use different business strategies that eventually impact their innovation performance.

(Audretsch et al., 2013) explored open innovation in UK SMEs using panel data from 2002 to 2014.Their findings reiterated further that collaborations with local and foreign partners significantly promote innovation, pointing to the relevance of strategic partnerships and geographical considerations.

Additionally, (Cirera and Muzi, 2020) reacted to the measurement of innovation by pointing to discrepancies in self-reported innovation rates because of variability across survey instruments or due to cognitive biases. In that respect, they recommended high-quality interviews and a clear understanding of innovation, which may obviate mismeasurement—insights transferable across the globe.

Furthermore, (Mei et al., 2019) investigated the inter-firm linkages and innovation activities of Chinese manufacturing SMEs through a questionnaire-based survey. Their study of 420 valid samples demonstrated how inter-firm collaborations and technological adoption considerably enhance innovation performance, thus putting innovation drivers into light while offering a broader view of factors influencing process innovation.

1.4 Overview of the Current State of SME Innovation in Egypt

SMEs play a significant role in economic growth, employment generation, and innovation. The status of SMEs in innovation in Egypt is characterized by dynamism and continuous development, affected by a wide variety of influential factors: economic, social, and technological (Rezk et al., 2016). SMEs represent 80% of Egypt's GDP, 75% of the private sector employment, and 99% of the non-agriculture employment. The government realized the contribution of SMEs in sustaining resilience and diversification; accordingly, it implemented a package of policies to spur innovation and entrepreneurship in this sector. Another strong incentive for innovation by SMEs in Egypt is the increased accessibility to digital technologies.

The importance of Business Model Innovation (BMI) in driving the success of SMEs, particularly in developing regions like the Middle East and Africa, cannot be overstated. Business Model Innovation (BMI) is identified as a critical resilience strategy that enables small businesses to adapt to challenges and uncertainties. BMI mediates the relationship between business ties and business performance(Miroshnychenko et al., 2021). (ElNaggar and ElSayed, 2023) shed light on an under-investigated area of Business Model Innovation in micro and small enterprises within emerging economies. They treat BMI as one of the strategies for resilience. The study assumes that drivers of BMI in smaller enterprises differ considerably from the drivers in large firms in developed economies. Contrarily, the reasons for such divergence turn around dynamic capabilities and efficacy of reconfiguration of resources towards the formation of market value by the owner. It provides an innovative explanation of BMI and its consequences in turbulent business times by linking social capital with market orientation. The study assesses the effect of managerial ties, business owner market orientation, and perceived environmental uncertainty on BMI, together with the consequent impact of BMI on business performance. This will be done by analyzing the responses of 426 micro and small enterprise owners Using Smart PLS 3.3; findings show that BMI mediates the relationship between business ties and business performance. Interestingly, the results show that the owners who have a reactive market orientation contribute more to the BMI compared with those having a proactive market orientation. The insight reached in this regard has brought into the limelight the role played by micro and small business owners in driving BMI and further illustrated that BMI relies on dynamically deploying and configuring resources to generate value

Complementing this, (Rezk et al., 2016) analyze the innovation landscape of private SMEs in Egypt. Drawing on data from the 2015 Egyptian National Innovation Survey, conducted on 2,700 SMEs operating in different manufacturing activities, their study assesses innovation practices and the role of cooperation toward improving SME performance. The results show that 35.5% of Egyptian SMEs had implemented at least one form of innovation, product, or process. It can be noticed that the innovation activity increases with the firm size as measured by the number of employees. Most of the innovative SMEs belong to the manufacturing sector. Only 2.1% of them are dependent on universities, the government, or public research institutions as significant sources of information for the development of products or process innovations. These results underline the role of firm size and extramural cooperation in driving innovation within the Egyptian SME sector.

Adding to this body of knowledge, (Becheikh and Bouaddi,2022) investigated the determinants of innovation in developing economies by examining SMEs at a macro level, wherein they examine significant obstacles they faced in Egypt, Morocco, and Tunisia. Their study thus embodies an effort to bridge the gap in the literature on the determinants of innovation in such contexts. Through the regression analysis, the researchers show that investment in R&D significantly and positively affects the likelihood that firms will innovate and successfully produce a diversity of products and process innovations. This study also revealed that other factors are involved, combinations of which uniquely affect innovation in each of the three countries. Notably, these findings underscore the critical role of R&D in driving innovation but indicate how SME policy support toward innovation should consider country-specific elements when formulated in developing economies.

Further exploring the innovation landscape in Egypt, (Elshamy, 2021) examines the relationship between innovation activity and firm size in Egyptian SMEs. To do this, a questionnaire-based survey, developed from the CIS, is used to collect quantitative and qualitative data about innovation activities related to manufacturing firms from different parts of Egypt: the north, middle, and Upper Egypt, covering 2010-2012. Estimating the relationship between innovation activity and firm size uses a logit model. Still, an ordered logit model is used to estimate the relationship between sources of innovation and performance. The results show a positive link between firm size and innovation activity. This study shows that, from a policy point of view, innovation policies have to define their main objectives and target groups if they wish to support the innovation activity of firms efficiently.

Providing a contemporary examination of the innovation process, According to (El-Gharbawy et al., 2024), considering the role of process innovation capability; this review is done to see the influence of information technology on innovative performance within Egyptian SMEs. Their study assessed how IT flexibility, integration, and agility influenced the creative performance and capabilities of SMEs in a stimulating technology-based business environment. The present research is done to further understand the relationships among IT, process innovation capability, and innovative performance by examining these variables from different perspectives. Based on the previous literature, using data analyzed with the Amos 25 software package for running structural equation modeling, it was found that IT measured by three dimensions—IT flexibility, IT integration, and IT agility—exhibits a significant favorable influence on innovative performance mediated by process innovation capability. The available evidence from Egyptian SMEs stresses the role of IT in enhancing innovation and growth.

Collectively, these studies contribute to providing nuanced insights into innovation and business model innovation of SMEs within the context of emerging economies. They underpin dynamic capabilities, social capital, market orientation,

R&D investment, firm size, external collaboration, and IT as drivers of innovation and business performance. Indeed, these insights must be helpful for policymakers and business leaders seeking to support and improve the innovative capacities of SMEs from developing regions. SME innovation in Egypt is characterized by a dynamic, growing ecosystem driven by digital transformation, supportive policies, international collaborations, and, more recently, regional co-creation. While the conditions are complex, the efforts of too many stakeholders secure the eventual path of an innovative, more resilient SME sector pushing to contribute massively to economic development and global competitiveness in Egypt.

1.5 Barriers to Innovation:

Innovation, in the dynamic business and technological landscape, becomes a significant promoter of growth and competitiveness. However, numerous factors act against the capability of firms to innovate. According to OECD & Eurostat, 2018, an innovation barrier is "a factor preventing non-innovative enterprises from innovating or hindering innovation-active enterprises from introducing particular types of innovation.". Most of the barriers are surmountable, and they may increase costs or cause some technical problems. The definition, therefore, underlines the complex nature of innovation barriers, referring to a wide range of issues that firms must overcome in the process of successful innovation.

Historical perspectives were provided by (Arundel, 1997; Baldwin & Lin, 2002) on the innovation barriers, which are likely to prevent firms from conducting their (R&D&I) activities. These issues are not exclusive to anyone's organization, they affect anyone trying to innovate. Starting from early studies on the topic, literature consistently groups these barriers in an homogeneous system. This systemized plan helps in providing a holistic view for us to be able to understand the common hurdles that companies face during their innovation process. Research by (D'Este et al.,2012 and Pellegrino, 2018) explores in more detail the effects of management hurdles on propensity to innovate. These results suggest that other management-related constraints, which may be higher in firms operating at lower quality levels, will deter innovation. But still then most of the businesses will face some sort of bottlenecks that can hinder them in carrying out their projects for innovation and achieving innovative results. The three barriers that are most widespread in this regard are related to costs, know-how and market. These barriers represent real threats but also driving chances to companies who can formulate solutions to address the same in order.

(Oduro,2020) investigated the barriers of open innovation adoption among SMEs in Ghana. In the year 2020, he conducted a study that used an exploratory sequential design to establish that the identified barriers that prevent OI include collaboration and organizational strategic barriers, while on the other hand, financial and knowledge factors act as driving forces rather than barriers without giving front to the prior literature views about the barriers of SMEs in implementing open innovation. This finding underpins the need for context-specific factors that shape the barriers to innovation.

(Fanelli, 2021) investigated how new technologies can enhance the competitiveness and productivity of SMEs within eight European countries. The main disruptive technologies were identified by this work, and policy recommendations were addressed to set up the most favorable conditions and incentives for the setting of innovative solutions within SME business models. The findings showed that what it takes is supportive policies and resources so it is possible to promote innovation in SMEs. While (Vitola, 2019) focused on challenges and opportunities to introduce innovation in rural SMEs across six European countries. Among others, major challenges were inappropriate policy, environmental deficiencies, knowledge and skill gaps, problems with the hiring of skilled labor, and low competitiveness. Having described the above challenges, in this study a number of policy recommendations for cooperation, networking, training, programs for innovation support, marketing, and workforce availability were proposed.

The obstacles to innovation were assessed among Indonesian manufacturing firms (Hartono and Kusumawardhani, 2018). The barriers were classified into market and institution barriers, employee and organization attitude barriers, finance and risk barriers, and knowledge and cooperation barriers. While employee and organization attitude barriers had a positive effect on innovation, financial, risk, knowledge, and cooperation barriers had a negative effect. This demonstrates how different kinds of barriers may variously influence innovation activity. (Adeyeye et al.,2018) investigated how each barrier individually affects innovative performance and external search strategies of firms in Nigeria. It was found that organizational rigidities lower innovativeness, while regulatory constraints increase innovativeness. Knowledge and infrastructure barriers have been observed to hinder innovation implementation and external knowledge search; therefore, it has been realized that infrastructure and the role of regulation could, in fact, play a very important role in fomenting innovation.

Together, these studies show how multifaceted and context-specific the consequences of innovation barriers may be, and how crucial it is to have supportive policies and resources in place in order to overcome them. Gaining an understanding of these obstacles could help businesses become more innovative and advance towards growth and competitiveness.

2. METHODOLOGY

The study adopts a descriptive research design to identify and understand the various barriers to innovation in non-innovative SMEs in Egypt. The authors conducted an in-depth analysis of data from the Community Innovation Survey (CIS) based on the Oslo Manual (OECD, 2018). The survey designed and implemented by Egyptian Science, Technology and Innovation Observatory affiliated to academy of Scientific Research and Technology and in cooperation with Central Agency for Public Mobilization and Statistics (CAPMAS).

2.1. Survey Sample Design

A stratified systematic sampling method is used to ensure representative coverage of the diverse sectors and regions within the Egyptian SME landscape, The survey sample was 2,197 firms. The manufacturing sector was targeted in 23 Egyptian governorates, The companies were divided into small and medium enterprises according to the number of employees, and the size of the company is classified into small enterprises (10-49 employees) and medium enterprises (50-249 employees) in order to allow comparison of results globally and rely on the European Union (EU) definition of small and medium enterprises (Turner & Ledwith, 2016).

Table 1 shows the sectoral distribution of firms in the sample by ISIC Rev.4 The small and medium-sized firms have a very representative spread across all manufacturing sectors that gives adequate coverage of the vast industrial terrain in Egypt. The largest sector in the sample is the manufacture of other non-metallic mineral products, with a representation of 374 firms and 17.0 percent of the total sample. After this, the manufacture of wearing apparel comprises 14.6% with a total of 320 firms, followed by the manufacture of food products with 11.4% and a total of 250 firms. For firm size, generally, in most firms in every sector, the majority are small firms. For instance, under the manufacture of food products sector, there are 159 small firms compared to 91 medium-sized firms. Similarly, while the manufacture of wearing apparel includes 178 small firms, it also contains 142 medium firms, thus showing some near balance in the distribution of small to medium-sized firms within the sector.

Table 1. Distribution of Sample in CIS based on ISIC Rev4.

Sector	Small Firm Sample	Medium Firm Sample	No of Firm	% of Firm in Sample
Manufacture of food products	159	91	250	11.4
Manufacture of beverages	3	7	10	.5
Manufacture of tobacco products	3	1	4	.2
Manufacture of textiles	133	77	210	9.6
Manufacture of wearing apparel	178	142	320	14.6
Manufacture of leather and related products	33	12	45	2.0
Manufacture of wood and of products of wood and cork, except furniture	18	7	25	1.1
Manufacture of paper and paper products	58	44	102	4.6
Printing and reproduction of recorded media	32	22	54	2.5
Manufacture of chemicals and chemical products	89	70	159	7.2
Manufacture of pharmaceuticals, medicinal chemical and botanical products	9	11	20	.9
Manufacture of rubber and plastics products	96	78	174	7.9
Manufacture of other non-metallic mineral products	119	255	374	17.0
Manufacture of basic metals	26	15	41	1.9
Manufacture of fabricated metal products, except machinery and equipment	105	41	146	6.6
Manufacture of computer, electronic and optical products	4	6	10	.5
Manufacture of electrical equipment	29	47	76	3.5
Manufacture of machinery and equipment	18	17	35	1.6
Manufacture of motor vehicles, trailers and semi-trailers	13	11	24	1.1
Manufacture of other transport equipment	6	1	7	.3
Manufacture of furniture	54	33	87	4.0
Other manufacturing	12	12	24	1.1
Total	1197	1000	**2197**	**100.0**

2.2. Data Collection and Analysis

The Community Innovation Survey (CIS) questionnaire, The Community Innovation Survey (CIS) includes sections on general information includes sections on general information about small and medium-sized firms, types of innovations, innovation activities and performance, and the size of employees. It includes a set of questions that identify innovative and non-innovative companies and the barriers to innovation. Data is collected through face-to-face interviews using CAPMAS teams, IBM SPSS Statistics 26 was used in Analysis of data.

The study differentiates between innovative and non-innovative SMEs by following the standards methods by the Oslo Manual (OECD, 2018). Where Innovative SMEs are those that have introduced new or significantly improved products, services, or processes within the last three years. This includes introducing new-to-market innovations or adopting existing innovations in a new way for the firm. While non-innovative SMEs, conversely, have not made any significant advancements in these areas. This criterion is commonly used in international studies on innovation (OECD, 2018).

3. RESULTS AND DISCUSSIONS

3.1. Innovative and non- innovative Firms

It can be seen in the analysis that 745 out of 2197 are innovative firms, accounting for 34%, and the rest are non-innovative accounting for 66%. That crops up is that the manufacture of other non-metallic products tops in terms of number at 124, representing roughly 33% of the firms in the given sector. In contrast, manufacturing of wearing apparel, with 233 firms accorded the connotation of non-innovative, tops the list of the most non-innovative sectors. It is followed by manufacturing of food products, where an initial look at Figure 1 indicates a huge disparity between the number of innovative at 89 and non-innovative firms at 161.

Figure 1. the Innovative and non-innovative firms by sector of manufacturing

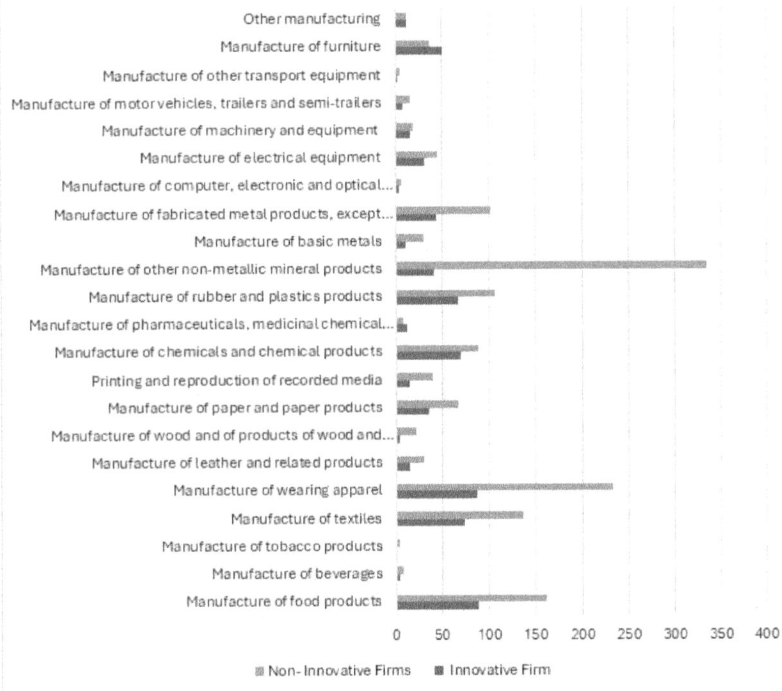

The analysis across innovative and non-innovative firms brings out very large differences in innovation activities among various manufacturing sectors in Egypt. 'Manufacturing of other non-metallic mineral products' shows up as the most innovative sector, with the largest number of enterprises that arc innovative at 124 out of a total of 374 firms, which works out to a share of 33%. At the same time, though small in terms of number, the manufacturing of computer, electronic, and optical products is very prominent, with an innovation rate of 6 out of 10 firms, which accounts for 60%. The manufacturing of motor vehicles, trailers, and semi-trailers is also relatively well balanced, as out of the 24 firms involved in this activity, 11 of them are innovative, which translates to 46%. The least innovative sectors are the manufacture of tobacco products, where no innovative activity is reported at all; in fact, none of the 4 firms were involved in innovation activities.

The second lowest in terms of innovative activity is the manufacture of beverages: here, only 3 out of 10 firms (30%) are innovative. Meanwhile, the leather and related products manufacturing sector needs increased innovation activity, rating 14 out of 45 as innovative. Although this is a bit higher compared to beverage manufacturing, 21 still somewhat reflects a low proportion warranting innovation encouragement and support. Table 2: Distribution of Innovative and Non-Innovative Firms Across

Various Governorates in Egypt. Cairo has 71 innovative firms with 255 non-innovative firms. Alexandria has 25 innovative firms, while non-innovative ones number 146. Port Said has 5 innovative firms, as opposed to the 34 non-innovative firms. In Suez, only 1 firm showed innovation, as opposed to non-innovative firms, of which there were 8. Damietta, on the other hand, had an increased rate of innovation, where out of 52 firms, 25 were innovative.

Table 2. Innovative and Non innovative firm By Governorate

Governorate	Innovative Firms	Non-Innovative Firms
Cairo	71	255
Alexandria	25	146
Port Said	5	34
Suez	1	8
Damietta	25	27
Dakahlia	9	83
Sharqia	154	118
Qalyubia	126	207
Kafr El-Sheikh	9	8
Gharbiya	75	81
Menoufia	43	50
Beheira	19	61
Ismailia	5	27
Giza	80	306
Bani Sweif	11	18
Fayoum	1	27
Minya	7	14
Asyut	6	15
Sohag	8	16
Qena	2	4
Aswan	2	1
Luxor	2	2
New Valley	1	2
Total	687	1510

Damietta is the most innovative governorate in Egypt, with a very high percentage of innovative firms. Of 52 firms in Damietta, 25 were innovative, hence an innovation rate of 48.1%. This very high percentage does indicate an actionable environment for innovation in Damietta. Suez comes as the least innovative governorate, with only 1 firm out of 9 firms being innovative, which gives an innovation rate of 11.1%.

3.1. Innovative and non- innovative SMEs

The distribution of innovative and non-innovative firms based on their size showed that showed that small enterprises have 314 innovative firms out of 1197, while medium have 373 innovative firms out of 1000. Once more, dispersion portrayed that the medium enterprise is generally more innovative than the small enterprise. In general, the innovation rate across all firms is 687 out of 2197, approximately 31.3%.While 356 small enterprises were non-innovative, only 120 small enterprises tried to be innovative in their business activities. On the other hand, 186 medium enterprise groups were non-innovative, while 134 others had the urge for innovative ideas in their business fields. This indicates that innovative enterprises are relatively well-balanced between small and medium enterprises, where medium ones, at 134, stand slightly higher than the small ones at 120. This is an indication that there is a strong involvement of both small and medium enterprise groups in embracing innovative ways of doing things. On the contrary, for non-innovative enterprises, there is a remarkable skewed attraction to small enterprises, with a majority of 356 in comparison with medium-sized ones at 186. This disparity may indicate that smaller enterprises are going through more formidable barriers to implementing innovation, probably because of limitations in resources or higher innovation risks.

Figure 2. breakdown of Innovation in SMEs

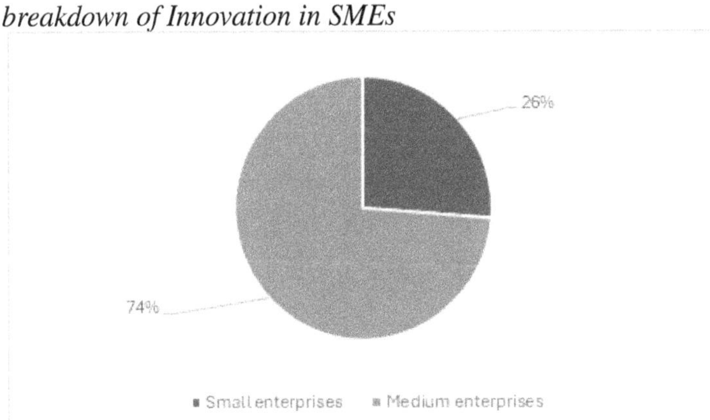

Table3 presents enterprises by manufacturing sectors. If the obtained results are taken into account, it is possible to put forward the sector of "Manufacture of Food Products," which has a great number of innovative enterprises (89) and non-innovative enterprises (161). On the other side, a sector such as "Manufacture of Tobacco Products" has very low counts for both innovative enterprises with 0 and non-innovative enterprises with 4, reflecting a poor competitive framework in which businesses could differentiate themselves by being innovative. While the number of noninnovative enterprises in the sector "Manufacture of Textiles" is high, at 137, the number of innovative enterprises is also considerably high, as it stands at 73. The same case applies to the "Manufacture of Wearing Apparel" sector, whereby the count of innovative enterprises stands at 87, with the number of non-innovative enterprises at 233, showing a fair degree of spread in innovation adoption in this sector.

An analysis by the manufacturing sector shows that in food products there is quite a strong tendency towards innovation, as manifested in the large number of 89 innovative enterprises in comparison to 161 non-innovative ones. While textiles and wearing apparel manifest a healthy mix of innovative and non-innovative enterprises. Textiles have 73 innovative versus 137 non-innovative enterprises, while wearing apparel has 87 innovative versus 233 non-innovative.

Table 3. Innovative and non- innovative enterprises by Manufacturing Sector

Manufacturing Sector	Innovative			Non- Innovative		
	Small enterprises	Medium enterprises	Total	Small enterprises	Medium enterprises	Total
Manufacture of food products	51	38	89	108	53	161
Manufacture of beverages	1	2	3	2	5	7
Manufacture of Tobacco Product	0	0	0	3	1	4
Manufacture of textiles	31	42	73	102	35	137
Manufacture of wearing apparel	37	50	87	141	92	233
Manufacture of leather and related products	8	7	15	25	5	30
Manufacture of wood and of products of wood and cork, except furniture	3	1	4	15	6	21
Manufacture of paper and paper products	17	18	35	41	26	67
Printing and reproduction of recorded media	6	9	15	26	13	39
Manufacture of chemicals and chemical products	37	33	70	52	37	89
Manufacture of pharmaceuticals, medicinal chemical and botanical products	4	8	12	5	3	8
Manufacture of rubber and plastics products	26	41	67	70	37	107

continued on following page

Table 3. Continued

Manufacturing Sector	Innovative			Non- Innovative		
	Small enterprises	Medium enterprises	Total	Small enterprises	Medium enterprises	Total
Manufacture of other non-metallic mineral products	13	27	40	106	228	334
Manufacture of basic metals	5	6	11	21	9	30
Manufacture of fabricated metal products, except machinery and equipment	19	24	43	86	17	103
Manufacture of computer, electronic and optical products	0	4	4	4	2	6
Manufacture of electrical equipment	7	24	31	22	23	45
Manufacture of machinery and equipment n.e.c.	5	11	16	13	6	19
Manufacture of motor vehicles, trailers and semi-trailers	4	4	8	9	7	16
Manufacture of other transport equipment	2	0	2	4	1	5
Manufacture of furniture	32	18	50	22	15	37
Other manufacturing	6	6	12	6	6	12
Total	314	373	687	883	627	1510

The difference between the number of innovative and non-innovative enterprises, 254 and 542, respectively, reflects a huge gap in adoption that characterizes this manufacturing sector. The non-innovative enterprises outnumber innovative ones by more than two to one, and there is a substantive share of the industry probably missing out on the competitive advantage that stems from innovation. Innovative Enterprises: This balance of small and medium-sized enterprises from 120 to 134 shows that both sizes approach each other in innovativeness and, therefore, are almost equally likely to be innovative. Medium enterprises have a slight advantage, probably since they have an easier way of accessing resources and more consolidated processes that can facilitate the push for innovative initiatives. For Non-Innovative Enterprises, the huge bias toward small enterprises at 356 compared with medium ones at 186 shows that the smaller enterprises are the ones that more often do not innovate. This may be due to several reasons: limited available finances, a lack of skilled personnel, and greater risk perceptions associated with innovative activities.

Table 4 and Figure 3 show the analysis of innovative and non-innovative firms based on small and medium enterprises across several governorates; hereby, regarding innovative small enterprises, Cairo prevails with 34 enterprises, while the Alexandria governorate came next with 10. When it came to innovative medium enterprises, Qalyubia prevailed with 73 enterprises, and the Sharqiyah governorate came next with 103. This distribution shows that Cairo, being an economic hub par excellence, may be said to have nourished a fairly large number of innovative small enterprises and that Qalyubia does in the case of medium enterprises. The largest number of non-innovative small enterprises is found in Cairo, with 165 enterprises, followed by of Qalyubia with 137. Likewise, for non-innovative medium enterprise categories, Cairo again leads the table with 90 enterprises, while Qalyubia comes next with 68. This is a trend in Cairo's wide economic base, accommodating both innovative and non-innovative enterprises in large numbers. The mean is 26.17, and the median is 6.50, indicating that half of the governorates have six or fewer of these enterprises.

The standard deviation of 63.34 tells that they are very highly variable in their distribution, ranging from 0 to 314. At the same time, innovative medium enterprises have a mean of 31.08 and a median of 3.50. This suggests an even greater variability with a standard deviation of 77.48 and a range from 0 to 373. On the other hand, non-innovative small enterprises show a mean of 73.58 and a median of 19.50, already indicating that, in general, non-innovative enterprises are more numerous. The standard deviation of 178.14 does show a good deal of dispersion, ranging from 1 to 883 counts. There was an average of 52.25 and a median of 9.00 for non-innovative medium enterprises, with a standard deviation of 130.65 and a range of 0 to 627. These statistics underpin how much more dominant and variable the non-innovative enterprises are compared with their innovative counterpart enterprises.

Table 4. Innovative and non-innovative enterprises by Governorate

Governorate	Innovative Small enterprises	Innovative Medium enterprises	Non- Innovative Small enterprises	Non- Innovative Medium enterprises
Cairo	34	37	165	90
Alexandria	10	15	97	49
Port Said	1	4	25	9
Suez	1	0	3	5
Damietta	19	6	18	9
Dakahlia	7	2	55	28
Sharqiyah	51	103	68	50
Qalyubia	53	73	137	70
Kafr El-Sheikh	7	2	4	4
Gharbiya	37	38	47	34
Menoufia	21	22	29	21
Beheira	12	7	52	9
Ismailia	2	3	21	6
Giza	30	50	96	210
Bani Sweif	6	5	9	9

continued on following page

Table 4. Continued

Governorate	Innovative Small enterprises	Innovative Medium enterprises	Non- Innovative Small enterprises	Non- Innovative Medium enterprises
Fayoum	0	1	14	13
Minya	5	2	9	5
Asyut	6	0	11	4
Sohag	5	3	14	2
Qena	2	0	4	0
Aswan	2	0	1	0
Luxor	2	0	2	0
New Valley	1	0	2	0
Total	314	373	883	627

Small innovative enterprises form the largest percentage in Fayoum and Luxor, with 100% indicating that activities are solely on innovation bases. Many governorates indicate a high percentage of small enterprises being cited as innovative, which depicts a mixed economic environment with different levels of innovation. The percentage of non-innovative small enterprises is highest in Qena, Aswan, Luxor, and New Valley, with 100%. The predominance of non-innovative small enterprises indicates that these governorates have both big challenges and, therefore, potential areas for incubating innovation. This stands to reason, as the percentage of innovative medium enterprises is highest in Giza at 62.5%, with considerable variability across governorates. Likewise, the percentage of non-innovative medium enterprises varied, pointing toward smaller populations of enterprises in some governorates being more non-innovative.

Figure 3. Innovative and non-innovative enterprises for both small and medium enterprises by Governorate

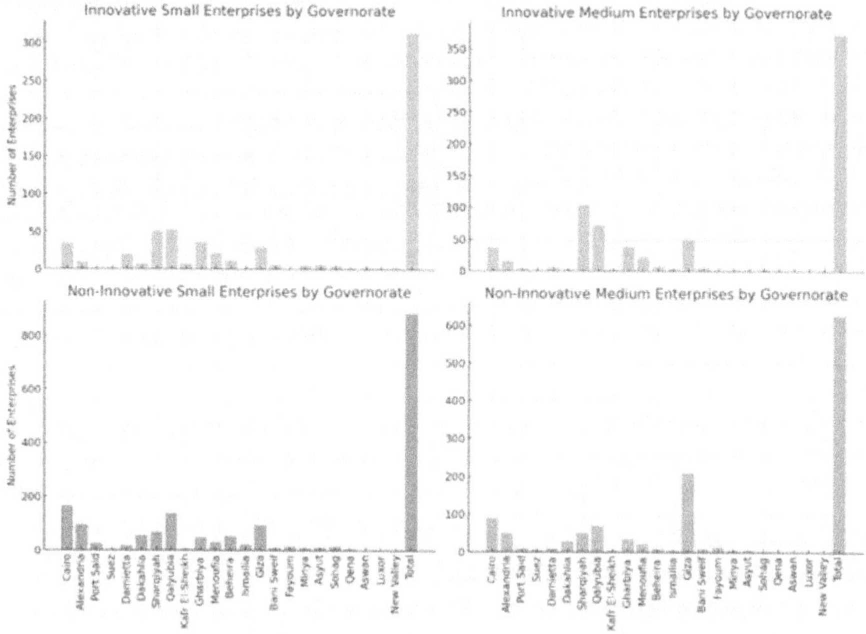

3.2. Barriers to Innovation

Data relating to barriers to innovation were collected on a Likert scale, with certain barriers receiving responses as either High, Medium, Low or Not a Constraint. Concerning the barrier relating to a lack of internal finance for innovation, 36.68% rated it as a high barrier. It represents a very high proportion and explains that a financial constraint is very pervasive amongst firms. Whereas 26.06% of the firms do not think that is a barrier, a variation in access to internal finance amongst different organizations could exist. In the case of no credit or private equity, more than half of the respondents do not consider it a barrier, which implies that many firms either have enough credit or some other source of funding. Nevertheless, this remains a serious problem area for a significant minority of 18.20 percent, strongly pointing to the need for access to external funding as particularly critical for some firms.

With regards to problems in getting public grants or subsidies, 47.16% of the respondents indicated no barrier in that aspect. The implication here is that a good number of enterprises do not find it very difficult to access mechanisms for public funding. However, 26.49% of firms found it high in this barrier. This means there are discrepancies in access to public funding between different parties. Among others, on high costs, 61.57% of the respondents rated it as a high barrier. This suggests that it is a case of innovation efforts being too often brought down by cost, making it the single biggest factor hamstringing innovation efforts for far too many firms. The low percentage of 10.92% that have not considered this to be a barrier paints an accurate picture of how great this constraint is. The lack of skilled employees in your enterprise has rather a broadly scattered response, with 25.91% rating it as high, 28.68 percent medium, and 28.82 percent as not a constraint. This could create the impression of mixed experiences regarding the availability of skilled labor. Correspondingly, while some firms encounter major problems finding skilled employees, others cope relatively well with the problem.

This high cost is the most prominent barrier, identified as a major issue by 423 firms, and indicates that the financial constraint has indeed been a large, pervading problem in making innovation too costly for many firms. The second most cited barrier is market competition, rated as a major issue by 284 firms. Strong competitors may deter firms from investing in innovation if they perceive that the potential gains are at risk and can be easily replicable by others. The second critical barrier is a lack of internal financing for innovation—perceived to be high in the case of 252 firms. It shows that most of these have internal financial constraints in many companies and, therefore, call for better streams or better financial management practices through internal funding mechanisms that will help facilitate innovation efforts. A notable barrier is the internal misalignment of priorities, which is pointed out by 226 firms. This, therefore, implies that the two concepts can ultimately limit innovation efforts within organizations due to conflicting goals and failure to focus on common strategies of innovation.

While only a few can easily find their way through public funding mechanisms, access to public funding, therefore, remains a challenge for 182 firms: lack of skilled employees —for which activities CRIs can and should invest in employee training and development programs—and poorer access to external knowledge, which is a barrier for 172 firms—to enhance knowledge transfer. Facilitating such interactions can enhance the innovation capability of firms due to better exploitation of external expertise and resources. Market uncertainty affects 123 firms, thus showing how unpredictability in the market demand can deter innovation. Extensive market research and the adoption of flexible innovation strategies may help firms bargain with such uncertainty and reduce the associated risks.

For instance, for 113 firms, this barrier is represented by the lack of appropriate collaboration partners. Public policy aimed at encouraging partnerships and networks will, therefore, facilitate better use of external expertise and resources and foster a more cooperative innovation environment among firms. Communication barriers with institutions of research bodies hinder innovation efforts for 110 firms. The problems related to access to markets amount to 67 firms; it foretells difficulty in reaching possible customers. Better market access through improved distribution channels and marketing strategies can help firms grow despite this constraint. For 62 firms, the lack of good ideas for innovations is a barrier. Such firms would need a more creative and open innovation culture that can facilitate the generation of many new ideas for improvement and continuous development. Some firms (62) do not see any need to innovate due to past innovations, thus showing satisfaction with the status quo. This could be beneficial in the short term but may result in complacency.

Figure 4. The number of non-innovative firms that rated various barriers to innovation as "High"

Figure 5: innovation barriers rated "High" by SMEs: From the analysis, high costs appeared as the most important barrier to both small and medium enterprises, where 208 small enterprises and 215 medium enterprises rated it as a high barrier. This shows that issues of cost were common in both firm sizes, and therefore, innovative ways of addressing these financial constraints are highly required. A large proportion of small- as well as medium-enterprises appear to suffer from internal financial constraints and, therefore, feel a need for improvement in internal funding mechanisms in both groups. Another important barrier is an internal misalignment of priorities, with 107 responses in small enterprises and 119 in medium-sized. In availing public grants or subsidies, there are more difficulties faced by medium-scale enterprises than small enterprises, with 98 medium enterprises as opposed to small enterprises with 84 identities that report as a barrier. As regards the attitude of employees, the second most significant barrier to small and medium enterprises is the lack of skilled employees, wherein 86 of the former and 92 of the latter firms reported this.

Small enterprises indicate more significant problems related to searching for external knowledge compared with medium enterprises—93 small against 79 medium enterprises. For market uncertainty, this barrier was very important for medium enterprises since it affected 69 firms, as opposed to 54 small enterprises. Finally, for both small and medium enterprises alike, nearly an equal share point out that the most important obstacle—the one which was selected by 58 and 55 of small and medium enterprises, respectively—is searching for appropriate collaboration partners. Communication with research bodies is, however, much more painful for medium enterprises, with 61 firms affected as compared to 49 small enterprises. Access to markets, though, is a hindrance equally faced by both enterprises- 33 small enterprises and 34 medium enterprises reported difficulties in getting to potential customers. This can be improved upon through better market access achieved via increased channels of distribution and marketing. Both small and medium enterprises also complained that it is hard to come up with good ideas for innovations. On this effect, 29 of the small enterprises and 33 of the medium enterprises pointed this out as one of the major problems. Just a few companies or firms of both categories feel content with the past innovations and, therefore, see no need to innovate more. For this question, 22 small enterprises and 30 medium enterprises responded that there was no need for more innovation.

Figure 5. The number of non-innovative firms by size of firm that rated various barriers to innovation as "High"

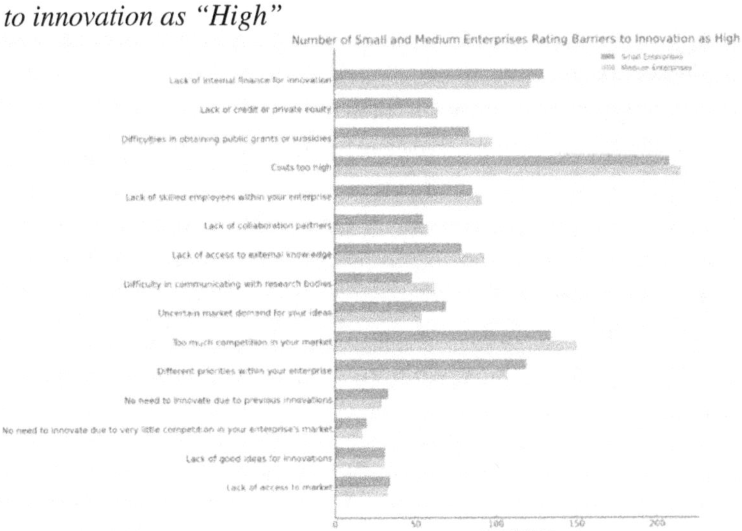

4. CONCLUSION

The analysis of innovative and non-innovative firms across various sectors and governorates in Egypt showed that, out of 2,197 firms, only 34% of them are innovative—a case of a wide gap in innovation adoption. This is observable across different dimensions of a manufacturing sector and firm sizes; hence, there is a requirement for targeted strategies to foster innovation.

This sector represents the highest innovative level since it has a large percentage of firms conducting innovative activities. Since all this falls on the other extreme, industries such as wearing apparel and food products manufacturing seem to have a higher proportion of non-innovative firms. This just means that although some sectors embrace innovation, others are somewhat left behind, probably because of the nature of the sector at play or the changes in the competition level or issues of market demand.

Geographically, this analysis portrayed Damietta as the most innovative governorate because of a large percentage of innovative firms, pointing to a vibrant local innovation environment. Other governorates, such as Suez, have very minimal innovation activity, which points to regional disparities that need to be addressed for there to be balanced economic development in the country. Among barriers to innovation, it comes out that very major ones are high costs and internal financial constraints, together with market competition. While widespread, these barriers rank differently across SMEs. Medium enterprises show a slightly higher openness to innovation than small ones, probably due to easier access to resources and process maturity in support of innovation.

Size-based analysis further reveals that the propensity for failure to innovate among innovative and non-innovative firms is higher in small enterprises due to a lack of resources and relatively higher perceived risks. This disparity, therefore, puts forth the need for special support to small enterprises to build up their innovative capacity. A robust innovation ecosystem in Egypt may only be fostered through a balanced approach that addresses sector-specific and regional challenges. This would cement the overall innovation landscape of the country, playing off its strengths of innovative sectors and regions while giving special attention to supporting the developing ones.

Recommendations

To overcome the barriers to innovation identified in the paper among SMEs in Egypt, several targeted strategies are recommended:

Improving access to finance: This is critical for innovation activities for many SMEs due to financial constraints. Therefore, the Egyptian government and financial institutions should enhance access to grants, subsidies, and low-interest loans dedicated to financing innovation activities.

Promoting business model innovation: Policymakers and business leaders need to create an enabling environment through training on implementing innovative business models for SMEs to adapt to innovative business models.

Investing in skills development: The second most important barrier to innovation comes from the lack of skilled labor in the market. Therefore, this requires investment in workforce development programs by the private sector and the state, which would, over time, provide SMEs with the skills they seek.

Promoting public-private collaboration: Collaboration between SMEs and academic institutions can foster innovation by enabling access to knowledge and resources. Public-private partnerships, especially in strategic sectors such as ICT and manufacturing, maybe the solution needed to boost Egypt's innovation capacity. In addition to implementing the strategies in place, addressing these barriers would go a long way in creating a better ecosystem for SMEs in Egypt. In this way, SMEs can operate with greater strength in an increasingly competitive global market.

ACKNOWLEDGMENT

We would like to extend our sincere gratitude to the Academy of Scientific Research and Technology (ASRT) for providing the invaluable data that made this study possible. Their support and contribution have been instrumental in enabling a comprehensive analysis of the innovation landscape across various sectors and regions in Egypt. Without access to this detailed data, our research would not have been as thorough and insightful. We appreciate the efforts of Nile University, Egypt and ASRT in promoting and facilitating research that contributes to the understanding and advancement of innovation in Egypt.

REFERENCES

Adeyeye, D., Egbetokun, A., Opele, J., Oluwatope, O., & Sanni, M. (2018). How Barriers Influence Firms' Search Strategies And Innovative Performance. *International Journal of Innovation Management*, 22(02), 1850011. DOI: 10.1142/S1363919618500111

Ahmed Alhaji, A. (2022). Financing as a Major Problem to Small and Medium Agriculture Enterprises in Nigeria. A Fact within the Literature. SSRN Electronic Journal. DOI: 10.2139/ssrn.4084771

Apa, R., De Marchi, V., Grandinetti, R., & Sedita, S. R. (2020). University-SME collaboration and innovation performance: The role of informal relationships and absorptive capacity. *The Journal of Technology Transfer*, 46(4), 961–988. DOI: 10.1007/s10961-020-09802-9

Arundel, A. (1997). Enterprise strategies and barriers to innovation. In *Innovation Measurement and Policies* (pp. 101–108). European Commission.

Audretsch, D. B., Belitski, M., Caiazza, R., & Phan, P. (2023). Collaboration strategies and SME innovation performance. *Journal of Business Research*, 164, 114018. DOI: 10.1016/j.jbusres.2023.114018

Bain, D., & Kleinknecht, A. (Eds.). (1993). *New concepts in innovation output measurement*. Springer., https://books.google.com.eg/books?hl=ar&lr=&id=Bv6-DAAAQBAJ&oi=fnd&pg=PA1&ots=bMFH1NjP9g&sig=ryPMFcPZN9W9hQFOdX6NxDtaGIs&redir_esc=y#v=onepage&q&f=false

Baldwin, J., & Lin, Z. (2002). Impediments to advanced technology adoption for Canadian manufacturers. *Research Policy*, 31(1), 1–18. DOI: 10.1016/S0048-7333(01)00110-X

Barbieri, L., Bragoli, D., Cortelezzi, F., & Marseguerra, G. (2019). Public Funding and Innovation Strategies. Evidence from Italian SMEs. *International Journal of the Economics of Business*, 27(1), 111–134. DOI: 10.1080/13571516.2019.1664834

Becheikh, N., & Bouaddi, M. (2022). Macro barriers to business and innovation in developing economies: Evidence from Egyptian, Moroccan, and Tunisian SMEs. Canadian Journal of Development Studies. *Canadian Journal of Development Studies*, 44(2), 151–171. DOI: 10.1080/02255189.2022.2102464

Bigliardi, B., & Galati, F. (2016). Which factors hinder the adoption of open innovation in SMEs? *Technology Analysis and Strategic Management*, 28(8), 869–885. DOI: 10.1080/09537325.2016.1180353

Božić, L., & Rajh, E. (2016, January). The factors constraining innovation performance of SMEs in Croatia. *Ekonomska Istrazivanja*, 29(1), 314–324. DOI: 10.1080/1331677X.2016.1168040

Calantone, R. J., Cavusgil, S., & Zhao, Y. (2002, September). Learning orientation, firm innovation capability, and firm performance. *Industrial Marketing Management*, 31(6), 515–524. DOI: 10.1016/S0019-8501(01)00203-6

Cirera, X., & Muzi, S. (2020). Measuring innovation using firm-level surveys: Evidence from developing countries. *Research Policy*, 49(3), 103912. DOI: 10.1016/j.respol.2019.103912

Cricelli, L., Greco, M., & Grimaldi, M. (2016). ASSESSING THE OPEN INNOVATION TRENDS BY MEANS OF THE EUROSTAT COMMUNITY INNOVATION SURVEY. *International Journal of Innovation Management*, 20(03), 1650039. DOI: 10.1142/S1363919616500390

D'Este, P., Iammarino, S., Savona, M., & von Tunzelmann, N. (2012). What hampers innovation? Revealed barriers versus deterring barriers. *Research Policy*, 41(2), 482–488. DOI: 10.1016/j.respol.2011.09.008

El-gharbawy, M. S. H., A. Ragab, A., A. Ragheb, M., & Farouk, M. (2024). The Effect of Information Technology on Innovative Performance with Mediation Role of Process Innovation Capability: Evidence from Egyptian SMEs. Journal of Business and Management Sciences, 12(2), 67–88. https://doi.org/10.126

ElNaggar, R. A., & ElSayed, M. F. (2023). Drivers of business model innovation in micro and small enterprises: Evidence from Egypt as an emerging economy. *Future Business Journal*, 9(1), 4. Advance online publication. DOI: 10.1186/s43093-022-00180-2

. Elshamy, H. (2021). The Impact of Firm Size on Innovative Activity: An Analysis Based on Egyptian Firm Data. 40(2), 60–75. DOI: 10.21608/caf.2021.154782

Eurostat (2015). Statistics explained - Community innovation survey (CIS). Available online at https://ec.europa.eu/eurostat/statistics-explained/index.php?Title= Glossary:Community_innovation_survey_(CIS)

Fanelli, R. M. (2021). Barriers to Adopting New Technologies within Rural Small and Medium Enterprises (SMEs). *Social Sciences (Basel, Switzerland)*, 10(11), 430. DOI: 10.3390/socsci10110430

Hartono, A., & Kusumawardhani, R. (2018). Innovation Barriers and Their Impact on Innovation: Evidence from Indonesian Manufacturing Firms. *Global Business Review*, 20(5), 1196–1213. DOI: 10.1177/0972150918801647

. Herr, H., & Nettekoven, Z. M. (2017). The role of small and medium-sized enterprises in Development. What can be learned from the German experience.

Hervás-Oliver, J. L., Parrilli, M. D., Rodríguez-Pose, A., & Sempere-Ripoll, F. (2021). The drivers of SME innovation in the regions of the EU. *Research Policy*, 50(9), 104316. DOI: 10.1016/j.respol.2021.104316

Hoelscher, M., & Schubert, J. (2015). Potential and Problems of Existing Creativity and Innovation Indices. *Creativity Research Journal*, 27(1), 1–15. DOI: 10.1080/10400419.2015.992656

. Joseph, A. (1934). Schumpeter, The theory of economic development: An inquiry into profits, capital, credit, interest, and the business cycle.

Latifi, M. A., Nikou, S., & Bouwman, H. (2021). Business model innovation and firm performance: Exploring causal mechanisms in SMEs. *Technovation*, 107, 102274. DOI: 10.1016/j.technovation.2021.102274

Lesáková, U., Gundová, P., Kráľ, P., & Ondrušová, A. (2017). Innovation Leaders, Modest Innovators and Non-innovative SMEs in Slovakia: Key Factors and Barriers of Innovation Activity. *Organizacija*, 50(4), 325–338. DOI: 10.1515/orga-2017-0024

Mei, L., Zhang, T., & Chen, J. (2019). Survey data on inter-firm linkages and innovation activities of Chinese manufacturing SMEs. *Data in Brief*, 27, 104671. DOI: 10.1016/j.dib.2019.104671 PMID: 31720326

Menna, A., Walsh, P. R., & Ekhtari, H. (2019). Identifying enablers of innovation in developed economies: A National Innovation Systems approach. *Journal of Innovation Management*, 7(1), 108–128. DOI: 10.24840/2183-0606_007.001_0007

Miroshnychenko, I., Strobl, A., Matzler, K., & De Massis, A. (2021a). Absorptive capacity, strategic flexibility, and business model innovation: Empirical evidence from Italian SMEs. *Journal of Business Research*, 130, 670–682. DOI: 10.1016/j.jbusres.2020.02.015

Support to implementation of strategies to foster Micro, Small and Medium Enterprises (MiSMESIS) development in Egypt. (n.d.). EEAS. https://www.eeas.europa.eu/node/43122_en

Chapter 6
Digital Transformation of Non-Governmental Organizations in Egypt: Harnessing Technology for Social Impact

Asmaa Moaz
Nile University, Egypt

Mohamed Mamdouh Awny
Nile University, Egypt

ABSTRACT

This study explores the drivers and challenges of digital transformation in Non-Governmental Organizations (NGOs) in Egypt, aiming to provide insights and best practices for decision-makers to enhance the success of digital initiatives in this sector. Utilizing a quantitative research methodology, the online survey gathered responses from 187 participants to collect comprehensive data. Key internal drivers identified include data-driven decision-making (24%), operational efficiency (23%), and employee empowerment (14%). A significant external driver is the competitive environment, noted by 31% of respondents. Major challenges include limited financial resources (52%), inadequate technology infrastructure (43%), and resistance to change (35%). This research proposes a framework for digital transformation in NGOs, featuring five main pillars and 12 sub-pillars. Additionally, a Digital Readiness Score (DRS) tool has been introduced to assess the readiness of NGOs for digital initiatives, offering a structured approach to navigate the transformation process effectively.

DOI: 10.4018/979-8-3373-1132-6.ch006

INTRODUCTION

"Digital Egypt" is a national initiative that aims to transform Egypt into a digital society by leveraging technology to drive economic growth and social development. The initiative seeks to provide access to digital services and opportunities for all Egyptians, including those in rural and underserved areas. It involves a range of projects and initiatives, such as building a digital infrastructure, expanding e-commerce, improving digital literacy, and promoting innovation and entrepreneurship. Through "Digital Egypt", the government is working to create a more competitive and dynamic economy, while also improving the quality of life for citizens. The initiative is an important step towards achieving Egypt's vision of becoming a leading digital economy and a hub for technology and innovation in the region. The digitalization of various government and public services is anticipated to boost productivity, reduce red tape, and improve data accessibility and sharing. These will contribute to rationalizing the decision-making process and result in further alignment, and better resource allocation (Kamel, 2021).

Digital transformation contributes to sustainable development in Egypt and helps the process of improving access to services, enhancing economic growth, enhancing education and healthcare, enhancing innovation and infrastructure development, enabling effective governance, supporting environmental sustainability, facilitating data-based decision-making, and promoting comprehensive development by encouraging international cooperation (Elgohary, 2022).

NGOs in Egypt have been instrumental in contributing to various aspects of development, including health, education, poverty alleviation, and other areas. They are playing a critical role in improving health outcomes, particularly in marginalized communities. They provide access to education for children who would otherwise be excluded from the formal education system, particularly in rural areas. They also assist families living in poverty, including access to food, shelter, and other basic needs (Ghazi, 2021).

In March 2021, the Egyptian Ministry of Social Solidarity (MOSS), launched, for the first time, an Integrated Electronic System (ngos.eg) for Civil Society Organizations (CSOs) to manage their operations nationally. According to the Ministry, the new system provides 36 services for CSOs, including case records, complaints, and beneficiaries' management. Additionally, it seeks to offer an interactive database of the activities and outcomes of the CSOs in Egypt. All the CSOs in Egypt are obliged to register on the new electronic system. According to the law, the CSOs who will not register will not be allowed to work in Egypt.

The Egyptian Ministry of Social Solidarity seeks to achieve digital transformation and automation of all the services provided by it for CSOs to ensure the quality and efficiency of service delivery. Based on that the CSOs in Egypt must adopt the digital transformation process to ensure the quality of services they are providing, accelerate the decision-making process, effectively use the available resources, and improve the overall performance.

LITERATURE REVIEW

The Role of NGOs in Egypt's Development

Non-governmental organizations (NGOs) target a range of social, economic, and political issues which consequently contribute significantly to Egypt's growth. Improving healthcare services and Egyptians' general quality of life is a major focus of many non-governmental organizations in Egypt. Their goal is to offer medical examinations, healthcare, and assistance to disadvantaged populations, especially those living in poor villages (Mohsen, 2018). Some NGOs engage in projects related to development and capacity building. Their objective is to improve the socio-economic conditions, knowledge, and skills of individuals and communities by offering training, resources, and support. NGOs acknowledge the significance of education in enabling people and societies. They concentrate on projects like encouraging students to get an adequate education, supporting educational institutions and schools, offering scholarships, and pushing for changes in the educational system. Sustainability and environmental conservation are main objectives of many NGOs. They aim to promote renewable energy, fight climate change, safeguard natural resources, and increase the general public's awareness of environmental issues. NGOs often focus on programs for economic growth and poverty reduction. To improve areas of disadvantage, NGOs focus on initiatives related to microfinance, entrepreneurship, job creation, vocational training, and community development.

The National Alliance for Civil Society Development in Egypt is a collaborative network of civil society organizations, non-governmental organizations, and other entities involved in development work within Egypt. According to the State Information Service (SIS) affiliated with the Presidency of the Republic (2024), this coalition was established in 2022 and aims to enhance the effectiveness of civil society initiatives, promote sustainable development, and coordinate efforts to address social, economic, and environmental challenges in Egypt. The National Alliance for Civil Society Development is composed of 34 developmental and service entities in Egypt. The National Alliance operates in 27 governorates in Egypt, including border and remote areas like Al-Wadi Al-Jadid, North Sinai, and South Sinai. It provided numerous services to the most vulnerable groups, benefiting more than 20 million citizens. It was able to deliver various services to the most vulnerable groups, reaching over 25 million beneficiaries with food assistance of all kinds, 5 million beneficiaries with healthcare services, and providing 40,000 job opportunities for people with disabilities. Additionally, it supported financially disadvantaged university students, paid off debts of indebted women, launched the "Plant" initiative targeting 100,000 small farmers to cultivate 150,000 Feddans, and rebuilt several homes for the needy. All of this was achieved at a cost exceeding 12 billion pounds. Furthermore, the alliance made significant relief efforts within Egypt and abroad to support our brothers in Palestine since the onset of the Israeli aggression on the Gaza Strip, and previously supported our Sudanese brothers due to the ongoing civil war in Sudan (Alsharkawy, 2024). In May 2024, President Abdel Fattah El-Sisi issued Presidential Decree No. 149 of 2024 to establish the National Alliance for Developmental Civic Work. Overall, NGOs in Egypt contribute to various aspects of development, including democracy promotion, healthcare improvement, human rights advocacy, capacity building, and collaboration for socio-economic progress.

NGOs are involved in addressing a wide range of Sustainable Development Goals (SDGs), including poverty eradication, quality education, gender equality, clean water and sanitation, climate action, and more. They work towards these goals by collaborating with governments, private sector entities, and local communities to create sustainable solutions and promote social and environmental change (Koshy, 2022).

Challenges and Success Factors of Digital Transformation

For organizations, digital transformation is not a straightforward process to complete; it may take years or multiple attempts before the transformation is implemented successfully. According to McKinsey (Martin, 2018), years of transformations studies have revealed that the success rate for digital transformation initiatives is typically low (e.g. fewer than 30% success). Even digitally aware sectors like high tech, media, and telecom are struggling. The success percentage in these sectors is no higher than 26%. Digital transformations are more difficult in more traditional sectors, including infrastructure, oil and gas, automotive, and pharmaceuticals, where success rates range from 4 to 11% (Martin, 2018).

Senior managers are working extremely hard to achieve true digital transformation, yet the outcomes fall far short of the goals and expectations of top leaders. CEOs and senior executives identified digital transformation risk as their top issue for 2019. However, "70% of all digital transformation initiatives do not reach their goals," and "of the $1.3 trillion that was spent on digital transformations last year, it was estimated that $900 billion went to waste" (Tabrizi et al., 2019).

Van Veldhoven and Vanthienen (2023) stated that projects for digital transformation are difficult and frequently unsuccessful. Organizations have difficulties establishing effective digital transformation initiatives. These projects' size and complexity make them challenging from a strategic standpoint as well (Schneider and Kokshagina, 2021). However, they are also accompanied by problems in terms of technology (Fuchs and Hess, 2018), culture (Diener and Špaček, 2021), and behavior (Saarikko, et al., 2020). Many digital transformations efforts fail, with an estimated failure rate of between 66% and 87.5% (Libert et al., 2016).

Many of the challenges experienced by the organizations may be attributed to the high percentages of digital transformation failures. Regarding the digital transformation property of cross-functional processes, a wide range of barriers, such as those listed in Table 1, are likely to be involved.

Table 1. Barriers to Success of Digital Transformation in Cross-Functional Processes

Barrier	Reference
Lack of urgency	(Fitzgerald et al., 2013)
Lack in digital talent	(Baskin, 2018)
Lack of digital culture	(Hartl and Hess, 2017), (Fitzgerald et al., 2013), (Baskin, 2018)
Poor communication	(Baskin, 2018)
Constant competition	(Baskin, 2018)
Not enough funding	(Fitzgerald et al., 2013)
Roles and responsibilities are not clear	(Fitzgerald et al., 2013)
Regulatory concerns	(Fitzgerald et al., 2013)
Unrealistic expectations	(Baskin, 2018)
Business units implementing independently in silos	(Fitzgerald et al., 2013)
Limitations of IT systems	(Fitzgerald et al., 2013)
Lack of vision	(Fitzgerald et al., 2013)
Unclear business case	(Fitzgerald et al., 2013)

Von Leipzig et al. (2017) addressed the following barriers faced by organizations during the phases of digital transformation, high implementation risks and costs, inadequate IT systems, a lack of technical competence, and unsuitable business processes. Even organizations that are at the most advanced stage of their digital transformation process in terms of activities and workflows are not yet completely ready to take on the challenges of digital transformation (Matt and Rauch, 2020).

The following are the key areas that must be prioritized for organizations to achieve the level of digital excellence (Legner et al., 2017):

§ Digital leadership and digital authorization
§ Data-driven agility

§ Stakeholder participation
§ Digital platform management
§ Business model innovation
§ IT architecture transformation
§ Process digitization and automation
§ Digital security and compliance

According to McKinsey's research, the twenty-one keys to success are a set of elements that may increase the success rate of a digital transformation. These variables can be categorized into the following five groups (Martin, 2018):

- having the right, digital-savvy leaders in place
- building capabilities for the workforce of the future
- empowering people to work in new ways
- giving day-to-day tools a digital upgrade
- communicating frequently via traditional and digital methods

Best Practices within the Digital Transformation in NGOs

Digital transformation is dramatically changing how NGOs operate. This change helps to improve the efficiency, reach out, and social impact of NGOs. Despite the difficulty in finding an abundance of published literature concerning best practices for digital transformation in NGOs, in the following part, the researcher suggests and clarifies best practices within digital transformation in NGOs, highlighting how these organizations benefit from technology to achieve their missions.

Sheombar and Ravesteijn (2022) discussed that Northern-based NGOs in the Netherlands are increasingly turning to digital social entrepreneurship, leveraging digital technologies such as the Internet and social media in their ICT for Development (ICT4D) projects. This shift allows them to become more independent from government funding and explore new ways of providing services. Digital transformation helped these organizations to be less dependent on traditional funding sources, such as government grants. By adopting digital technologies, they can diversify their sources of funding, engage in social enterprise activities, and generate revenue through innovative methods. On the other hand, Digital transformation allows Northern-based NGOs in the Netherlands to improve their service delivery models. They leverage digital technologies to reach more beneficiaries, deliver services more efficiently, and measure their impact more effectively.

In the health sector, Tanzania has a digital mapping platform for NGOs to strengthen civil society engagement, in pursuit of universal health coverage. This platform is an interactive website that gathers, analyzes, and displays information about NGOs in Tanzania from many sources using a variety of interactive charts, graphs, and maps. Technology offers countless potential for stakeholders, particularly communities, as it makes it possible to capture and analyze data in new ways and to use mapping as a tool for visibility, information sharing, fostering collaboration, and improving transparency (Mathenge et al., 2023). According to Bhuiyan & Haque (2024), the COVID-19 pandemic brought widespread familiarity with telehealth and telemedicine services. NGOs provided all kinds of health care, which were made accessible to the poor in Dhaka, Bangladesh, using mobile networks.

According to Omdena (2024), there are many NGOs (like Oxfam International, Care International, and the International Committee of the Red Cross) that support people through the distribution of tangible items like food, clothing, and many other things. These NGOs efficiently manage their distribution and logistics processes by utilizing digital tools. They use many well-liked digital technologies, such as GoFreight, Soloplan Carlo, and Linbis (Software for logistics that offers a broad range of functions for supply chain management and freight transportation. It provides strong automation tools and extensive customization).

Concerning the fund opportunities, there is a digital platform – Tasharuky- specialized in crowdfunding designed to be a safe environment to provide smart solutions using the latest financial technology methods and a portal that brings together trusted humanitarian institutions with funders. Tasharuky is distinguished in applying the principle of participation since it operates in an international scale and provides supporters with multiple options in presenting the organizations' projects in one place. Moreover, it is compatible with the principles of Islamic Sharia. Tasharuky allows receiving money for projects all over the world through secure payment methods. This digital platform was established in November 2022, registered in Indonesia, headquartered in Jakarta (Tasharuky, 2022).

In Egypt, some NGOs use different technologies for various purposes. For example, Misr El-Khier Foundation used an ERP system to automate and speed up the internal process, delivering its service to the beneficiaries in a shorter time. Egyptian Food Bank used SMS verification code technology to control fraud and ensure that the service was received by the right beneficiaries.

RESEARCH METHODOLOGY

Research Problem

Although digital technologies hold great potential, previous research, indicates that in general NGOs have not fully adopted digital technology, with a significant emphasis on communication. Digital technologies can support NGOs in achieving and maintaining sustainability and fulfilling their missions more efficiently and effectively (Lynn et al., 2022). To date, no academic research has developed a digital transformation framework specifically for NGOs in Egypt. In the present digital age, digital transformation of institutions has become a necessity. There is no choice, as today's world has become increasingly dependent on digital technology. Therefore, there is an urgent need to understand the challenges and drivers associated with the adoption and integration of digital technologies by NGOs in Egypt.

Hence, the problem addressed in this research is: "Understanding the challenges and drivers with the aim of developing a comprehensive digital transformation framework specifically designed for the non-profit sector in Egypt." This framework is expected to provide practical guidance, strategies, and best practices to help NGOs effectively navigate, benefit from, and communicate with other entities in this digital transformation journey. This should be done by addressing the specific barriers the NGOs face and leveraging digital technologies to enhance their operations, improve stakeholder engagement, and achieve their social mission more quickly and efficiently.

Research Objectives

The ultimate objective of this research is to develop a digital transformation framework specifically suiting Egyptian NGOs addressing the challenges they face in leveraging digital technologies effectively, and efficiency, and highlighting their transparency. Hence, generating a favorable impact on Egyptian NGOs.

Research Design

This study employs a quantitative research methodology, utilizing an online questionnaire as the primary data collection method. Quantitative research focuses on the systematic investigation of phenomena through statistical, mathematical, or computational techniques, enabling researchers to gather numerical data and analyze it for patterns and trends. While the study also includes a literature review to contextualize the research questions, it is essential to clarify that this review serves to inform the study's objectives rather than contribute qualitative data. Summarizing existing literature is crucial for mapping the research field, as it highlights gaps that the current study aims to address (Snyder, 2019). The researcher developed the online questionnaire by conducting a comprehensive literature review and consulting with experts from NGOs in Egypt. This collaboration helped identify key themes and variables pertinent to the research questions. The insights gained from the literature and foucus group discussions with these experts informed the design of the questionnaire, ensuring it effectively addresses the study's objectives.

The online survey facilitates the collection of data from a broad and diverse population, allowing for a more robust analysis of the research questions. This quantitative approach provides advantages such as generalizability and precision in the results, focusing on measurable outcomes and ensuring statistical validity.

Quantitative Research (Online Questionnaire)

The researcher has taken into account several guidelines while developing the online survey (Regmi et al., 2017). The online questionnaire is a popular and efficient method for collecting research data from the target audience. The researcher designed an online questionnaire based on the insights from the available literature review, expert focus group discussions, and her own experience. The online questionnaire aims to collect primary data from the experienced participants related to the study about the issues raised from the literature reviews. The questionnaire is to be distributed electronically (email, LinkedIn, etc.), with the aid of an online survey platform, to targeted people in the non-profit sector in Egypt. Participants will be given a defined period to respond, and reminders will be sent to maximize response rates (anonymously).

Sampling

Collecting primary data through online questionnaires is a comprehensive approach that allows the researcher to gather insights from different perspectives. The online questionnaires allow researchers to reach participants from different geographical locations. This helps in obtaining diverse perspectives and reduces the limitations associated with physical proximity. A random sampling technique was employed to select a diverse group of individuals closely connected to the non-governmental organizations in Egypt. The random sampling is a method of selecting a sample from a population in such a way that each individual in the population has an equal chance of being included in the sample. This approach helps to reduce selection bias and ensures that the sample is representative of the entire population. A sample size of about 200 participants was targeted to ensure statistical significance. The announcement for the online survey was disseminated across professional networks, emails, and social media. Data collection started on 22 February 2024, and the survey was available for 30 days. Clear communication of the research objectives and the voluntary nature of participation was emphasized. The researcher targeted 200 responses; however, the actual collected responses were 187.

Data Analysis

The Statistical Packages for the Social Sciences (SPSS) version 20 is used to analyze the data gathered from the respondents. To figure out the frequency and percentage of the total population, the researcher conducts both descriptive and inferential analysis.

RESULTS AND DISCUSSIONS

This study's primary research tool was an online survey intended to gather input from individuals working in the non-governmental organization (NGOs) sector in Egypt. The data that had been collected was examined using SPSS (Statistical Package for the Social Sciences). The utilization of SPSS facilitated the computation of frequencies and percentages, hence enabling a more comprehensive analysis of the collected data. Highlighting the primary findings from the survey are the tables and statistics that follow.

Sample Description

The following points show the basic information of the sample.

Age Categories of the Respondents

The age distribution of respondents in the survey provides a glimpse into the potential readiness and enthusiasm among individuals within non-governmental organizations (NGOs) to lead digital initiatives. Despite varying levels of the current digital engagement, the survey indicates a diverse representation across different age groups, reflecting a spectrum of experiences and perspectives. Notably, 49% of respondents fall within the age range of 36 to 45 years, which may signify a cohort of mid-career professionals primed to take on leadership roles in driving digital transformation efforts. Moreover, the significant presence of individuals aged between 20 to 35 years, comprising 30% of respondents, suggests a keen interest and aptitude among younger generations to leverage digital technologies for organizational advancement. Additionally, the involvement of seasoned professionals aged 46 to 60 years (19% of respondents) and individuals over 60 years (2% of respondents) underscores the wealth of knowledge and expertise that can be harnessed to inform and guide digital initiatives. This wide range of age groups emphasizes the opportunity for multigenerational cooperation and idea exchange to help NGOs adopt digital transformation with passion and efficiency.

Gender Representation of the Sample

The gender breakdown of respondents in the survey offers insight into the gender representation within non-governmental organizations (NGOs) regarding digital transformation initiatives. The data reveals a significant majority of male respondents, comprising 75% of the total, indicating a predominant presence of men in roles related to digitalization within NGOs. Conversely, female respondents accounted for 25% of the total, reflecting a notable but comparatively smaller representation in the digital transformation landscape. While this gender disparity is evident, it also underscores the importance of fostering gender diversity and inclusivity in driving digital initiatives within NGOs. Addressing gender imbalances and promoting gender equity in leadership and decision-making positions can enrich perspectives, enhance innovation, and contribute to more effective and inclusive digital transformation strategies within the sector. By actively engaging and empowering women in digital roles and leadership positions, NGOs can harness the full potential of diverse talent and perspectives to drive meaningful change and innovation in the digital age.

The Positions (role) of the Respondents

Among the surveyed respondents, officers represented the largest proportion at 41% as shown in Table 2, indicating a significant presence of frontline staff who are preparing for digital transformation within their respective organizations. Managers followed closely behind, constituting 33% of respondents, suggesting a substantial managerial workforce involved in overseeing and planning digital initiatives. Additionally, coordinators accounted for 14% of respondents, highlighting the importance of individuals responsible for facilitating collaboration and communication across different departments and stakeholders to ensure smooth transitions into digital frameworks. Directors, while representing a smaller proportion at 7%, play a critical strategic role in setting the vision and direction for digital transformation efforts within their organizations. Their leadership and guidance are instrumental in fostering a culture of innovation and digital readiness. Moreover, the presence of individuals categorized under "Others" was at 6%, underscoring the diverse range of roles and expertise required to navigate the complexities of preparing for digital transformation within NGOs. Collectively, the distribution of roles among respondents reflects the collaborative and multifaceted nature of preparing for digital transformation within NGOs, emphasizing the importance of cohesive teamwork, effective communication, and strategic leadership in driving organizational readiness for the digital age.

Table 2. Sample description (positions)

No.	Positions	Frequency	Percent
1.	Director	13	7%
2.	Manager	61	33%
3.	Officer	76	41%
4.	Coordinator	26	14%
5.	Others	11	6%
Total		187	100%

The Organization Size

The size distribution of organizations represented in the survey provides valuable insights into the diversity of the non-governmental sector's digital transformation landscape. Among the surveyed respondents, organizations with more than 500 employees constituted the largest proportion, accounting for 56% of the total, as shown in Table 3. This suggests a significant presence of larger NGOs actively engaged in preparing for digital transformation, leveraging their resources and scale to drive comprehensive digital initiatives. Conversely, organizations with 10 to 50 employees represented 33% of respondents, indicating substantial participation from smaller NGOs in digital readiness efforts. Despite potentially limited resources and capacity, these organizations demonstrate a commitment to embracing digital transformation and maximizing their impact in an increasingly digital world.

Additionally, organizations with 51 to 100 employees accounted for 6% of respondents, while those with 101 to 250 employees and 251 to 500 employees each represented 2% of the total. Overall, the size distribution of organizations in the survey underscores the inclusive nature of digital transformation efforts within the NGO sector, with organizations of varying sizes actively preparing for the challenges and opportunities presented by the digital age. Regardless of size, the collective commitment to embracing digitalization reflects a shared recognition of its potential to enhance organizational efficiency, effectiveness, and impact in advancing social and humanitarian causes.

Table 3. Sample description (organization size)

No.	Organization size	Frequency	Percent
1.	From 10 to 50 employee	62	33%
2.	From 51 to 100 employee	12	6%
3.	From 101 to 250 employee	4	2%
4.	From 251 to 500 employee	4	2%
5.	More than 500 employee	105	56%
Total		187	100%

Current Status of Technological Landscape on the Egyptian NGOs

The following presents the findings of the survey concerning the status of the technological landscape in Egyptian NGOs.

Technological Infrastructure of the Egyptian NGOs

Technological infrastructure serves as the backbone of digital transformation initiatives within non-governmental organizations (NGOs), laying the foundation for the seamless integration of digital tools and technologies into organizational workflows. The survey findings reveal a diverse range of technological infrastructural capabilities among Egyptian NGOs. A notable 25% of respondents rated their technological infrastructure as excellent. Furthermore, 48% of respondents rated their infrastructure as good. Conversely, 19% of respondents described their technological infrastructure as fair. Additionally, 8% of respondents rated their infrastructure as weak. Addressing these weaknesses and enhancing technological infrastructure will be essential for NGOs to unlock the full potential of digital transformation, drive operational efficiency, and maximize social impact.

Overall, the survey findings underscore the importance of investing in robust technological infrastructure to support the digital transformation journey within Egyptian NGOs.

Digital Skills Development in Egyptian NGOs

The survey findings provide valuable insights into the perspectives of employees within non-governmental organizations (NGOs) regarding the prioritization of digital skills development by their respective organizations. A significant majority of respondents, comprising 68%, indicated that their organizations are keen on enhancing their employees' digital skills. 18% of respondents reported that their organizations are not actively prioritizing the enhancement of employees' digital skills. From the employees' viewpoint, this may indicate potential challenges or missed opportunities within their organizations, such as limited resources, competing priorities, or a lack of recognition regarding the importance of digital skills development. Employees may perceive this as a barrier to their growth and advancement within the organization, prompting potential concerns about future career prospects and job satisfaction. Moreover, 13% of respondents expressed uncertainty about their organization's stance on raising employees' digital skills. This uncertainty may stem from a lack of clear communication or strategic direction within the organization regarding digital skills development initiatives. From the employees' perspective, this ambiguity may create feelings of disengagement or insecurity about their organization's commitment to supporting their professional growth and development in an increasingly digital-centric landscape.

Digitalization in the Egyptian NGOs

The survey findings reflect the perspectives of employees within non-governmental organizations (NGOs) regarding the digitalization or automation of processes within their scope of work. A significant majority of respondents, constituting 60%, expressed that they do have a plan to convert all processes to digital or automated. 18% of respondents indicated that they do not have a plan to convert all processes to digital or automated. This perspective may stem from various factors such as organizational inertia, resource constraints, or a lack of awareness regarding the benefits of digitalization and automation. Addressing these barriers and fostering a culture of innovation and continuous improvement will be essential to empower employees to embrace digital transformation initiatives and drive organizational change. Moreover, 22% of respondents expressed uncertainty about whether there is a plan to convert all processes to digital or automated within their scope of work.

Overall, the survey findings emphasize the importance of employee perspectives in driving digital transformation initiatives within NGOs. By empowering employees to champion digitalization and automation efforts within their respective roles, organizations can harness the collective expertise and enthusiasm of their workforce to realize the full potential of technology in achieving organizational objectives.

Drivers of Digital Transformation in the NGOs

The following presents the findings of the survey concerning the internal and external drivers of digital transformation in Egyptian NGOs.

Internal Drivers

The survey findings underscore a significant internal driver for digital transformation among employees within non-governmental organizations (NGOs). Boosting the data-driven decision-making processes represents the largest proportion at 24% as shown in Figure 1. This percentage highlights the importance of harnessing data analytics to inform strategic decision-making, reflecting a recognition of the transformative power of data in enhancing organizational agility, effectiveness, and responsiveness to challenges and opportunities. In alliance with the previous driver, increasing operational efficiency was cited by 23% of respondents. Strong data-driven decision-making processes will increase the operational efficiency of the organization. These two drivers are vital for non-governmental organizations (NGOs) because of the limitation of the resources. Empowering employees, identified by 14% of respondents. Improving the experience of stakeholders, cited by 13% of respondents, reflects a customer-centric approach aimed at enhancing engagement, satisfaction, and loyalty among donors, partners, beneficiaries, and other stakeholders. Changing organizational culture was similarly mentioned by 13% of respondents. Increasing innovation, cited by 12% of respondents, highlights a focus on fostering creativity, experimentation, and entrepreneurial thinking to drive breakthrough solutions and propel the organization toward greater social impact and sustainability.

External Drivers

The survey findings shed light on the external drivers for digital transformation perceived by employees within non-governmental organizations (NGOs). Among the respondents, the most prominent external driver identified is competition, cited by 31% of participants as shown in Figure 2. This reflects a heightened awareness among employees of the need for NGOs to remain agile and innovative in response to competitive pressures within the sector. The recognition of competition as a driving force for digital transformation underscores the imperative for NGOs to leverage digital technologies to differentiate themselves, enhance service delivery, and maintain relevance in an increasingly crowded landscape.

Additionally, funding pressures and transparency emerged as significant external drivers, with 24% of respondents citing this driver. This finding suggests that NGOs are increasingly cognizant of the importance of demonstrating accountability, transparency, and efficiency in the use of funds to secure and maintain donor support. The integration of digital tools and platforms can facilitate improved financial management, reporting, and transparency, thereby enhancing donor confidence and supporting sustainable fundraising efforts.

Furthermore, crises such as the COVID-19 pandemic were identified as key external drivers for digital transformation, also cited by 24% of respondents. The disruptive impact of crises underscores the importance of organizational resilience and adaptability, prompting NGOs to accelerate digital transformation initiatives to ensure business continuity, remote collaboration, and service delivery in times of uncertainty.

Stakeholder expectations emerged as another significant external driver, with 21% of respondents highlighting this factor. NGOs are increasingly under pressure to meet the evolving needs and expectations of various stakeholders, including donors, beneficiaries, partners, volunteers, and communities. Embracing digital transformation can enable NGOs to enhance stakeholder engagement, thereby fostering stronger relationships and partnerships.

Challenges of Digital Transformation in NGOs

In the following section, the findings of the survey concerning the different challenges of digital transformation in Egyptian NGOs are presented.

Resource-Related Challenges of Digital Transformation in NGOs

The survey findings offer insights into the resource-related challenges perceived by employees within non-governmental organizations (NGOs) in the context of digital transformation. Among the respondents, the most prevalent challenge identified is limited financial resources, cited by a substantial 52% of participants. This underscores a common concern among employees regarding the financial constraints that may impede the implementation of digital initiatives within their organizations. The scarcity of financial resources can pose significant barriers to acquiring necessary technologies, investing in staff training, and undertaking digital transformation projects, hindering organizational progress toward digital readiness. Furthermore, a notable 34% of respondents highlighted the lack of competencies as a significant resource-related challenge. Additionally, uncertainty in funding emerged as a challenge, cited by a smaller proportion (11%) of respondents. Lastly, a minority of respondents (3%) cited other resource-related challenges not captured by the predefined options. Figure 3 illustrated the resource-related challenges of digital transformation in NGOs.

Organization-Related Challenges of Digital Transformation in NGOs

The survey results provide valuable insights into the organization-related challenges perceived by employees within non-governmental organizations (NGOs) in the context of digital transformation. Among the respondents, the most prevalent challenge identified is resistance to change, cited by 35% of participants as shown in Figure 4. Overcoming resistance to change requires effective change management strategies, clear communication, and engagement with stakeholders at all levels to foster a culture of openness and readiness for digital transformation. Furthermore, a significant proportion of respondents (19%) highlighted the lack of vision as a challenge. This finding suggests that employees perceive a gap in organizational leadership or strategic direction concerning digital transformation initiatives. Without a clear vision or roadmap guiding digital efforts, organizations may struggle to align priorities, allocate resources effectively, and sustain momentum towards achieving digital objectives. Clarifying organizational vision and goals for digital transformation is essential to provide employees with a sense of purpose and direction in their efforts. Additionally, respondents identified other organization-related challenges, including unclear roles and responsibilities (13%), diversity of stakeholder needs (11%), lack of support from high-level management (8%), and poor communication (5%). Addressing these challenges requires a holistic approach that involves clarifying roles and responsibilities, fostering stakeholder engagement, securing leadership buy-in, and enhancing communication channels to facilitate collaboration and alignment towards digital goals. Lastly, a minority of respondents (10%) cited other organization-related challenges not captured by the predefined options, indicating the presence of additional nuanced barriers specific to their organizational contexts.

Technical and Operational-Related Challenges of Digital Transformation in NGOs

The survey outcomes illuminate critical technical and operational challenges perceived by employees within non-governmental organizations (NGOs) regarding digital transformation endeavors. Notably, a significant proportion of respondents (43%) identified insufficient technological infrastructure as the foremost challenge. This underscores a fundamental obstacle hindering the effective adoption and implementation of digital solutions within NGOs. Inadequate infrastructure may impede the deployment of necessary digital tools, hinder seamless connectivity, and limit scalability, thereby constraining organizational efforts to leverage technology for innovation and efficiency gains. Furthermore, integration with legacy systems emerged as a notable challenge, cited by 24% of respondents as shown in Figure 5. Additionally, data security and privacy concerns were highlighted as significant challenges, with 20% of respondents expressing apprehension in this regard. The increasing volume and complexity of digital data pose heightened risks for NGOs, necessitating robust measures to safeguard sensitive information, mitigate cybersecurity threats, and ensure compliance with data protection regulations. Addressing these challenges is paramount to maintaining stakeholder trust, safeguarding organizational reputation, and mitigating potential legal and financial liabilities.

Moreover, a smaller proportion of respondents (8%) cited measuring impact and return on investment (ROI) as a challenge. This underscores the importance of establishing robust evaluation mechanisms and metrics to assess the effectiveness and efficiency of digital transformation initiatives for demonstrating the tangible benefits and outcomes of digital investments, informing strategic decision-making, and optimizing resource allocation. Lastly, a minority of respondents (4%) identified other technical and operational challenges not captured by the predefined options, indicating the presence of additional nuanced barriers specific to their organizational contexts. Exploring and addressing these unique challenges will be critical for designing tailored solutions and strategies to support successful digital transformation efforts within NGOs.

Elements of Digital Transformation in NGOs

The online questionnaire included a list addressing a number of the most important digital transformation elements. Each respondent was allowed to check one of four choices to reflect his/her point of view. The used statistical method is the weighted percentage. This technique helps to rank the elements by giving a score for each response (very important=4, important=3, neutral=2, not important=1) as shown in Table 4. Then the total score has been calculated for each element by summing up the frequencies multiplied by the given score. Then the weighted percentage has been calculated by dividing the total score by the highest score which is 748 (total number of respondents 187 * the highest given score 4).

Based on the respondents' perspectives and the weighted percentage technique, the elements of digital transformation in NGOs are ranked as follows:

1. **Technological Infrastructure**: This element ranks first, with 96% of respondents emphasizing its significance. Having a robust technological foundation is considered paramount for driving digital transformation initiatives within NGOs.
2. **Cybersecurity**: Following closely behind, cybersecurity is ranked second, with 96% of respondents acknowledging its importance. This underscores the critical need for robust security measures to safeguard sensitive data and protect against cyber threats.

3. **Digital Skills**: Digital skills are ranked third, with 95% of respondents highlighting their significance. This emphasizes the importance of equipping NGO employees with the necessary skills and knowledge to effectively leverage digital tools and technologies.
4. **Monitoring & Evaluation**: Ranking fourth, monitoring and evaluation frameworks are considered crucial by 93% of respondents. This highlights the need for assessing the impact and effectiveness of digital initiatives to ensure their success.
5. **Digital Organizational Culture**: Digital organizational culture ranks fifth, with 92% of respondents emphasizing its importance. Fostering a culture of innovation, collaboration, and adaptability is deemed essential for driving digital transformation within NGOs.
6. **Leadership**: Leadership is ranked sixth, with 92% of respondents recognizing its significance. Visionary leadership plays a pivotal role in championing digital initiatives and driving organizational change.
7. **Innovation**: Innovation ranks seventh, with 90% of respondents acknowledging its importance. Fostering creativity and embracing new ideas are crucial for driving digital innovation within NGOs. Innovation is a part of the digital organizational culture.
8. **Change Management**: Change management is ranked eighth, with 90% of respondents highlighting its significance. Effective change management strategies are essential for facilitating smooth transitions and mitigating resistance to digital transformation.
9. **Digital Strategy**: Digital strategy ranks ninth, with 89% of respondents emphasizing its importance. Aligning digital initiatives with organizational goals and objectives is crucial for driving successful digital transformation.
10. **Governance**: Governance is ranked tenth, with 89% of respondents recognizing its significance. Establishing clear governance structures and processes is essential for overseeing and guiding digital transformation initiatives.
11. **Competition**: Competition ranks eleventh, with 88% of respondents acknowledging its impact. Analyzing competitors and market trends is important for NGOs to differentiate themselves and stay relevant in a competitive landscape.
12. **Partnership**: Partnership ranks twelfth, with 86% of respondents emphasizing its significance. Strategic collaborations with technology providers, other NGOs, and stakeholders can amplify digital efforts and drive success.
13. **Stakeholder Engagement**: Lastly, stakeholder engagement is ranked thirteenth, with 83% of respondents highlighting its importance. Involving stakeholders throughout the digital transformation journey ensures alignment with their needs and priorities, enhancing overall success.

Overall, these elements collectively contribute to the successful implementation of digital transformation initiatives within NGOs, underscoring the multidimensional nature of the digital journey in the nonprofit sector. Table 4 summarizes the elements' frequency, percentage, and ranking.

Table 4. Ranking of digital transformation elements in NGO

Element	Very important		Important		Neutral		Not important		Total marks	Weighted %	Rank
	F	%	F	%	F	%	F	%			
Technological infrastructure	160	86%	27	14%	0	0%	0	0%	721	96%	1
Cyber security	155	83%	31	17%	1	1%	0	0%	715	96%	2
Digital skills	152	81%	33	18%	1	1%	1	1%	710	95%	3
Monitoring & Evaluation	138	74%	46	25%	3	2%	0	0%	696	93%	4
Digital organizational culture	134	72%	46	25%	7	4%	0	0%	688	92%	5
Leadership	134	72%	48	26%	3	2%	2	1%	688	92%	6
Innovation	117	63%	66	35%	3	2%	1	1%	673	90%	7
Change management	119	64%	60	32%	7	4%	1	1%	671	90%	8
Digital strategy	110	59%	71	38%	6	3%	0	0%	665	89%	9
Governance	117	63%	58	31%	10	5%	2	1%	664	89%	10
Competition	108	58%	68	36%	10	5%	1	1%	657	88%	11
Partnership	96	51%	78	42%	11	6%	2	1%	642	86%	12
Stakeholders engagement	89	48%	74	40%	18	10%	6	3%	620	83%	13

*F: frequency

By combining these data collection and analysis results, the study aims to provide a comprehensive and nuanced understanding of digital transformation among NGOs in Egypt, addressing both the quantitative and qualitative aspects of the research problem. In the upcoming section, the developed framework and recommendations are shown.

Framework for Digital Transformation in NGOs

After conducting a comprehensive literature review and analyzing the responses to the online survey, the researcher designed and developed a comprehensive framework to guide digital transformation in the non-governmental organizations in Egypt. This framework, shown in the following Figure 6, summarizes the essential components needed for an effective transition into the digital age, helping

NGOs navigate the digital age, and leverage digital tools and processes to enhance operational efficiency and increase social impact. The framework consists of five main pillars; Strategy and Leadership, Organizational Culture and Skills, Technological Infrastructure and Security, Monitoring and Governance, and External Collaboration and Competitiveness. Each main pillar has sub-pillars as explained in the following section.

1. Strategy and Leadership Pillar

Digital Strategy

According to Chanias et al. (2019), a digital strategy is responsible for organizing, prioritizing, and carrying out the organization's transformation initiatives. In the long run, it also overlooks the organization's journey towards achieving the intended future state of being digitally transformed. Developing and implementing a digital transformation strategy has become a top priority. It requires alternating between learning and doing, which is a very dynamic process.

Leadership

The success of an organization's digital transformation depends on digital leaders since they are in charge of performing organizational surgery, selecting and retaining the proper team, and buying-in the vision both internally and externally. Due to their adaptability, change-oriented nature, transformative vision, and forward-looking viewpoint, digital leaders are essential to bringing about the transformation that is needed (Benitez et al., 2022). Selecting the right strong leadership with a clear vision to inspire and guide teams is a prime key.

Change Management

Change management plays a crucial role in ensuring the sustainability of digital transformation initiatives. Digital transformation is not just about technology, but also about people. So employees are the key drivers of successful digital transformation. By involving employees in the change process and addressing their concerns, change management helps to overcome resistance and increase employee buy-in, which is essential for the long-term sustainability of digital transformation. Also, effective communication is a key for the digital transformation journey. Clear and transparent communication helps to raise awareness, manage expectations, and overcome any concerns or resistance (Pacolli, 2022). Start with WHY and let your team understand the reasons for change and are engaged in the process, which contributes to the sustainability of the transformation.

2. Organizational Culture and Skills Pillar

Digital Organizational Culture

Digital organizational culture consists of the attitudes, convictions, habits, and behavioral patterns that influence an organization's approach toward the digital environment and how it adapts to digital transformation actions. Effective digital transformation depends on an organization's ability to adapt to the dynamic digital landscape, enhance operational effectiveness, and introduce innovative concepts into its services and procedures (Salah El Din, 2023). Every NGO has to develop its digital organizational culture and be included in day-to-day activities.

Digital Skills

Digital skills not only include technical abilities; they also include other aspects like problem-solving and critical thinking (Audrin & Audrin, 2022). According to Audrin et al. (2024), digital skills have different dimensions that are necessary to adapt to a digital work environment. They include the following dimensions: use of technology, cybersecurity, content management, effective communication and collaboration, critical inquiry, responsibility, ensuring well-being, and protecting and developing one’s identity online.

3. Technological Infrastructure and Security Pillar

Technological Infrastructure

A robust and reliable digital infrastructure, including networks, servers, cloud computing, and connectivity, is essential for organizations to leverage digital technologies effectively. Without a solid infrastructure, it would be challenging to implement and support digital initiatives (Saarikko et al., 2020).

Cyber Security

Organizations should have a thorough understanding of the cybersecurity threats they may face during the digital transformation process. This includes being aware of potential interruptions due to malicious activities or unauthorized access by attackers aiming to alter, destroy, or extort sensitive information from users. Organizations should implement effective cybersecurity measures to protect their digital assets from cyber threats. This may include implementing encryption, authentication, and access control measures to protect data and networks from unauthorized access or malicious activities (Saeed et al., 2023).

4. Monitoring and Governance Pillar

Monitoring and Evaluation

Monitoring the implementation of digital initiatives through applying measurable success and key performance criteria to assess the progress and effectiveness of these initiatives. This monitoring helps identify any challenges or issues that may arise during implementation and allows for timely adjustments and improvements. Ex ante evaluation refers to the assessment of projects before their implementation, while ex post evaluation is conducted after the completion of projects. These evaluations help measure the impact and outcomes of strategic projects, identify lessons learned, and inform future decision-making (Meyerhoff Nielsen and Mahmoud Ali, 2021).

Governance

Effective digital governance ensures the security, transparency, and efficiency of digital processes. It involves establishing clear accountability, roles, and decision-making authority for managing digital tools and processes. Regular audits and assessments help in maintaining the integrity of digital activities.

5. External Collaboration and Competitiveness Pillar

Partnership

Partnerships are mutually beneficial, offering significant advantages for both sides. For NGOs, partnerships they provide access to technology, funding, and expertise that can drive digital transformation. For corporations, partnering with NGOs enhances their Corporate Social Responsibility (CSR) profile, increases employee engagement, and builds customer loyalty. Several NGOs have successfully partnered with corporations to advance their digital transformation efforts.

Stakeholder Engagement

NGOs have a diverse range of stakeholders that play crucial roles in their activities and operations. These stakeholders include customers who benefit from the organization's services, employees who contribute to its daily operations, donors who provide financial support, governmental organizations that collaborate and provide regulatory support, other NGOs that partner on common goals, communities affected by NGO initiatives, industry representatives relevant to their work, and even other NGOs. Engaging with these stakeholders is essential for transparency, accountability, and the effective delivery of the NGO's mission and programs, allowing them to maximize their impact and create positive change in society.

Competition

Digital transformation in NGOs significantly enhances their competitive edge and opens up new funding opportunities. By adopting advanced technologies and digital tools, NGOs can streamline their operations, improve efficiency, and deliver services more effectively. This modernization not only helps NGOs to stand out in a crowded sector but also enables them to provide transparent and real-time impact reporting, which is highly attractive to donors and funders.

The following Table 5 summarizes the main five pillars and the 12 sub-elements of the digital transformation framework for NGOs in Egypt proposed by this research.

Table 5. Digital transformation framework for NGOs in Egypt

<table>
<tr><td colspan="3">Digital Transformation Framework for NGOs</td></tr>
<tr><td colspan="3">Pillar 1: Strategy and Leadership</td></tr>
<tr><td>Digital Strategy</td><td>Leadership</td><td>Change Management</td></tr>
<tr><td colspan="3">Pillar 2: Organizational Culture and Skills</td></tr>
<tr><td>Digital Culture</td><td colspan="2">Digital Skills</td></tr>
<tr><td colspan="3">Pillar 3: Technological Infrastructure and Security</td></tr>
<tr><td>Technological Infrastructure</td><td colspan="2">Cyber Security</td></tr>
<tr><td colspan="3">Pillar 4: Monitoring and Governance</td></tr>
<tr><td>Monitoring & Evaluation</td><td colspan="2">Governance</td></tr>
<tr><td colspan="3">Pillar 5: External Collaboration and Competitiveness</td></tr>
<tr><td>Competition</td><td>Partnership</td><td>Stakeholder Engagement</td></tr>
</table>

Implementation Tips for Digital Initiatives in NGOs Based on the PDCA Cycle

PDCA or the Deming cycle is a management methodology that aims to continually improve processes. This cycle is based on four stages: Plan, Do, Check, and Act. To ensure successful implementation of digital initiatives in non-governmental organizations (NGOs), applying the PDCA cycle offers a structured approach. Here are the core steps (shown in Table 6) for each phase of the PDCA cycle:

Table 6. PDCA cycle for digital initiatives in NGOs

Plan **Define Objectives and Goals: Set clear, specific goals that align with the organization's mission and strategic objectives.** **Conduct Needs Assessment: Evaluate the current state and identify gaps.** **Establish KPIs: Define measurable indicators to monitor progress.**	*Do* **Implement Pilot Projects: Start with small-scale projects to test new technologies and processes.** **Allocate Resources: Ensure financial, human, and technical resources are available to support implementation.** **Execute the Plan: Follow the timeline to deploy digital tools, train staff, and update procedures.**
Act **Make Adjustments**: Based on evaluation, make necessary changes. **Document Lessons Learned**: Capture lessons learned and best practices for future initiatives. **Standardize Practices**: Integrate successful practices into standard operating procedures. **Plan for Continuous Improvement**: Continue the PDCA cycle to foster ongoing improvement.	***Check*** **Monitor Progress**: Regularly review progress against KPIs and milestones. **Gather Feedback**: Collect feedback from the different stakeholders. **Evaluate Results**: Analyze data to determine if objectives are being met and identify areas for improvement.

The Digital Readiness Score (DRS) Tool

The Digital Readiness Score is a valuable tool to guide digital transformation efforts in NGOs and ensure alignment with organizational goals and objectives. Based on the Digital Readiness Score (DRS), the organization can identify strengths and areas for improvement. DRS, shown in Table 7, can be conducted by individuals or teams within the NGOs who have expertise in digital transformation, technology, organizational development, and/or strategic planning. This depends on the size and structure of the organization, and this may include members of the IT department, the digital transformation team, senior management, or external consultants specializing in strategy and digital transformation. Evaluation by internal or external calibers depends on various factors such as the organization's resources, experience, and ability to conduct such evaluations. Internal evaluations may provide greater familiarity with the NGO's culture and operations, while external evaluations may provide a new perspective and specialized expertise. In both cases, the process of assessment of the digital readiness inside the NGOs must be real, honest, and reflect reality for measures to be taken that will help organizations move forward in the digital transformation journey.

Table 7. NGOs Digital Readiness Score Sheet

NGOs DRS Sheet	
Organization Name:	
Date:	
Instructions: Rate your organization's readiness for digital transformation on a scale from 1 to 5, where:	
1 = Strongly Disagree	
2 = Disagree	
3 = Neutral	
4 = Agree	
5 = Strongly Agree	
Digital Strategy	
Assessment Criteria	**Rating (1-5)**
Our organization has a clear digital strategy in place.	
The digital strategy aligns with our organizational goals and objectives.	
Key stakeholders are involved in the development of the digital strategy.	
Clear goals, timelines, and performance indicators are defined in the digital strategy.	
Leadership	
Assessment Criteria	**Rating (1-5)**
Our leadership demonstrates a vision for digital transformation	
Leaders are actively involved in driving digital initiatives	
Leadership provides support and resources for digital transformation efforts	
There is effective communication and alignment between leadership and staff regarding digital initiatives	
Digital Organizational Culture	
Assessment Criteria	**Rating (1-5)**
Our organizational culture encourages innovation and experimentation	
There is a willingness to embrace change and adapt to new technologies	

continued on following page

Table 7. Continued

NGOs DRS Sheet	
Digital initiatives are recognized and rewarded within the organization	
Employees feel empowered to contribute ideas and participate in digital initiatives efforts	
Digital Skills	
Assessment Criteria	**Rating (1-5)**
Employees have the necessary digital skills to perform their roles effectively	
There are opportunities for ongoing training and development in digital skills	
The organization invests in building digital capabilities among staff	
Employees feel confident in using digital tools and technologies.	
Technological Infrastructure	
Assessment Criteria	**Rating (1-5)**
Our technological infrastructure is robust and scalable	
Hardware, software, and network capabilities support digital initiatives effectively.	
There are systems in place for regular maintenance and updates of technological infrastructure	
The organization has invested in modern technology solutions to meet its digital needs	
Cyber Security	
Assessment Criteria	**Rating (1-5)**
Our organization has implemented strong cyber security measures to protect data and systems	
There are policies and procedures in place for data encryption and access controls	
Employees are trained on cyber security best practices and protocols	
There are mechanisms for monitoring and responding to cyber security threats	
Monitoring & Evaluation	
Assessment Criteria	**Rating (1-5)**
Our organization has systems in place for monitoring and evaluating digital initiatives.	
Key performance indicators and metrics are defined to measure the effectiveness of digital initiatives	
Regular reviews and assessments are conducted to track progress and identify areas for improvement.	
Feedback from monitoring and evaluation activities is used to refine digital strategies and initiatives.	
Total DRS	

Final Recommendations for Digital Transformation in Egyptian NGOs

In summary, this study highlights the key drivers and challenges of digital transformation in Egyptian NGOs. The main drivers include data-driven decision-making, improved operational efficiency, employee empowerment, and competition within the NGO sector. On the flip side, significant challenges involve limited financial resources, insufficient technological infrastructure, and resistance to change.

To address these issues, the researchers propose a tailored framework for digital transformation in Egyptian NGOs. This framework consists of five core pillars, divided into 12 sub-pillars, offering a structured approach for organizations to manage their digital initiatives effectively. Additionally, a Digital Readiness Score (DRS) tool has been introduced to aid in continuous improvement, allowing NGOs to monitor their transformation progress. These contributions aim to provide NGOs with practical guidance, strategies, and best practices to successfully implement digital transformation. By utilizing the proposed framework and DRS tool, organizations can boost their readiness for digital change and maximize their impact.

Key Recommendations:

1. **Invest in data analytics:** NGOs should prioritize developing data analytics capabilities. This investment enables data-driven decision-making, enhancing program effectiveness and overall social impact. By systematically collecting and analyzing relevant data, NGOs can gain insights that inform their strategies and operations.
2. **Enhance technological infrastructure:** To address inadequate technological infrastructure, NGOs must invest in reliable and scalable IT systems. Strengthening this foundation is crucial for supporting digital initiatives and ensuring efficient operations in a fast-evolving digital landscape.
3. **Foster a culture of innovation and digital literacy:** NGOs should cultivate a culture of innovation and digital literacy among employees. This can be achieved through awareness programs, training sessions, workshops, and knowledge exchange initiatives. Encouraging employees to adopt new technologies and innovative practices will facilitate a smoother transition to the digital age.
4. **Seek partnerships and collaborations:** Exploring partnerships with other organizations—including government agencies, private sector entities, and technology providers—is vital for NGOs. Collaborations can enhance resource capabilities, facilitate the sharing of best practices, and open up funding opportunities for digital transformation initiatives. Leveraging collective strengths will accelerate NGOs' digital journeys.

5. **Address resistance to change:** Implementing effective change management strategies is essential for NGOs. This includes fostering open communication, engaging stakeholders, and creating a supportive environment that encourages experimentation and learning. Proactively addressing resistance to change will help organizations transition more smoothly to digital practices.
6. **Continuously evaluate and adapt:** NGOs should regularly assess the impact of their digital transformation efforts and adjust based on evaluation results. Defining key performance indicators, gathering stakeholder feedback, and staying informed about emerging technologies and trends are essential for continuous improvement. This iterative process ensures NGOs remain responsive to evolving needs and opportunities.

In conclusion, decision-makers within NGOs must recognize that digital transformation is an ongoing journey requiring a strategic and holistic approach. Initiating pilot projects to identify best practices and scaling successful initiatives can facilitate broader adoption of digital technologies. By embracing the identified drivers, NGOs in Egypt can enhance operational effectiveness, improve stakeholder engagement, and fulfill their social missions with greater efficiency and impact.

REFERENCES

About Tasharuky. Tasharuky Platform. (n.d.). https://www.tasharuky.com/about-us

Alsharkawy, N. (2024, May 9). *Crowned by a Presidential Decree: Establishment of the National Alliance for Developmental Civic Work*. Al-Marsad. https://marsad.ecss.com.eg/81575/

Audrin, B., Audrin, C., & Salamin, X. (2024). Digital skills at work – Conceptual development and empirical validation of a measurement scale. *Technological Forecasting & Social Change. Technological Forecasting and Social Change*, 202, 123279. DOI: 10.1016/j.techfore.2024.123279

Audrin, C., & Audrin, B. (2022). Key factors in digital literacy in learning and education: A systematic literature review using text mining. *Education and Information Technologies*, 27(6), 7395–7419. DOI: 10.1007/s10639-021-10832-5

Baskin, K. (2018, August 16). *5 reasons companies struggle with Digital Transformation*. MIT Sloan. https://mitsloan.mit.edu/ideas-made-to-matter/5-reasons-companies-struggle-digital-transformation

Benitez, J., Arenas, A., Castillo, A., & Esteves, J. (2022). Impact of digital leadership capability on innovation performance: The role of platform Digitization Capability. *Information &amp. Information & Management*, 59(2), 103590. DOI: 10.1016/j.im.2022.103590

Bhuiyan, M. I., & Haque, A. (2024). Role of ngos in providing available and affordable health care services to the slum people in Dhaka. *Clinical Epidemiology and Global Health*, 25, 101478. DOI: 10.1016/j.cegh.2023.101478

Chanias, S., Myers, M. D., & Hess, T. (2019). Digital transformation strategy making in pre-digital organizations: The case of a financial services provider. *The Journal of Strategic Information Systems*, 28(1), 17–33. DOI: 10.1016/j.jsis.2018.11.003

Diener, F., & Špaček, M. (2021). Digital Transformation in banking: A managerial perspective on barriers to change. *Sustainability (Basel)*, 13(4), 2032. DOI: 10.3390/su13042032

Elgohary, E. (2022). The role of digital transformation in Sustainable Development in Egypt. *The International Journal of Informatics. Media and Communication Technology*, 4(1), 71–106. DOI: 10.21608/ijimct.2022.219953

Fitzgerald, M., Kruschwitz, N., & Welch, M. (2013). Embracing Digital Technology: A New Strategic Imperative, Capgemini Consulting Worldwide [Online]. *MIT Sloan Management Review*, 55(1), 1–13.

Fuchs, C., & Hess, T. (2018). Becoming Agile in the Digital Transformation: The Process of a Large-Scale Agile Transformation. *ICIS 2018 Proceedings*. 19. https://aisel.aisnet.org/icis2018/innovation/Presentations/19

Ghazi, R. A. (2021). *Civil Society Organizations' Support for Vulnerable Youth in Transition: a case study in Egypt* [Thesis, the American University in Cairo]. AUC Knowledge Fountain. https://fount.aucegypt.edu/retro_etds/2555

Hartl, E., & Hess, T. "The Role of Cultural Values for Digital Transformation: Insights from a Delphi Study" (2017). *AMCIS 2017 Proceedings*. 8. https://aisel.aisnet.org/amcis2017/Global/Presentations/8

Kamel, S. (2021). (working paper). *The Potential Impact of Digital Transformation on Egypt* (pp. 1–75). Cairo, Egypt: The Economic Research Forum (ERF).

Koshy, P. (2022, May 31). *The role of ngos in attaining the sdgs*. UN Today. https://untoday.org/the-role-of-ngos-in-attaining-the-sdgs/

Legner, C., Eymann, T., Hess, T., Matt, C., Böhmann, T., Drews, P., Mädche, A., Urbach, N., & Ahlemann, F. (2017). Digitalization: Opportunity and challenge for the business and Information Systems Engineering Community. *Business &amp. Business & Information Systems Engineering*, 59(4), 301–308. DOI: 10.1007/s12599-017-0484-2

Libert, B., Beck, M., & Wind, Y. (2016, July 14). *7 questions to ask before your next digital transformation*. Harvard Business Review. https://hbr.org/2016/07/7-questions-to-ask-before-your-next-digital-transformation

Lynn, T., Rosati, P., Conway, E., Curran, D., Fox, G., & O'Gorman, C. (2022). *Digital Technologies and Civil Society*. Essay, Palgrave Macmillan Cham. DOI: 10.1007/978-3-030-91247-5_5

Martin, F. (2018, October 29). *Unlocking success in digital transformations*. McKinsey & Company. https://www.mckinsey.com/capabilities/people-and-organizational-performance/our-insights/unlocking-success-in-digital-transformations#/

Mathenge, V. W., Sambaiga, R., Bitebo, A., Komba, M., Mushashu, A., Sang'anya, M., Shaweji, F., Mbwilo, G., Mayao, V., Mlembwa, J., Rangi, J., Onuekwe, C., Kileo, N., Mwengee, W., & Sagoe-Moses, C. (2023). The development of a digital NGO mapping platform in Tanzania, as a tool to strengthen civil society engagement, in pursuit of Universal Health Coverage. *The Pan African Medical Journal*, 46. Advance online publication. DOI: 10.11604/pamj.2023.46.115.42333 PMID: 38465003

Matt, D. T., & Rauch, E. (2020). SME 4.0: The role of small- and medium-sized enterprises in the digital transformation. *Industry 4.0 for SMEs*, 3–36. https://doi.org/DOI: 10.1007/978-3-030-25425-4_1

Meyerhoff Nielsen, M., & Mahmoud Ali, W. K. (2021). Governing and monitoring the Digital Transformation: Assessing the Qatari experience since 2003. *14th International Conference on Theory and Practice of Electronic Governance*. https://doi.org/DOI: 10.1145/3494193.3494227

Mohsen, M. (2018). *What are NGOs doing in Egypt and what can you do to help? | Egyptian Streets*. Egyptian Streets. https://egyptianstreets.com/2018/07/31/what-are-ngos-doing-in-egypt-and-what-can-you-do-to-help/

Omdena. (2024). *How can NGOs develop an effective data strategy for digital transformation*. https://www.omdena.com/blog/ngos-digital-transformation

Pacolli, M. (2022). Importance of Change Management in Digital Transformation Sustainability. *IFAC-PapersOnLine*, 55(39), 276–280. DOI: 10.1016/j.ifacol.2022.12.034

Regmi, P. R., Waithaka, E., Paudyal, A., Simkhada, P., & Van Teijlingen, E. (2017). Guide to the design and application of online questionnaire surveys. *Nepal Journal of Epidemiology*, 6(4), 640–644. DOI: 10.3126/nje.v6i4.17258 PMID: 28804676

Saarikko, T., Westergren, U. H., & Blomquist, T. (2020). Digital Transformation: Five recommendations for the digitally conscious firm. *Business Horizons*, 63(6), 825–839. DOI: 10.1016/j.bushor.2020.07.005

Saeed, S., Altamimi, S. A., Alkayyal, N. A., Alshehri, E., & Alabbad, D. A. (2023). Digital transformation and cybersecurity challenges for businesses resilience: Issues and recommendations. *Sensors (Basel)*, 23(15), 6666. DOI: 10.3390/s23156666 PMID: 37571451

Salah El Din, P. (2023). The Role of Digital Organizational Culture in Organizational Development In light of Digital Transformation adoption in Egypt. *Arab Journal of Administration*, 45(4), 1–20. DOI: 10.21608/aja.2023.173954.1353

Schneider, S., & Kokshagina, O. (2021). Digital Transformation: What we have learned (thus far) and what is Next. *Creativity and Innovation Management*, 30(2), 384–411. DOI: 10.1111/caim.12414

Sheombar, A., & Ravesteijn, P. (2022). Digital Transformation of development NGOs: The case of transitioning northern-based development NGOs. *Lecture Notes in Networks and Systems*, 600–613. https://doi.org/DOI: 10.1007/978-3-030-96293-7_50

Snyder, H. (2019). Literature review as a research methodology: An overview and guidelines. *Journal of Business Research*, 104, 333–339. DOI: 10.1016/j.jbusres.2019.07.039

Tabrizi, B., Lam, E., Girard, K., & Irvin, V. (2019, March 13). *Digital transformation is not about technology*. Harvard Business Review. https://hbr.org/2019/03/digital-transformation-is-not-about-technology

The National Alliance for Developmental Civil Work. SIS. (2024, January). https://shorturl.at/dveoT

Van Veldhoven, Z., & Vanthienen, J. (2023). Best practices for digital transformation based on a systematic literature review. *Digital Transformation and Society*, 2(2), 104–128. DOI: 10.1108/DTS-11-2022-0057

Von Leipzig, T., Gamp, M., Manz, D., Schöttle, K., Ohlhausen, P., Oosthuizen, G., Palm, D., & von Leipzig, K. (2017). Initializing customer-orientated digital transformation in enterprises. *Procedia Manufacturing*, 8, 517–524. DOI: 10.1016/j.promfg.2017.02.066

Chapter 7
Implications of Artificial Intelligence for Medication Therapy Management and Its Efficacy in Clinical Pharmacy

Hadeer Ehab Barakat
Department of Clinical Pharmacy and Pharmacy Practice, Ahram Canadian University, Egypt

Islam Tharwat Abdel Halim
School of Information Technology and Computer Science, Nile University, Egypt

ABSTRACT

Medication therapy management (MTM) is of importance for the assistance of healthcare providers to create medication plans that fit with each patient's specific requirements. MTM includes services provided by pharmacists or other qualified healthcare professionals to improve the treatment regimen for each patient. These services can involve assessment of the patients' health status, conduction of pharmaceutical care plan, and decrease the hospital stays.Artificial intelligence (AI) allows MTM to be conducted with higher efficiency and effectiveness in the applications of clinical pharmacy. AI has been indispensable in the medical field by assisting providers in reducing expenditures, enhancing patient outcomes, and increasing the overall efficacy of patient treatment through enhancing medication safety, predicting health risks and outcomes across diverse populations, reducing time and costs of

DOI: 10.4018/979-8-3373-1132-6.ch007

treatment, monitoring medication adherence, and scheduling follow-up healthcare services. As a result, healthcare providers are currently obligated to implement AI to assist with MTM services

INTRODUCTION

Clinical pharmacy is defined as "the area of pharmacy concerned with the science and practice of rational medication use" ("The Definition of Clinical Pharmacy," 2008). Clinical pharmacists optimize drug therapy and promote health, wellbeing, and illness prevention for patients. Clinical pharmacy integrates the therapeutic knowledge, experience, and judgement to provide optimal pharmaceutical care plan and patient outcomes. As a discipline, clinical pharmacy must generate new knowledge to improve health and quality of life. In order to accomplish the therapeutic objectives that are intended, the clinical pharmacists employ evidence-based therapeutic guidelines, emerging sciences, and evolving sciences, technologies, as well as pertinent legal, ethical, social, cultural, economic, and professional principles. Therefore, clinical pharmacists are responsible for Medication Therapy Management (MTM) in direct patient care settings, whether they are practicing independently or in consultation or collaboration with other health care professionals.

Clinical pharmacy services include medication order review (the process of evaluating the appropriateness of current medications on the basis of health conditions, indication, and therapeutic goals of each medication), medication reconciliation (the process of comparing patient's medication orders with all the medications the patient has taken in order to detect medication discrepancies and then communicate the newly reconciled list to the patient and the clinical care providers), and patient education (Ranchon et al., 2023). Application of clinical pharmacy services actively improve medication adherence, optimize drug therapy outcomes, avoid adverse drug events, shorten hospital stays, increasing access to healthcare, reduce readmission rates and mortality and maximize satisfaction of the healthcare team.

The expanding requirement for personalized treatment especially for geriatrics with multidrug regimen for multiple comorbidities, the dosage regimen, and numerous or complex dosage forms, and administration instructions emphasizes the critical role of clinical pharmacists. Therefore, it is imperative to make efficient use of cutting-edge technologies, which has started during the last years with the automation of the drug process, including manufacturing and administration of medications in community or hospital settings (Flynn et al., 2021). Further accelerating the demand for more automation and scalability is the growing digitization of healthcare data. The theory and creation of computer systems that carry out activities like learning, problem-solving, language processing, and visual perception that often call for human intelligence is known as Artificial Intelligence (AI), which can augment human intelligence and improve decision making and operational processes (Schutz et al., 2020).

Currently, AI is gaining traction across various sectors, and it is clear that healthcare is beginning to embrace its potential. These advancements promise to enhance patient care and simplify administrative tasks in hospitals, insurance companies, and pharmaceutical firms. Recent studies indicate that AI can match or even surpass human capabilities in certain healthcare functions. For instance, AI systems have proven to be more effective than radiologists at detecting cancerous tumors and are helping researchers assemble patient groups for expensive clinical trials. Nevertheless, despite these advancements, it will likely be several years before AI assumes broader roles in medical practice (Wen & Huang, 2022).

The chapter provides in-depth insight into the changing effects of AI on MTM practices and elucidates the relative robustness of AI over conventional methods in MTM services. Having considered the uses of AI in patient data collection and analysis, predictive analytics, drug regimen planning, as well as medication compliance documentation, we seek to understand how much MTM services can be improved in terms of volume and quality. It will also address the issues of the so-called 'unintended consequences' of these tools, in particular, the ethical and other risks of AI use. In addition, the chapter describes the current trends in the MTM and services that premises the increasing role of AI preventer in clinical pharmacy workforce health care responsibility. Overall, the objective of this chapter is to encapsulate how cutting-edge AI technologies will integrate into MTM systems and how patient benefits will increase through improved MTM systems.

The Role of MTM in Clinical Pharmacy

Medication Therapy Management is also referred to as MTM and is a distinctive caregiving style mainly provided by clinical pharmacists. MTM has been designed with the goal of improving patients' health for proper management of medications (Schussel et al., 2019). But, what does professional clinical pharmacist do in relation to MTM, and what sorts of problems do they find themselves facing? MTM encompasses a wide range of services, but it can be broken down into five key components (Burns, 2005):

- Medication Therapy Review (MTR)
- Personal Medication Record (PMR)
- Medication-Related Action Plan (MAP)
- Intervention and/or Referral
- Documentation and Follow-Up

Even though we do not need yet another acronym in healthcare, it is critical to realize that the use of the MTM optimizes the dosages which patients require for treatment of their health conditions and enhances the benefits offered by prescribed medications. MTM is also involved in the early determination and elimination of preventable problems like the adverse drug events that cost our healthcare system numerous billions of U.S. dollars annually.

However, an MTM program can benefit all patients on their medications, with the highest gains going to the chronic disease patient population and those taking multiple drug regimens. Usually, they are more often elderly patients receiving Medicare, and it can be realized how valuable and helpful the clinical pharmacist with compassionate and consultative approach is in order to help such patient to manage their health properly.

The Role of a Clinical Pharmacist in MTM

There should be a question arising in a mind, 'Why can't physicians advice patients about the medication they give?' Let's assume a patient who has been prescribed different medications by different providers. Many of them may be prescribed by their primary care physician, whereas others may be prescribed following a visit to the emergency room or urgent care. Because the various medication regimens are prescribed for various conditions, it becomes hard for any single physician to advise patients accurately on each drug.

This is where the clinical pharmacist comes into play time and again, patients have more than one chronic condition and are likely to be prescribed treatments from more than one specialty physician, for instance, a cardiologist for hypertension, a gastroenterologist for irritable bowel syndrome and a rheumatologist for autoimmune disorder. Thus, we must have a medication specialist to make sure that the medications being given are safe and effective. This is where role of clinical pharmacists themselves as the drug experts in delivery of high quality and comprehensive MTM services especially for patients with chronic diseases and multi-drug regimens.

Current Challenges in MTM

Although there are some professionals who are employed in academic institutions and hospitals, the majority of pharmacists continue to be employed in community pharmacies. Unfortunately, in these more traditional environments, the clinical pharmacists find themselves performing numerous roles that may encompass inventory control, accounting and sundry tasks. Not surprisingly, currently, pharmacy hours are an alarming problem and many pharmacists experience burnout (Dee et al., 2023).

Closely related to their other responsibilities, the performance of clinical pharmacists is expected to embrace the delivery of MTM consultations in community settings. If you have visited a pharmacy recently you will know how crowded it gets with long queues of people, customers waiting for their drugs to be dispensed and phones ringing all around. Therefore, under these circumstances, how can a pharmacist conduct standard one on one private medication counseling and education to the patient? It's simply not feasible.

MTM programs also encounter difficulties related to technology interoperability. Pharmacists are not integrated into the healthcare information technology (HIT) infrastructure as extensively as other healthcare professionals. Moreover, many community pharmacies lack the connectivity to Electronic Health Record (EHR) systems used by healthcare providers. This disconnect leaves them without crucial patient information, such as claims data, necessary for delivering high-quality consultations and improving patient outcomes.

Growing Demand for Personalized Medicine in Clinical Pharmacy

Personalized medicine is currently being challenged with rising improvements in technologies and human genomics to deliver personalized treatments. This transition from mass market pharmaceuticals to taking care of each patient as a separate organism is explained by the fact that every person is individual concerning his or her genetic factors, environmental conditions and daily routines would result in the body's response to a particular drug (Hassan et al., 2022).

Through pharmacogenomics and genomic biomarkers, the caregivers and clinicians can select patients who would likely benefit most from particular therapy or who are at peril of adverse effects of certain medications. Such a concept has the possibilities of increasing the variety of beneficial results for patients, increasing medication safety, and decreasing overall healthcare expenses. However, adoption of personalized medicine suffers several drawbacks; some of these are the following: • Data privacy • Ethical issues • General infrastructure • Technical know-how To achieve this goal, those outlined above challenges will have to be addressed in order to make the personalized medicine widely available and fairly distributed (Goetz & Schork, 2018).

How Technology Empowers Clinical Pharmacists

Currently, the use of technology in the delivery of health care services has taken a new dimension and so has the clinical pharmacists. Here are some ways technology is empowering them to deliver better patient care (H. et al., 2018; Mancini, 2019; Reis & Afonso, 2011):

1. Electronic Health Records (EHRs):
   - Centralized patient information: EHRs contain the longitudinal summary of patient's medical history, current medications and medication history, known allergies, laboratory and diagnostic data.
   - Improved communication: Through the use of EHRs, information sharing between the pharmacists and other care givers is made easy in order to enhance co-ordination.
   - Medication reconciliation: EHRs contribute to the documentation of medication order and prevent medication errors from being prescribed to a patient.
2. Medication Adherence Tools:
   - Reminders and alerts: Use of technology in nursing care for patients can produce alerts for the dosage times and also produce alerts if the patients have not taken the medication.
   - Tracking and monitoring: Exemplary tools also enable monitoring of medication taking and non-compliance, so that pharmacists can need intervention.
3. Clinical Decision Support Systems (CDSS):
   - Evidence-based recommendations: CDSSs offer pharmacists evidence supported decision making, particularly on the choice of a drug, its dose and the necessary monitoring.
   - Drug interaction alerts: These systems can establish whether there are likely to be interaction between the drugs you are providing and prevent adverse events.
4. Telepharmacy:
   - Expanded access to care: By telepharmacy, pharmacists are able to offer their services virtually leading to easy access for those patients in the rural and other unserved areas.
   - Virtual consultations: There are a number of other such areas where the presence of pharmacists can be employed effectively, e.g., pharmacists can make house calls and give medication consultations.
5. Mobile Apps:

- Patient education: Using the mobile apps, patients could find the necessary information about the medications they use, the possible side effects, contraindications, and the method of administration.
- Medication tracking: This is since patients can download different apps through which they can monitor the medications they use.

These technologies can therefore be utilized to improve efficiency of clinical pharmacists, deliver superior patient care and enable them to fill a more central role in delivering overall healthcare services.

GROUNDWORK AND HISTORICAL EVALUATION

Artificial Intelligence: A Historical Perspective

AI is an all-embracing discipline that examines the means of imitating human intelligence through computers without the use of human explanation (Hamet & Tremblay, 2017). Turner used the idea of testing Artificial Intelligence and tested it in 1950 calling it Turing test which sought to determine whether a machine could make a decision that a human being could make (Shieber, 2007). This challenge brought the abstract concept of AI into a concrete focus, thereby becoming a research objective. Many core concepts of AI are attributed to Turing but credit for the real start of AI is often given to the 1956 Dartmouth conference on artificial intelligence where data scientists, engineers, mathematicians discussed the applied potential of AI (Kline, 2011). These pioneers expected that AI will evolve, and will one day equal, if not surpass the human intellect in every endeavor.

The emergence of industrial AI is also in the 1950s with an aim of the manufacturing of devices that provided decision making and actions. For instance, in 1955, General Motors used automated manufacturing control to invent the initial robotic arm (Moran, 2007). Joseph Weizenbaum at MIT developed the world's first chatterbot, Eliza in 1964 which mimics a human therapist (Weizenbaum, 1966). It was also in 1960s that Shakey was created after which it become the first robot that could comprehend what a human was saying as well as respond to it. These early breakthroughs assured that the aspiration of an AI future was not equated to a science fiction dream, but in reality AI is a growing and developing domain.

AI started achieving great progress within the medical domain during the seventies . INTERNIST-1 was developed in 1971 and included the first decision support system for use in examining patients and making diagnoses (Miller et al., 1986). It was the first precedent which showed that AI can somewhat decrease the diagnostic burden of healthcare workers. The first AI in Medicine conference was organized shortly after in 1975, and encouraged cross-disciplinary cooperation that continued to progress medical AI (Kulikowski, 2019). MYCIN another AI system that was developed alongside it helped physicians in treatment by giving them antibiotics for infectious diseases was an example of the concrete ways that AI could make a difference in the healthcare field (Shortliffe, 1977). More progress was made in the 1980s with the advent of DXplain which helped clinicians in diagnosing illnesses out of symptoms. This system was built on the basis of INTERNIST-1, provided more numerous diagnostic suggestions, and formed the base for the medical information databases (Hoffer, 1992).

Come to the present age in which possessing such systems like IBM's Watson have brought several changes in the medical field and in our everyday life (Ferrucci et al., 2013). For example, in the beginning of the 2017 Watson conducted data search and analysis that led to identification of RNA-binding proteins linked to amyotrophic lateral sclerosis (Bakkar et al., 2018). Pharmbot initiated in 2015 was an assistant that could teach patients as well as their families about medicines along with treatment options.

Machine and Deep Learning in Health Care

As shown in Figure 1, within the broad field of AI is the sub-field of Machine Learning (ML) that primarily uses algorithms and data to make predictions so as to draw interpretations directly from the data without having to involve human beings (Higgins et al., 2023). Tasks of ML are mainly implemented with statistical learning techniques including logistic regression, linear regression and random forest among others for works including data preprocessing and prediction respectively. K-nearest neighbor is one of those methods which can be applied for data classification and regression (Secinaro et al., 2021). Of all the types of ML frameworks, deep learning and neural networks are the most well-known frameworks (Schmidhuber, 2015). Neural networking is an endeavor to mimic how neurons perform in the human brain so as to build artificial neurons also known as nodes (Hinton & Salakhutdinov, 2006). These networks enable information to be processed as in the human central nervous system. One of these networks is convolutional neural networks CNN which can handle complicated multidimensional signals hence help in image and language recognition (Indolia et al., 2018). CNNs, as an example, can identify diseases from medical images just like it's done by a doctor (Aljuaid & Anwar, 2022).

Figure 1. Hierarchy and classification of intelligent algorithms (Amin et al., 2024).

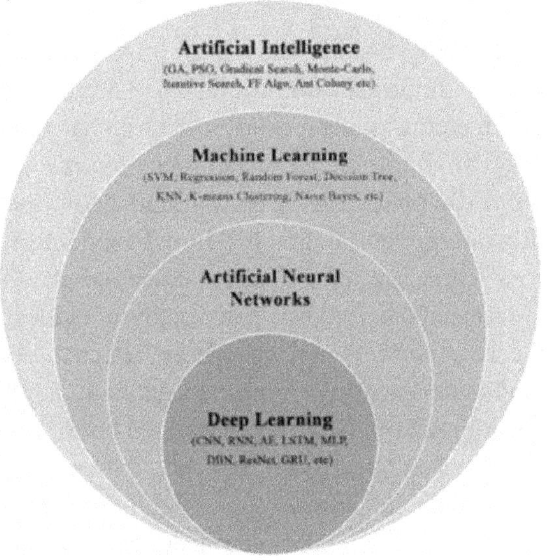

More specifically, a number of studies have predicted that ML has achieved great progress in the radiology area, where computer-assisted detection aid clinicians by drawing attention to specific irregular patterns in imaging scans, and possibly help decrease misdiagnosis (van Leeuwen et al., 2021). In dermatology, applications of ML algorithms are found in classification of skin diseases, analysis of pathology slides and identification and detection of skin cancers (Habehh & Gohel, 2021). CNNs, for instance, the Google Inception V3 model have been trained with thousands of dermatological images, and in some instances, the model makes a diagnosis as efficiently as dermatologists (Chan et al., 2020). These systems make it possible to show how effectiveness of the medical diagnosis can be improved through the use of ML.

Even though radiological image interpretation and dermatology are the most renowned examples of ML's medical applications, AI is much more vast. AI is also applied in the following areas, disease risks predicting, patient usage, and genetic studies. The possibility of realizing additional, as yet unimagined, applications continues to remain staggering given the future progression of AI and its progressively deeper integration into the healthcare environments.

AI Applications in Medication Therapy Management

As AI automates recurring, manual, and routine processes, pharmacists' time and attention can be diverted to intricate clinical duties that offer high-touch, compassionate patient care. Time-consuming responsibilities including regular monitoring, patient and drug safety monitoring, and data processing may be offloaded by AI systems. AI technology can operate in the background to give the pharmacist, at the right moment, readily interpreted and graphically digested information to help with patient care decisions (Chalasani et al., 2023; Damiani et al., 2023). Pharmacists will need to actively participate with this new supply of evidence-based data as it synthesizes. Healthcare paradigms may change as AI capability expands into the diagnostic domain and some diagnoses can be independently validated by AI-enabled equipment and applications. Eventually, patients might be under ongoing diagnostic monitoring. Especially for underserved patient populations, pharmacy leadership should concentrate on enhancing access to care through scalable models based on AI-enabled diagnostic surveillance and pharmacy drug management. Regarding to the novelty of this technology, systems should be built with a functional degree of autonomy that matches the degree of confidence users can have in the system. Healthcare paradigms could change as AI capability expands into the diagnostic domain and some diagnoses can be independently verified by AI-enabled equipment and applications (Abràmoff et al., 2018).

Beyond its effects on care delivery paradigms, AI is broadening the range of available medications. Software platforms made possible by advances in AI capabilities are emerging for use alongside, or in place of, pharmaceutical therapy in individuals with chronic illness conditions. Pharmacists ought to be involved in the design of AI-enabled digital medicines because of the expected trend of growing development and usage of digital therapies and a combination of chemical, biological, and digital therapies. For monitoring of safety and efficacy, pharmacists ought to have access to the data from AI-enabled digital therapies (Searchfield et al., 2021). Looking for chances to work with other healthcare organizations, pharmacy organizations can contribute to the development of standards and guidelines on how to use AI-enabled digital therapies into patient care.

Informatics

Because clinical trial design and evidence-based medicine are rigorous fields, AI models must be trained, assessed, rectified, and applied to data that closely resemble actual practice. Run on disproportionately represented data, poorly sourced data, or connected data thought to be causal, AI models may inadvertently exacerbate systemic prejudice or discrimination (Ramyap et al., 2022). Furthermore, since clinical practice and data inputs or distributions change over time, AI models need to be maintained and watched over. Especially pharmacist with an emphasis in informatics, have the perfect baseline skill sets for developing and implementing AI because of their experience in experimental design, research methods, and problem solving. Pharmacists should encourage the creation of interdisciplinary teams or dedicated positions for integrating AI solutions within the health system. They can also further specialize in AI and data science or form alliances with data scientists to create, test, and validate new AI models linked to medication management (Kiryu, 2022, 2023).

Clinical Applications

Incorporating AI and AI-enabled clinical decision support tools into conventional clinical workflows should be an open and eager adaptation by pharmacists to improve patient care. Drugs are a key component of therapeutic suggestions that AI applications are extending beyond diagnostics to encompass.

Pharmacists should be in the forefront of the research, development, application, and quality enhancement of therapy recommendation models because drugs are their main emphasis. To assess these models, improve patient care, lower costs, and provide patients complete medication management, pharmacy departments should work with healthcare systems to use pharmacists and new AI-enabled diagnostic and decision support tools (Akhlaghi et al., 2024).

Clinical Pharmacy Services

Among clinical pharmacy services (CPS), medication order review the most frequently targeted by AI-powered apps and tools, health product dispensing, and pharmaceutical interviews and therapeutic education (Ranchon et al., 2023).

Medication Order Review

In medication order review, AI models identify medication orders needing clinical pharmacist interventions using machine learning. Comprehensive medication order reviews or targeted reviews on specific medication classes, such as anti-infective agents could be conducted. A unique method for training the supervised algorithm involved previously analysing prescription orders from clinical pharmacists who needed pharmaceutical interventions for drug-related issues. In spotting patients with a drug-related issue, the decision support algorithm performed better than traditional systems. Notably, in each of these experiments, improper prescriptions found by machine learning algorithms needed to be reevaluated by clinical pharmacists to confirm the occurrence of the error.

Health Products Dispensing

Dispensing of medications and medical devices can be performed using various AI methods (machine learning, deep learning and conversational AI). A model was developed to predict early preparation of health product dispensing before outpatient arrival at the hospital pharmacy to reduce patient wait times (Alhorishi et al., 2021). Another model was developed for blister-packaged drug identification to assist hospital pharmacists in medication dispensing (Ting et al., 2020).

Pharmaceutical Interviews and Therapeutic Education

Kessler et al. discussed the use of a commercially available AI platform (https://www.surveyorhealth.com/) for telephone medication management interventions conducted by clinical pharmacists in collaboration with physicians. This platform uses Bayesian probabilistic inference to assess the risk of each medication for potential adverse events. It is a cloud-based application that incorporates demographic, laboratory, and medication fill data to identify high-risk Medicaid patients and guide pharmacists' interventions (Kessler et al., 2021).

By examining wireless signals in the patient's residence, without making physical contact, Zhao et al. established AI-powered tools based on neural network models to identify drug self-administration problems. Volunteers who had received training from pharmacists on how to use insulin pens and inhalers observed others self-administer. Through comparison of the model's predictions with human annotations, it was determined whether patients used these devices correctly. Such findings allow medical specialists to customize instruction on the correct use of medicine devices (Zhao et al., 2021).

Pharmacy Operations

From an operational perspective, AI platforms can enhance inventory management, facilitate product verification, and allow pharmacists to utilize their skills more fully (Khan et al., 2023). As AI becomes more reliable, standard pharmacy operations will become increasingly automated, enabling pharmacists to focus on high-value patient-care activities. It is crucial for pharmacy executives to not only adopt AI but also lead efforts to shape the future of pharmacy and the role of pharmacists in an AI-driven environment. Pharmacy leadership should prioritize the development and deployment of operational AI systems that promote personalized, continuous, and preventive care.

Education and Engagement

All facets of pharmacy practice need exposure to and education on artificial intelligence (Hamid et al., 2023). Throughout their PharmD programme, pharmacy students should be exposed to the fundamentals of data science and artificial intelligence through a health informatics curriculum. Pharmacists should get chances to learn about AI through ongoing education. For individuals who would like additional practical experience in AI creation, governance, and application, there could be data science or pharmaceutical residencies with an AI concentration. The pharmacy education system needs to be flexible as new technologies develop quickly to guarantee that the profession is prepared to handle these shifts in patient care.

ENHANCING MTM EFFICACY WITH AI

Personalized Care

The Electronic Medical Records and Genomics (eMERGE) Network has discovered previously unidentified relationships between genetic information and clinical data in a number of therapeutic domains, including ophthalmology and cardiovascular diseases, by combining patients' '-omics' data (e.g., genomics, proteomics, microbiomics) with EHR information.

Healthcare Management and Resource Optimization

Real-time predictive analytics of clinical data in EHRs can be performed to maximise resource management and carry out cost-benefit studies. Among the predictive results of EHR data analysis are the identification of risk indicators for readmissions, decompensation, triage, and high-cost patients.

Advancing EHR Studies

Use multicentric, multilanguage approaches to access larger datasets, diverse populations, and enable cross-center comparisons. Validate clinical NLP system outputs against expert-reviewed clinical notes for sensitivity and recall. Ensure data confidentiality and security in line with ethical, national, and international regulations.

Applications and Challenges in Hospital Pharmacy

While system pharmacies have been reluctant to incorporate AI, pharmacovigilance has been greatly enhanced by the use of NLP and machine learning to examine unstructured EHR data. This involves tracking treatment adherence, comorbidities and risk factors-based prescription errors, and hitherto unidentified drug-related adverse events, disparities in patient prescriptions. Important drug pricing information found in EHRs can be used to assess treatment alternatives and lessen patients' financial burdens. Because it affects treatment adherence and total healthcare costs, drug costs disclosure in EHRs has grown in significance. Real-time assessments of drug efficacy, safety, and comparative effectiveness made possible by NLP and machine learning analysis of EHR data can support decision-making in pharmacy and therapeutics (P&T) Committees.

AI in Hospital Pharmacy

Modern approaches in the fields of artificial intelligence (AI) and big data analytics to take use of current huge medical databases have been encouraged by the digitalization of health and medicine and the increasing availability of electronic health records (EHRs) by clinical researchers and healthcare professionals. Within the Hospital and Health System (Del Rio-Bermudez et al., 2020)

Natural language processing (NLP) and machine learning applied to access and analyse the unstructured, free-text data (e.g., medication safety, patients' medication history, adverse drug reactions, interactions, medication errors, therapeutic outcomes, and pharmacokinetic consultations) captured in millions of EHRs may become a crucial tool to enhance patient care and carry out real-time assessments of the efficacy, safety, and comparative effectiveness of available drugs in pharmacies. The potential of this method to facilitate share-risk agreements and direct decision-making in pharmaceutical and therapeutics (P&T) Committees is immense.

Ethical and Other Risks of AI Use in MTM

Despite its potential, the implementation of AI in MTM is not without challenges. One significant barrier is the resistance to change among healthcare providers who may be hesitant to trust AI systems over their clinical judgment. Additionally, there are concerns regarding data privacy and security, as the use of AI requires access to sensitive patient information. Moreover, integrating AI technology into existing healthcare systems can be complex and costly. Addressing these challenges is crucial to ensure the successful adoption of AI in medication therapy management.

Ethical Risks

- Data Privacy and Security: AI systems work based on a significant amount of patient information, and as such, issues such as data violation arise.
- Bias and Discrimination: This means that it is possible to obtain new results from the AI algorithms that could promote discrimination in the prescriptions of medication.
- Autonomy and Accountability: Health care providers are concerned that heavy dependence on artificial intelligence deprives them of control and can lead to ambiguous responsibility for negative outcomes.

Practical Risks

- Technical Limitations: However, there are several drawbacks of AI algorithms; the algorithms could not learn about some unusual or extraordinary cases, which means that they would give wrong recommendations to the doctors. Also, there are several limitations of AI algorithms in handling complex patient cases
- Lack of Transparency: AI is very black box in nature and thus the models lack the explanation of why a certain decision has been made in the process.
- Dependence on high-quality data for accurate AI predictions: Patient data is often fragmented across different healthcare providers and systems, making it difficult to collect and integrate. Also, data may contain errors, inconsistencies, or biases that can affect the accuracy of AI predictions.
- Cost and Accessibility: The adoption of AI for MTM solutions can also be costly, and this might reduce its reach to the few people with adequate monetary capabilities.

Addressing the Risks

To mitigate these risks, it is essential to:

- Develop robust data privacy and security measures: The other preventive measure that need to be installed include high level of encryption, restrictions on user access to file and data anonymizing techniques.
- Ensure algorithmic fairness: Make a point to feed the AI models with datasets that are free from bias in order to ensure that the AI models themselves are unbiased.

- Maintain human oversight: It is this healthcare provider's ultimate responsibility to make the final decisions and make the decisions counter to what AI suggests if needed.
- Promote transparency and explainability: Train decision making AI models that can recount why they reached the decision they made.
- Address cost and accessibility: Strive to democratize AI-driven MTM solutions serving the patients.

By and large, when functioning well, AI allows machines to perform tasks that appear highly intelligent to us, such as translating human languages. It might seem like AI gives machines human-like intelligence, but in reality, even with AI, machines do not possess human-like intelligence. Recognizing this, we must always consider the safety and effectiveness of AI. Tutt has suggested creating an "FDA for Algorithms" to ensure evidence supports the safety and effectiveness of various AI implementations before they are used in practice (Tutt, 2017). Whether or not such an agency is established, pharmacists have a crucial role in generating the evidence needed to guide decisions on the widespread implementation of AI in routine pharmacy practice.

Future Directions and Innovations of AI in MTM

As AI advances over the years, its application in MTM is becoming broader and broader providing new thrilling approaches to provide the best care for the patients and increase the effectiveness of the clinical pharmacy services. This synthesis reveals the following social promise that AI is expected to introduce in MTM in the future: Several innovative changes that can transform how medication therapy is managed, monitored, and individualized will be implemented by AI in the future.

One of them is the continued development of AI-based predictive analytics as a critical area. The same big amount of data provided by patients allows AI to foresee outcomes that any particular patient is likely to exhibit, distinguish between potential negative effects of medications, as well as determine an optimal course of treatment. Such specific details will specifically enable pharmacists to develop intricate and highly individualized medication schedules especially considering the compartmentalized and specialized natures of most contemporary diseases. These can reduce symptoms and enhance the administration of medications for the benefit of the patients and greatly reduce the over-expenditure in medical costs.

Furthermore, the use of Decision Support Systems (DSS) in clinical pharmacy practice supported by Artificial intelligence is expected to grow significantly in the future. These systems can give the current recommendations based on the current clinical practice protocols or info on drug interactions, or patient's medical record data. Application of AI in EHR systems means that pharmacists will be able to easily explain patient information and carry out more effective and efficient decisions in care delivery.

AI application for automating frequent tasks in MTM is another perspective direction. AI can save time for many routine tasks involved in medication reconciliation, data entry and detection of medication related issues in order to help pharmacists perform more patient-oriented work. Automation can also increase efficiency and effectiveness in that it reduces the amount of time that is spent on paperwork and tightens interaction with patients.

Other developing IT solutions such as natural language processing (NLP) as well as machine learning (ML) are also expected to improve PT communication as well as education in MTM. For example, AI chatbots and virtual assistants can help patients get medication-related information, help patients remember when to take their medications and other health education. Cohort: In large patient database, these tools can be quite useful in overcoming the communication barriers and supporting the patients by replying to their queries periodically.

Further, continuing advances in Artificial Intelligence are anticipated to augment its deployment with telepharmacy and remote patient monitoring systems. Telehealth expands the use of AI where it can help monitor a patient's progress from distance, and report signs of noncompliance to medications or the first signs of new complications. This will go a long way to increasing the coverage of MTM services, say for populations that do not often get reached or for those in remote areas.

Therefore, the application and development of AI in MTM carry a number of potential innovations that can enhance privacy, speed, and comprehensiveness of medication management services. Using artificial intelligence technologies at their fingertips, clinical pharmacists would be in a position to manage clinical care appropriately thus improving on patient care, disease treatments and positive healthy status.

REFERENCES

Abràmoff, M. D., Lavin, P. T., Birch, M., Shah, N., & Folk, J. C. (2018). Pivotal trial of an autonomous AI-based diagnostic system for detection of diabetic retinopathy in primary care offices. *NPJ Digital Medicine*, 1(1), 39. Advance online publication. DOI: 10.1038/s41746-018-0040-6 PMID: 31304320

Akhlaghi, H., Freeman, S., Vari, C., McKenna, B., Braitberg, G., Karro, J., & Tahayori, B. (2024). Machine learning in clinical practice: Evaluation of an artificial intelligence tool after implementation. *EMA -. Emergency Medicine Australasia*, 36(1), 118–124. Advance online publication. DOI: 10.1111/1742-6723.14325 PMID: 37771067

Alhorishi, N., Almeziny, M., & Alshammari, R. (2021). Using machine learning to predict early preparation of pharmacy prescriptions at PSMMC - a comparison of four machine learning algorithms. *Acta Informatica Medica*, 29(1), 21. Advance online publication. DOI: 10.5455/aim.2021.29.21-25 PMID: 34012209

Aljuaid, A., & Anwar, M. (2022). Survey of Supervised Learning for Medical Image Processing. *SN Computer Science*, 3(4), 292. Advance online publication. DOI: 10.1007/s42979-022-01166-1 PMID: 35602289

Amin, A. A., Sajid Iqbal, M., & Hamza Shahbaz, M. (2024). Development of Intelligent Fault-Tolerant Control Systems with Machine Learning, Deep Learning, and Transfer Learning Algorithms: A Review. In *Expert Systems with Applications* (Vol. 238). DOI: 10.1016/j.eswa.2023.121956

Bakkar, N., Kovalik, T., Lorenzini, I., Spangler, S., Lacoste, A., Sponaugle, K., Ferrante, P., Argentinis, E., Sattler, R., & Bowser, R. (2018). Artificial intelligence in neurodegenerative disease research: Use of IBM Watson to identify additional RNA-binding proteins altered in amyotrophic lateral sclerosis. *Acta Neuropathologica*, 135(2), 227–247. Advance online publication. DOI: 10.1007/s00401-017-1785-8 PMID: 29134320

Burns, A. (2005). Medication Therapy Management in Community Pharmacy Practice: Core Elements of an {MTM} Service (Version 1.0): American Pharmacists Association and National Association of Chain Drug Stores Foundation. *Journal of the American Pharmacists Association*, 45(5), 573–579. DOI: 10.1331/1544345055001256

Chalasani, S. H., Syed, J., Ramesh, M., Patil, V., & Pramod Kumar, T. M. (2023). Artificial intelligence in the field of pharmacy practice: A literature review. In *Exploratory Research in Clinical and Social Pharmacy* (Vol. 12). DOI: 10.1016/j.rcsop.2023.100346

Chan, S., Reddy, V., Myers, B., Thibodeaux, Q., Brownstone, N., & Liao, W. (2020). Machine Learning in Dermatology: Current Applications, Opportunities, and Limitations. In *Dermatology and Therapy* (Vol. 10, Issue 3). DOI: 10.1007/s13555-020-00372-0

Damiani, G., Altamura, G., Zedda, M., Nurchis, M. C., Aulino, G., Heidar Alizadeh, A., Cazzato, F., Della Morte, G., Caputo, M., Grassi, S., & Oliva, A. (2023). Potentiality of algorithms and artificial intelligence adoption to improve medication management in primary care: A systematic review. *BMJ Open*, 13(3), e065301. Advance online publication. DOI: 10.1136/bmjopen-2022-065301 PMID: 36958780

Dee, J., Dhuhaibawi, N., & Hayden, J. C. (2023). A systematic review and pooled prevalence of burnout in pharmacists. In *International Journal of Clinical Pharmacy* (Vol. 45, Issue 5). DOI: 10.1007/s11096-022-01520-6

Del Rio-Bermudez, C., Medrano, I. H., Yebes, L., & Poveda, J. L. (2020). Towards a symbiotic relationship between big data, artificial intelligence, and hospital pharmacy. *Journal of Pharmaceutical Policy and Practice*, 13(1), 75. Advance online publication. DOI: 10.1186/s40545-020-00276-6 PMID: 33292570

Ferrucci, D., Levas, A., Bagchi, S., Gondek, D., & Mueller, E. T. (2013). Watson: Beyond jeopardy! In *Artificial Intelligence* (Vols. 199–200). DOI: 10.1016/j.artint.2012.06.009

Flynn, A. J., Fortier, C., Maehlen, H., Pierzinski, V., Runnebaum, R., Sullivan, M., Wagner, J., & Stevenson, J. G. (2021). A strategic approach to improving pharmacy enterprise automation: Development and initial application of the Autonomous Pharmacy Framework. In *American Journal of Health-System Pharmacy* (Vol. 78, Issue 7). DOI: 10.1093/ajhp/zxab001

Goetz, L. H., & Schork, N. J. (2018). Personalized medicine: motivation, challenges, and progress. In *Fertility and Sterility* (Vol. 109, Issue 6). DOI: 10.1016/j.fertnstert.2018.05.006

H., R., J., C., M., M., & A., O. (2018). Assessing pharmacists' acceptance of prescription dispensing data infographic to initiate pharmacist-prescriber collaboration. In *Journal of the American Pharmacists Association* (Vol. 58, Issue 3).

Habehh, H., & Gohel, S. (2021). Machine Learning in Healthcare. *Current Genomics*, 22(4), 291–300. Advance online publication. DOI: 10.2174/1389202922666210705124359 PMID: 35273459

Hamet, P., & Tremblay, J. (2017). Artificial intelligence in medicine. *Metabolism: Clinical and Experimental*, 69, S36–S40. Advance online publication. DOI: 10.1016/j.metabol.2017.01.011 PMID: 28126242

Hamid, H., Zulkifli, K., Naimat, F., Che Yaacob, N. L., & Ng, K. W. (2023). Exploratory study on student perception on the use of chat AI in process-driven problem-based learning. *Currents in Pharmacy Teaching & Learning*, 15(12), 1017–1025. Advance online publication. DOI: 10.1016/j.cptl.2023.10.001 PMID: 37923639

Hassan, M., Awan, F. M., Naz, A., Deandrés-Galiana, E. J., Alvarez, O., Cernea, A., Fernández-Brillet, L., Fernández-Martínez, J. L., & Kloczkowski, A. (2022). Innovations in Genomics and Big Data Analytics for Personalized Medicine and Health Care: A Review. In *International Journal of Molecular Sciences* (Vol. 23, Issue 9). DOI: 10.3390/ijms23094645

Higgins, O., Short, B. L., Chalup, S. K., & Wilson, R. L. (2023). Artificial intelligence (AI) and machine learning (ML) based decision support systems in mental health: An integrative review. In *International Journal of Mental Health Nursing* (Vol. 32, Issue 4). DOI: 10.1111/inm.13114

Hinton, G. E., & Salakhutdinov, R. R. (2006). Reducing the dimensionality of data with neural networks. *Science*, 313(5786), 504–507. Advance online publication. DOI: 10.1126/science.1127647 PMID: 16873662

Hoffer, E. P. (1992). DxPlain: Diagnostic consultation by computer for the practicing physician. *The Journal of Medical Practice Management*, 7(4), •••.

Indolia, S., Goswami, A. K., Mishra, S. P., & Asopa, P. (2018). Conceptual Understanding of Convolutional Neural Network- A Deep Learning Approach. *Procedia Computer Science*, 132, 679–688. Advance online publication. DOI: 10.1016/j.procs.2018.05.069

Kessler, S., Mebine, P., Von Schweber, E., Von Schweber, L., Desai, M., McConnell, W., Jai, E. M., Nguyen, J., Kiroyan, C., & Ho, D. (2021). Economic and utilization outcomes of medication management at a large medicaid plan with disease management pharmacists using a novel artificial intelligence platform from 2018 to 2019: A retrospective observational study using regression methods. In *Journal of Managed Care and Specialty Pharmacy* (Vol. 27, Issue 9). DOI: 10.18553/jmcp.2021.21036

Khan, O., Parvez, M., Kumari, P., Parvez, S., & Ahmad, S. (2023). The future of pharmacy: How AI is revolutionizing the industry. *Intelligent Pharmacy*, 1(1), 32–40. Advance online publication. DOI: 10.1016/j.ipha.2023.04.008

Kiryu, Y. (2022). Medical Big Data Analysis Using Machine Learning Algorithms in the Field of Clinical Pharmacy. *Yakugaku Zasshi*, 142(4), 319–326. Advance online publication. DOI: 10.1248/yakushi.21-00178-1 PMID: 35370185

Kiryu, Y. (2023). Development of a Medical Big Data Analysis System Utilizing Artificial Intelligence Analytics in Clinical Pharmacy. In *Yakugaku Zasshi* (Vol. 143, Issue 6). DOI: 10.1248/yakushi.22-00179-4

Kline, R. R. (2011). Cybernetics, automata studies, and the dartmouth conference on artificial intelligence. *IEEE Annals of the History of Computing*, 33(4), 5–16. Advance online publication. DOI: 10.1109/MAHC.2010.44

Kulikowski, C. A. (2019). Beginnings of Artificial Intelligence in Medicine (AIM): Computational Artifice Assisting Scientific Inquiry and Clinical Art - with Reflections on Present AIM Challenges. *Yearbook of Medical Informatics*, 28(1), 249–256. Advance online publication. DOI: 10.1055/s-0039-1677895 PMID: 31022744

Mancini, J. (2019). E-health Records: A slow marathon to get online. *Health Science Inquiry*, 5(1). Advance online publication. DOI: 10.29173/hsi179

Miller, R. A., McNeil, M. A., Challinor, S. M., Masarie, F. E., & Myers, J. D. (1986). The INTERNIST-1/QUICK MEDICAL REFERENCE Project - Status report. *The Western Journal of Medicine*, 145(6). PMID: 3544509

Moran, M. E. (2007). Evolution of robotic arms. *Journal of Robotic Surgery*, 1(2), 103–111. Advance online publication. DOI: 10.1007/s11701-006-0002-x PMID: 25484945

Ramyap, P., Abrahamp, S., Pattasserilp, M. B., Mp, M. M., Mohan Kp, N., & Kp, S. (2022). Role of Artificial Intelligence in the Pharmaceutical Industry. In *IJISET-International Journal of Innovative Science* (Vol. 09). Engineering & Technology.

Ranchon, F., Chanoine, S., Lambert-Lacroix, S., Bosson, J. L., Moreau-Gaudry, A., & Bedouch, P. (2023). Development of artificial intelligence powered apps and tools for clinical pharmacy services: A systematic review. In *International Journal of Medical Informatics* (Vol. 172). DOI: 10.1016/j.ijmedinf.2022.104983

Reis, D., & Afonso, R. (2011). PDSS: Pharmacy Decision Support System - Concept approach. *Proceedings of the IADIS International Conference E-Health 2011, Part of the IADIS Multi Conference on Computer Science and Information Systems 2011, MCCSIS 2011.*

Schmidhuber, J. (2015). Deep Learning in neural networks: An overview. In *Neural Networks* (Vol. 61). DOI: 10.1016/j.neunet.2014.09.003

Schussel, K. E., Forbes, S., Taylor, A. M., & Cooley, J. H. (2019). Implementation of an interprofessional medication therapy management experience. *American Journal of Pharmaceutical Education*, 83(3), 6584. Advance online publication. DOI: 10.5688/ajpe6584 PMID: 31065160

Schutz, N., Olsen, C. A., Mclaughlin, A. J., Yi, W. M., Nelson, S. D., Kalichira, A. L., Smith, A. H., Miller, K. A., Le, T., Chaffee, B. W., Worthy Woodbury, C. D. R. K., & Patel, H. (2020). ASHP Statement on the Use of Artificial Intelligence in Pharmacy. In *American Journal of Health-System Pharmacy* (Vol. 77, Issue 23). DOI: 10.1093/ajhp/zxaa249

Searchfield, G. D., Sanders, P. J., Doborjeh, Z., Doborjeh, M., Boldu, R., Sun, K., & Barde, A. (2021). A State-of-Art Review of Digital Technologies for the Next Generation of Tinnitus Therapeutics. In *Frontiers in Digital Health* (Vol. 3). DOI: 10.3389/fdgth.2021.724370

Secinaro, S., Calandra, D., Secinaro, A., Muthurangu, V., & Biancone, P. (2021). The role of artificial intelligence in healthcare: A structured literature review. *BMC Medical Informatics and Decision Making*, 21(1), 125. Advance online publication. DOI: 10.1186/s12911-021-01488-9 PMID: 33836752

Shieber, S. M. (2007). The turing test as interactive proof. *Noûs (Detroit, Mich.)*, 41(4), 686–713. Advance online publication. DOI: 10.1111/j.1468-0068.2007.00636.x

Shortliffe, E. H. (1977). MYCIN: A KNOWLEDGE-BASED COMPUTER PROGRAM APPLIED TO. *Infectious Diseases*.

The definition of clinical pharmacy. (2008). In *Pharmacotherapy* (Vol. 28, Issue 6). DOI: 10.1592/phco.28.6.816

Ting, H. W., Chung, S. L., Chen, C. F., Chiu, H. Y., & Hsieh, Y. W. (2020). A drug identification model developed using deep learning technologies: Experience of a medical center in Taiwan. *BMC Health Services Research*, 20(1), 312. Advance online publication. DOI: 10.1186/s12913-020-05166-w PMID: 32293426

Tutt, A. (2017). An fda for algorithms. In *Administrative Law Review* (Vol. 69, Issue 1). DOI: 10.2139/ssrn.2747994

van Leeuwen, K. G., Schalekamp, S., Rutten, M. J. C. M., van Ginneken, B., & de Rooij, M. (2021). Artificial intelligence in radiology: 100 commercially available products and their scientific evidence. *European Radiology*, 31(6), 3797–3804. Advance online publication. DOI: 10.1007/s00330-021-07892-z PMID: 33856519

Weizenbaum, J. (1966). ELIZA-A computer program for the study of natural language communication between man and machine. *Communications of the ACM*, 9(1), 36–45. Advance online publication. DOI: 10.1145/365153.365168

Wen, Z., & Huang, H. (2022). The potential for artificial intelligence in healthcare. *Journal of Commercial Biotechnology*, 27(4). Advance online publication. DOI: 10.5912/jcb1327

Zhao, M., Hoti, K., Wang, H., Raghu, A., & Katabi, D. (2021). Assessment of medication self-administration using artificial intelligence. *Nature Medicine*, 27(4), 727–735. Advance online publication. DOI: 10.1038/s41591-021-01273-1 PMID: 33737750

Chapter 8 The Revolutionary Biocomputing Technology of Organoid Intelligence (OI) as the Forefront of AI and Medicine

Al-Hassan Soliman Wadan
https://orcid.org/0009-0009-0843-8975
Galala University, Egypt

ABSTRACT

Recent advancements in human stem cell-derived brain organoids have the potential to replicate critical molecular and cellular aspects of learning, memory, and cognition in vitro. The term "organoid intelligence" (OI) is proposed to encapsulate these advancements, aiming to establish a multidisciplinary field dedicated to OI. Researchers can now manufacture standardized, 3D myelinated brain organoids with heightened cell density, increased levels of glial cells, and gene expression essential for learning. Advanced 3D microelectrode arrays allow precise spatiotemporal electrophysiological signaling and recording. This lets scientists study the molecular processes that make learning and memory possible. Emerging technologies enable innovative biocomputing models through stimulus-response training and organoid-computer interfaces. The vision involves networked interfaces where brain organoids interface with real-world sensors and output devices, and eventually with sensory organ organoids.

DOI: 10.4018/979-8-3373-1132-6.ch008

INTRODUCTION

Experimental animals, along with 2D lines of cells, proved to be effectively employed in biomedical studies for the last 100 years. Their contributions have enhanced our comprehension of cellular signaling networks, facilitated the formulation of criteria for designing new molecules, revealed prospective targets for drug development, and illuminated the fundamental pathogenic mechanisms of disorders. The global application of model systems in contemporary biomedical research highlights their significant importance (Tang et al., 2022). Therefore, it is essential to improve how we study biological processes. We must better connect what happens at the tiny cell level with what happens in larger tissues. This will help make our lab experiments more accurate (Tang et al., 2022). The advancement of human pluripotent stem cell technology and adult stem cell methods of culture has facilitated the creation of individual-specific models in laboratory settings. These models can be used to study genetic variations linked to human diseases and to generate large quantities of cellular materials for compound screening. In 2006, Kazutoshi Takahashi and Shinya Yamanaka achieved the successful conversion of mice fibroblast cell types into pluripotent stem cells through the introduction of four gene transcription factors (TFs), namely *Oct4, Sox2, Klf4*, and *c-Myc*. The cells, which exhibited similar gene expression and developmental potential as ESCs, were referred to as induced pluripotent stem cells (iPSCs). The inception of hiPSC technology in 2007 marked a transformative milestone (Takahashi et al., 2007; Yu et al., 2007). Since then, it has been widely employed to establish human "disease-in-a-dish" models, offering the potential for personalized medicine initiatives. Furthermore, the advent of clustered regularly interspaced short palindromic repeats (CRISPR)/Cas9 endonuclease technology (Drost et al., 2015, 2017; Schwank et al., 2013) has enabled the creation of genetically modified hiPSC-based disease models. Initially, research primarily focused on utilizing iPSCs to develop 2D cell cultures, which lacked essential features such as cell-cell interactions, cellular diversity, and microcircuitry. Recently, there has been a notable shift towards advancing in vitro human 3D organ culture techniques, often called organoids. These techniques have become popular because they can create structures similar to real organs and simultaneously include different types of cells.

Organoids produced from primary tissues, ESCs, or iPSCs, can regenerate themselves, organize themselves, and demonstrate the exact characteristics of multicellularity and functionality as organs in living organisms. The term "organoid" was initially used in oncology as a synonym for "teratoma," first documented in 1946 (Smith & Cochrane, 1946). During the 1960s, the term "organoids" was coined to refer to organogenesis, which involved conducting tests where cells were dissociated and then reaggregated by developmental biologists (Weiss & Taylor, 1960).

Significantly, these organoids effectively replicated the structural arrangement of tissues in living organisms and comprised a complete set of stem cells, progenitor cells, and fully developed cell types. Researchers can quickly modify the 3D culture system by adjusting growth factors and cell isolation processes. This allows them to create many types of normal human and malignant organoids, including the brain (Lancaster et al., 2013a; Watanabe et al., 2017), colon (van de Wetering et al., 2015), stomach (Bartfeld et al., 2015), and liver organoids (Broutier et al., 2017; Huch et al., 2015). Scientists use organoids because they are small, repeatable models that show how complex human tissues work during growth, normal function, and disease. Many labs can use these models. They are easy to change, study, and test without the problems found in live animal experiments. When grown from human cells, they help move research from animals to humans while following ethical rules (Lehmann et al., 2019). Organoids are characterized by utilizing a sustainable tissue source from stem cells or primary tissue. This tissue is then cultured in a controlled environment, allowing it to self-organize into a structure resembling the modeled healthy or diseased tissue. Organoids also incorporate various aspects of the cellular complexity in the modeled tissue. Furthermore, organoids can be propagated and shared either as a culture or through a specific population of stem or progenitor cells (Lehmann et al., 2019).

Human brain organoids have emerged as an innovative model for studying neurodevelopment and simulating neurological disorders. These simplified 3D culture methods provide unparalleled opportunities to study the initial phases of brain structure and functioning in a laboratory setting (Hartung et al., 2023). The old term 'mini-brain' is no longer used. Organoids show great potential for basic research, testing drugs, personalized medicine, cell treatments for injuries and brain diseases, and even new ways of computing with biology (Smirnova et al., 2023). The future holds immense possibilities as organoids are being transplanted into animal models to enhance their functionality or incorporated into interfaces between brains and computers. Nevertheless, the characteristics that contribute to the scientific reliability of brain organoids also give rise to distinctive ethical dilemmas (Hartung et al., 2023) They are not detected in organoids of other tissues and necessitate meticulous deliberation (**fig. 1**) (Hartung et al., 2023).

Figure 1 displays a comprehensive outline of the proposed scenario. Brain organoids are cultivated in an organ-on-chip device in the standard perspective, allowing for drug testing for individualized therapy (upper branch). In the given situation, brain organoids are connected to a MEA (lower branch) to control the input and output of electrical impulses using an artificial intelligence (AI) model. This setup is utilized in learning and memory activities, which opens up the potential for creating personalized learning models. All illustrations have been generated with BioRender.com.

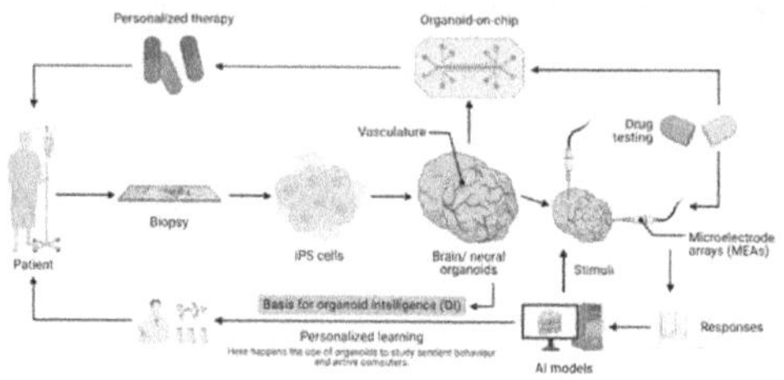

Current State, Prospects, and Possibilities of Brain Organoid Research

Our approach to OI is made possible by two significant advancements in biological technology: the pioneering research on converting human somatic cells into iPSCs. Over the last decade, there has been a notable movement in neural cell cultures, moving away from traditional single-layer cultures towards advanced 3D cultures that closely mimic organs, referred to as brain organoids. Each organoid has an outer diameter under 500 millimeters and comprises fewer than 100,000 cells. Each organoid is around one 3-millionth of the human brain's size, potentially corresponding to memory storage that equals 800 Megabytes. Other research teams have documented the presence of brain organoids that typically measure 3 and 5 millimeters in diameter and have been cultured for over a year (Hartung et al., 2023; Smirnova et al., 2023).

Brain organoids produced from human pluripotent stem cells can spontaneously arrange themselves into structured neural tissue, displaying distinct brain regions and a variety of different cell types (Pa ca, 2018). Every organoid resembles the nascent human brain, albeit with a less complex arrangement, cellular makeup, and dimensions. Various techniques enable the production of organoids specific to different regions, such as the forebrain, midbrain, or hypothalamus (Lancaster et al., 2013b; Pa ca, 2018). Lancaster et al. use brain organoids as a means to simulate neurodevelopmental principles and elucidate processes of neurological diseases (Lancaster et al., 2013b).

Continual progress improves organoids' maturity, functioning, and structural development. Future advancements, including integrating additional cell types, 3D patterning, and orthotopic transplantation, can potentially enhance the development and validation of this technology using human clinical information. The use of animal models will be crucial in demonstrating its practicality. OI is creating intelligent systems inspired by biology, employing human brain organoids and other biological elements (Shi et al., 2024). OI arises from the convergence between biological computation and human-machine interface technology. Its primary objective is to create computational representations with 3D cultures of human neurons, specifically brain organoids (Smirnova et al., 2023). This field combines progress in the biology of stem cells, tissue engineering, biomaterials, electrophysiological sensors, microfluidics, and ML (Bai et al., 2024). The main objective of OI is to utilize the computational capacities of biological neural network structures to generate novel types of intelligence and biocomputing. Furthermore, researchers are advancing microfluidic perfusion systems to function as substitutes for blood vessels in brain organoids, ensuring the delivery of essential nutrients and removal of waste crucial for organoid stability and survival (Duzagac et al., 2021). Special 3D sensors measure electrical signals from brain organoids. This helps scientists study how these organoids process information and learn (Huang et al., 2022). Many researchers try to copy the complex brain networks that control how we think and act. Brain connections form when neurons grow, move, and link up nearby and far apart. Scientists can now turn special stem cells into organoids that mimic specific brain areas. By joining these organoids, they can study how neurons move and connect (Pa ca, 2018). Assemblids utilize self-organization to facilitate complex interactions between cells, the development of circuits, and the maturation process in long-term cultures (Miura et al., 2022).

Engineering has sought to replicate brain-like functioning throughout the computer era, primarily by pursuing AI (Smirnova et al., 2023). However, we have not yet reached the point where we can replicate brain activity using computer simulations (Smirnova et al., 2023). OI is a developing field combining various disciplines to create biological computing systems employing 3D cultures of brain organoids and brain-machine interface technologies (Smirnova et al., 2023). In terms of functionality, the brain operates distinctly compared to a computer. Human brains can function more effectively with limited information through intuitive thinking and readily incorporate new information (Smirnova et al., 2023). Development of 3D organ equivalents using bioengineering (Alépée et al., 2014), the utilization of iPSCs has significantly transformed the field of biomedical research (Roth, 2021), offering progressively more applicable model systems, including those of the brain (Koo et al., 2019). Organoids and organ-on-chip models are two distinct technologies under microphysiological systems (MPS). These systems involve several bioengineering advancements that aim to replicate the structure and function of organs in laboratory experiments.

Advancements in stem cell technology have created various human models and testing systems, making experimental tools more accessible through global collaborations. For studying cognitive functions of the brain, neuronal cultures aim to mimic human organ structure and function. The development of human stem cells and microphysiological systems (MPS) bioengineering allows for the potential achievement of learning and memory capabilities in organoids and organ-on-chip systems. Using models that simulate basic cognitive abilities, known as "intelligence- or cognition-in-a-dish," researchers can explore these abilities with unconventional therapies and quick assessments. This marks a significant milestone in research. Brain organoids are increasingly incorporating glial cells, such as microglia, and show self-organization, which enhances understanding of synaptic plasticity and long-term memory (Xu & Wen, 2021) (fig.2).

Figure 2 illustrates the progression from 2D neuronal culture to 3D organoid brain culture.

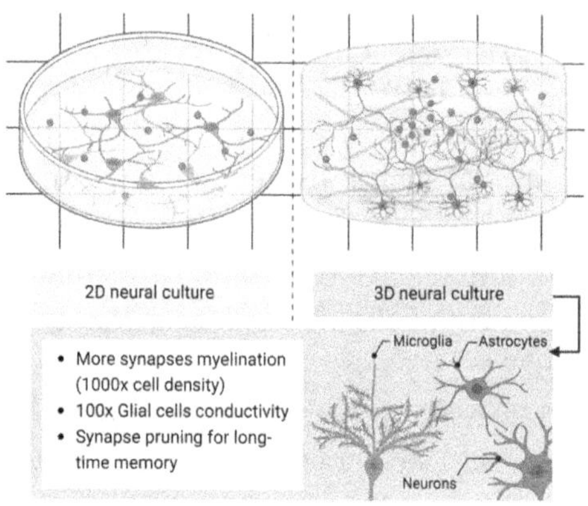

By creating disease models for neurodevelopmental disorders and neurodegeneration, we can enhance the efficiency and efficacy of chemical safety evaluations and therapy development for some of the most costly and challenging diseases. Examine the potential of experimentally studying the human body's complex physiology, namely in understanding, diagnosing, and treating different brain diseases in both healthy and pathological states. One can investigate inquiries concerning the precise elements that contribute to the disturbance of brain development, the impairment of brain function, or the hastening of brain function decline. Furthermore, it is possible to explore other approaches for addressing these issues. These problems can be handled using engineered MPS models, providing cost-effective and time-efficient testing. Conduct research on artificial intelligence models in a controlled setting to study the basic principles of human cognitive abilities. Additionally, these models will be employed to improve the detection of harmful substances that cause neurological disorders and explore potential remedies for such conditions. Lastly, the aim is to achieve biological, computational capabilities complementing traditional computing methods (Smirnova et al., 2023).

Bioengineering and stem cell technology have created organoids and organ-on-chip systems that mimic complex organs' structural and functional organization. The ultimate goal of these advancements is to recreate cognitive capabilities and intelligence in the human brain, comprehend and imitate the learning process, and create functional biological computing systems that work harmoniously with artificial intelligence. Researchers effectively showcased learning elements by training human iPSC-derived neuronal cells to play a basic computer game called *Pong* (Smirnova & Hartung, 2022). A feedback loop strategy was employed, wherein the individual was subjected to white noise (incomprehensible input) as a penalty if the controlled paddle failed to hit the ball. These findings indicate that the improvements in gameplay were enough, and further trials with careful control showed a correlation between the type of feedback given and the observed learning effects. Brain organoids derived from iPSCs have created 3D neural cultures that replicate specific characteristics of brain cellular makeup, structure, and function (Lancaster et al., 2013b; Lancaster & Knoblich, 2014). Cognition is the ultimate measure of functionality for the brain. Researchers have created the term OI to refer to combining organoids with AI. These two technologies and electrophysiology as part of sensor technologies are believed to be crucial in implementing cognitive capabilities. Utilize the spontaneously formed apparatus of 3D human brain cell cultures to store and process information. OI is a platform that allows for investigating the fundamental processes involved in learning and development. It provides valuable information about human neuroscience and has similar compensatory mechanisms that make interpretation challenging. Additionally, OI is a cost-effective alternative for drug development. The current constraint on brain organoids, which is their limiting size of approximately 500 μm, is mainly caused by restricted oxygen and nutrition supply to the core. Consequently, we cannot anticipate their substantial cognitive capacities. Researchers must devise innovative techniques to cultivate brain organoids, increasing their size from 50,000 to a minimum of 1 billion cells, approximately 1 centimeter in diameter. Support structures, such as the vasculature and a blood-brain barrier, are required (Smirnova et al., 2023).

Interacting with the Organoid - Gathering Data at Multiple Scales from Organoids

The interface includes electrodes for input-output and feedback, essential for training brain organoids. This process involves analyzing an organism's physical traits using advanced imaging and mapping neural connections. Current research focuses on developing real-time electrophysiological systems and imaging experiments that respond to changes instantly, operating at various spatial and temporal resolutions. Shell electrodes are used for measuring electrophysiological activity (Huang et al.,

2022), MEAs (Kagan et al., 2022), microelectrode meshes (McDonald et al., 2023), and neuropixels (Jun et al., 2017) Are being explored. The primary method of input in the experiments will predominantly include electrical stimulation. However, chemical signals (such as neurotransmitters or their pharmaceutical agonists) or optogenetics may also be used (fig. 3) (Lee et al., 2020).

Figure 3 illustrates the fundamental principle of organoid intelligence (OI). In addition to electrodes, neurotransmitters, and optogenetics can be used as alternative methods to input information into a brain organoid. Connecting with artificial intelligence and receiving corresponding input can create a mapping of responses.

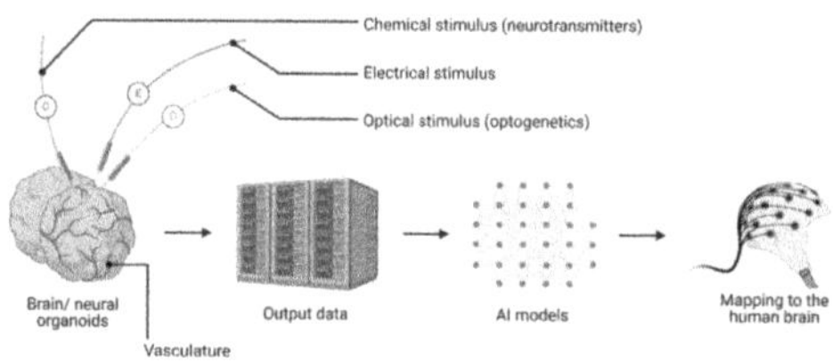

The next phase will consist of organoid crosstalk, which involves combining the brain organoid alongside sensory data from such assembloids, starting at the retina. This will provide complex inputs as well as outputs to control other organs. By employing conceptual frameworks and closed-loop system experiments, one can analyze fundamental aspects of cognition, such as the formation and recall of memories, plasticity, developmental routes, and predication. This approach allows for the investigation of the behavior of brain organoids when subjected to various input stimuli. Research has shown that brain organoids derived from stem cells taken from patients with Alzheimer's dementia display specific characteristics linked to the disorder. In a recent study, researchers observed a more significant toxic impact on the brain organoid system when a specific gene associated with autism exposure and a chemical risk element were combined. This finding demonstrates synergistic or combinatorial toxicity (Modafferi et al., 2021). An alternative and more realistic possibility is that knowledge about the brain's processes can improve computer architecture by advancing the field of neuromorphic computer science (Schuman et al., 2022). Neuromorphic computing is a computer architecture methodology that designs components by emulating the brain's mechanisms and the nervous system. Integrating three novel technologies, namely stem cell-based bioengineering, sensor technologies, and AI, can potentially exceed what we expect in advancing OI. Biological computing, or biocomputing, can outperform silicon-based computing and AI concerning speed, efficiency, and power while costing considerably fewer resources (Smirnova et al., 2023).

How to Build a Biocomputing Model

Organoids possess diverse characteristics that amplify their capacity for biocomputing. Firstly, the cell densities in these three-dimensional models are comparable to that found in existing organisms and considerably higher than in single-layer cell cultures. In 3D models, the ratio between the volume of cells and the volume of the surrounding fluid is significantly more noticeable compared to single layers. Moreover, most brain organoids have innate electrophysiological functions and react to stimulation from electricity, as evidenced by elicited field potential (Pamies et al., 2017), verifying the existence of functioning synapses. Furthermore, the organoids exhibit a significant amount of myelination in their axons. Pamies et al. pioneered creating a three-dimensional model of the human brain that demonstrated substantial axon myelination (Pamies et al., 2017). Approximately 40% of the axons present in the brain organoids were found to be myelinated (Chesnut et al., 2021a; Chesnut et al., 2021b), which approximates the 50% observed in the human brain (Buyanova & Arsalidou, 2021; Williamson & Lyons, 2018). Myelination has been replicated in different brain organoids (Monzel et al., 2017; Smits et al., 2020). Myelin decreases the capacitance of the axonal membrane and facilitates saltatory conduction between Ranvier's adjacent nodes. As the process of myelination enhances electrical conductivity by about 100 times, it holds the potential to enhance the performance of biological computing significantly. However, its actual influence in this model has yet to be proven. Ultimately, these organoid cultures have the potential to be enhanced with many cell types that play a role in biological learning, specifically oligodendrocytes, microglia, and astrocytes. Glia cells are vital in eliminating synapses during biological learning (Huang et al., 2020; Raiders et al., 2021; Xin & Chan, 2020). However, these levels have not yet been observed in brain organoid models at biologically significant concentrations. Initial research conducted in our organoid model has demonstrated the capacity for astroglia cell proliferation to reach biologically significant levels (Smits et al., 2020). Memory implies that it is more appropriate to examine these processes from a perspective that focuses on the relationship between neurons and glial cells rather than the often-employed approach that focuses solely on the interaction between neurons (Raiders et al., 2021). Additionally, modifying cell culture conditions to boost the production of immediate early genes (IEGs) is expected to improve brain organoids' learning and memory abilities. IEGs are vital for learning and are only expressed in neurons involved in memory formation.

Microfluidic Perfusion Systems

Brain organoids can replicate the spatial and temporal patterns of molecules and gene expression networks (Aparna et al., 2019). However, certain histoarchitectures, such as the cortex patterning and the neuron phenotypes in the human brain, do not accurately represent the regional architecture and intricate neural circuitry required for higher-order brain function (Hongjun et al., 2019). The complexity of the human brain is partially attributed to its substantial size and the intricate network of blood arteries that support its growth (Sansom et al., 2009). There is ongoing development of models for the brain's vasculature (Yuichiro et al., 2021); most brain organoid models do not have blood arteries and rely on passive diffusion to transport nutrients. The average diffusion range is around 300 mm before central necrosis, resulting from food scarcity, occurs (Zhengpenget al., 2021). Hence, the absence of a vasculature that fluids may penetrate is a significant constraint in enhancing the intricacy of biological systems and achieving functionality similar to that observed in living organisms (Zhengpeng et al., 2021). Microfluidic devices that replace blood vessels, enabling precise delivery of oxygen, nutrients, and growth stimulants while eliminating waste products, will be essential for the efficient and long-lasting cultivation of brain organoids (Atsushi et al., 2017). These substances will help maintain the stability and survival of the organoids, enabling a more natural and advanced development toward a model that closely resembles the complexity and characteristics of living organisms. Microfluidics, with the ability to fold itself and adapt quickly, can already transport chemicals with precise control over their three-dimensional location and timing (Jamal et al., 2011); recent advancements in 3D printing using sacrificial materials have the potential to produce scaffolds that can be penetrated by fluids in order to support the growth of organoids (Grigoryan et al., 2019). Additionally, these microfluidic technologies will provide chemical communication with organoids. The significance of spatiotemporal chemical patterns in encoding information is widely recognized in neuroscience and behavioral research (Mei et al., 2022). Recent advancements in micropatterning and microfluidics have significantly improved the control over the chemical environment around neurons using 2D chips. Tree-like microfluidic gradient generators are commonly employed to create precise chemical patterns, facilitating various biological and chemical research applications (Dertinger et al., 2001). 3D spatiotemporal microfluidic interfaces enable precise delivery and replication of chemical environments that contain neurotransmitters, neuropeptides, and other neurochemicals (Berger et al., 2018; Chen et al., 2017; Huang et al., 2022). ss

Machine Learning Models and Complex Biological Systems

Recently, the development of advanced organoid culture techniques has allowed for the large-scale manufacture of several types of organoids (Shi et al., 2024). Microfluidic technology has become an essential component in this field. This method achieves large-scale improvement in quality control by enclosing suspension-cultured self-assembling stem cell organoids in hydrogels, effectively reducing heterogeneity (Velasco et al., 2020). However, this significant progress in scalability also presents challenges regarding sample characterization and comprehensive data analysis. These are now essential components in managing the enormous volume of information generated by high-throughput capabilities platforms. Traditional analytical instruments have demonstrated the capability to generate dependable results (Broguiere et al., 2018); however, they are today confronted with the challenging task of managing the vast data derived from these samples. The field must confront the difficulty of reconciling the advantages of high-throughput screening with the requirement for precise and comprehensive data processing as it approaches the era of organoid intelligence. AI systems have shown remarkable ability in analyzing large amounts of genomic, proteomic, and clinical data, detecting patterns and correlations that may be difficult for humans to notice. ML algorithms are computational models that utilize mathematical principles to analyze input data and identify hidden patterns in fresh data. This is achieved by iteratively optimizing various cost functions to reach the global minimum (Kegeles et al., 2020). These methods have been extensively employed in biomedical research, namely, to lower the dimensionality of high-dimensional information to generate more straightforward embedded data that may be readily seen. Abdelmoula et al. employed mass spectrometry imaging in conjunction with a t-distributed stochastic neighbor embedding (t-SNE) method to distinguish between patients who have cancer of the breast and others suffering from gastric tumors. They classified them into separate subgroups and revealed the existence of variability within tumors (Abdelmoula et al., 2016). In a separate investigation conducted in 2017, Lynch et al. utilized diverse supervised machine models to analyze a dataset comprising information on cancer patients' surveillance, epidemiology, and outcomes. The researchers achieved a prediction accuracy rate exceeding 90% in determining the survival rates of lung cancer patients (Lynch et al., 2017). The integration of stem cell biology, bioengineering, and AI has initiated a novel phase of organoid research (fig. 4). OI revolutionizes our comprehension of developmental biology, disease causes, and treatment approaches by utilizing stem cell technology, precision engineering, and data-driven insights (Schuman et al., 2022).

Figure 4 illustrates the amalgamation of organoid conceptual technologies and AI to examine organoid attributes, simulate human development, and facilitate drug screening processes.

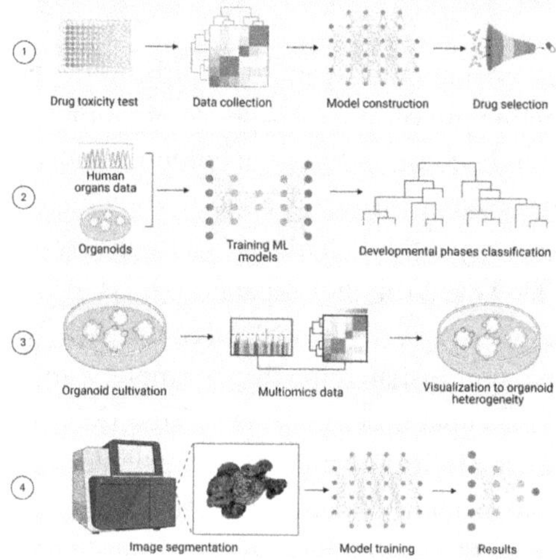

Enhancing Learning and Memory in Organoids: Utilizing Biological Feedback, Big Data, and AI-Machine Learning for Training

It is essential to comprehend the learning capabilities of brain organoids to ascertain their potential for computational use by leveraging the benefits of biological learning. At this level, learning is recognized as a heightened occurrence of displaying and memorizing a response pattern to a stimulating pattern. The researchers intend to utilize the various versions of technology mentioned before to connect organoids and computers to launch simulations involving supervised learning (Smirnova et al., 2023). In order to accomplish this objective, it is necessary to subject the brain organoids to specific patterns of electrical and chemical stimulation, which occur in space and time. The generated data will then accurately comprehend the relationships between the inputs and the outputs. A feedback loop is essential for training a learning system. The modifications in the configuration and functional capacities (including connections between neurons and electrophysiological connections) of brain organoids that arise from these learning cycles can then be analyzed. Synaptic plasticity, essential for memory and learning, is influenced by two key factors. The responses of brain organoids to electrical or chemical stimuli will indicate whether learning can occur. A reliable method for encoding and interpreting signals in human brain tissue is necessary. Recent advancements in neuroscience, mainly using intracortical microstimulation (ICMS) and MEAs, have shown promising results in decoding motor and sensory cortices in human participants (Osborn et al., 2021).

However, additional progress in magnitude and clarity will be required. Organoid-MEAs will produce extensive recording datasets that necessitate analysis using statistics and machine learning approaches. Considering the density and volume of the recordings, a new and specialized infrastructure for handling big data and a supercomputing capacity specifically designed to meet the complex requirements of this type of modern biological data will be required (Kagan et al., 2022; Lee et al., 2020). Integrating organoids with diverse types of advanced input and output prompts and recording crossing points (Bozhkov Petia, 2019) will enable us to comprehend the capacity for instantaneous management (Smirnova et al., 2023).

How Scientists Have Been Trying to Model the Human Brain and Organ Connections in A Dish

Retinal organoids, a complex sensory organ that has been modeled, could be linked to a brain organoid (Wahlin et al., 2017). In due course, networks of organoids will be interconnected to facilitate the implementation of more intricate OI (Kallman et al., 2020). The organoid will be connected to machines that can sense electrical and fluidic signals. It will get feedback on a cellular level, allowing the brain organoid to manage the machines by receiving information about the outcomes of its actions. By linking retinal and brain organoids, we can assess the transmission of signals between these distinct neuronal organoids and examine how the brain organoid interprets this external biological signaling. This process establishes the initial foundation for communication across organoids. Retinal organoids have been created that consist of layered retinas with the formation of structures resembling outer segments and synaptic ribbons (Kallman et al., 2020). Synapse development between two organoids is complicated and recently established in assembloids (Miura et al., 2022). Retinal and brain organoids can be linked via electrodes or direct connection. Due to the preliminary and very inefficient nature of methods for generating mature, naturally occurring light-sensing retinas, designed photosensitive ion channels are utilized. These channels are generated using a photoreceptor-specific promoter (Rost et al., 2017) and can serve as a proxy for light-sensitive photoreceptors. A recent publication has described optimizing growth conditions for retinal organoids to enhance signaling efficacy (Cowan et al., 2020).

Advancements in Medical Research and Innovation Driven by OI

An in-depth protein expression analysis is crucial to fully comprehend the signaling cascades and synaptic plasticity required for learning the organoid's biochemistry, particularly concerning neurotransmitters (Smirnova et al., 2023). Furthermore, OI research will enable the investigation of variations in neurodevelopment and neurodegeneration among stem cell donors, in addition to its groundbreaking utilization of human brain organoids for computation and learning (Soliman & Abdellatif, 2023; Wadan et al., 2024; Tolba et al., 2023; Wadan & Liaquat, 2024). Research on AD and other dementias may be considered a crucial goal. The success rates of clinical trials for new AD treatments have been extremely low. This is partly due to the premature application of successful results from animal models, which only mimic some characteristics of the disease in humans (Drummond Thomas, 2016). To thoroughly investigate memory, learning, and cognition, as well as the effects of neurodegeneration on these processes, it is necessary to have neuron-to-glia ratios comparable to physiological conditions and high degrees of biological complexity and connectivity between different brain regions. In addition to assembloids (Miura & Pa ca., 2019), characteristics of the lower and upper regions in the absence of exogenous substances that influence physical development (Renner et al., 2017) or the process of recapitulating growth factors (Kim et al., 2019) This has been replicated, underscoring an inherent capacity for self-organization in both instances. Recently, Cederquist et al. demonstrated that integrating an early organoid with a cluster of cells that secrete sonic hedgehog (SHH) led to the development of positional axes along the dorsal-ventral and anterior-posterior directions (Cederquist et al., 2019). Nevertheless, organoids lack consistent anatomical structure and defined spatial arrangement and typically fail to exhibit distinctive developmental asymmetry (Cederquist et al., 2019; Miura & Pa ca., 2019).

The Ethical Impacts in OI Applications

1. Moral Status and Consciousness

A critical ethical concern is the moral status of organoids, particularly brain organoids, and whether they could develop a consciousness that would warrant moral consideration (Hartung et al., 2023a). As brain organoids become more complex and resemble the developing human brain, there are questions about what level of structural complexity or cognitive abilities like consciousness would grant them moral status (**fig. 5**). Factors like the ability to experience emotions or sentience are debated as determinants of moral status (de Jongh et al., 2022). There are suggestions to develop sensitive, objective tests to screen brain organoids for advanced cognitive capabilities and consciousness. However, scientists emphasize that brain organoids will unlikely ever develop human-like consciousness. Nonetheless, if consciousness could potentially develop, there are questions about how to prevent unethical experimentation and what rights these entities would have (Bai et al., 2024; de Jongh et al., 2022; Morales Pantoja et al., 2023; Shi et al., 2024).

Figure 5 illustrates cognitive capacities, serving as a framework to guide the regulation of research on animal models and human brain organoids, ensuring ethical and scientific rigor in neurobiological studies.

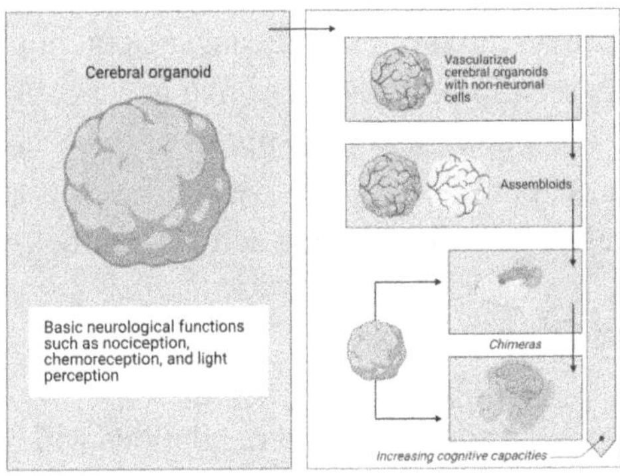

2. Informed Consent and Donor Rights

Another major ethical issue is informed consent for organoid research and the rights of tissue donors. As organoids are seen as beings with moral status, this raises questions about donor consent and control over their 'clones.' There are challenges in developing informed consent models that account for the unpredictability of research directions and the possibility of organoids developing consciousness. Concerns include whether donors can withdraw consent if organoids develop consciousness, what rights donors have if organoids are commercialized, and whether donors should be informed if research reveals something relevant to their health. To address these concerns, there are debates about the adequacy of current consent procedures and the need for unique research ethics review boards. These boards, specifically designed to handle the complex ethical issues of organoid research, can provide reassurance and confidence in the oversight process (Hartung et al., 2023a).

3. Commercialization and Intellectual Property

The commercialization of organoid research and intellectual property rights are other ethical flashpoints. There are questions about whether ownership rights can be enforced if an organoid-machine hybrid is sentient and at least partially human.

Privacy concerns also arise around what an organoid might reveal about a cell donor (Hartung et al., 2023b).

Societal Impacts of Organoid Intelligence Research

1. Public Perception and Engagement

Organoid intelligence research has significant societal implications that require public involvement. As seen in past instances like genetically modified crops, the concerns about public acceptance and perceptions of this technology highlight the importance of transparent communication. We can foster trust and understanding by ensuring that the public is well-informed and part of the conversation.

2. Governance and Regulation

Appropriate governance frameworks and regulations will be crucial to ethically steering organoid intelligence research. As research progresses rapidly, policymakers must work with scientists, ethicists, and the public to ensure ethical and regulatory questions are fully explored. This includes developing tailored ethical guidance, oversight, and proportionate regulation for organoid research.

3. Societal Benefits and Risks

On the positive side, organoid intelligence research holds great promise for advancing our understanding of the brain and brain disorders, potentially leading to new treatments. However, technology also carries risks of misuse, such as in AI systems. Responsible innovation will require balancing the societal benefits and risks through an interdisciplinary framework incorporating scientific, ethical, legal, and public perspectives.

Practical Application of OI in Medicine and AI

1. Applications of OI in AI

Open Innovation (OI) is a paradigm that encourages organizations to use external ideas, technologies, and collaborations to advance their innovation processes. This approach is particularly relevant in AI and medicine, where rapid advancements and complex challenges require diverse expertise and resources. Organizations often engage in collaborative research initiatives to develop AI algorithms that can improve decision-making processes. For example, IBM's Watson has collaborated with various healthcare institutions to develop AI systems capable of diagnosing diseases. Through partnerships with institutions like the Cleveland Clinic, IBM leverages external data and expertise to refine its AI models, enhancing the accuracy of medical diagnoses (Silver et al., 2015).

Additionally, crowdsourcing is a powerful tool in AI development, allowing organizations to tap into various ideas and solutions. Kaggle, a platform for data science competitions, exemplifies this. Companies can challenge data scientists globally, seeking innovative solutions to specific problems. For instance, in the Netflix Prize competition, data scientists worldwide collaborated to develop algorithms that improved the accuracy of movie recommendations, demonstrating how open innovation can drive AI advancements. Large corporations often collaborate with startups to leverage innovative AI solutions. For example, Google's acquisition of DeepMind enabled Google to enhance its AI capabilities significantly. DeepMind's work on reinforcement learning and neural networks has influenced various Google products, including Google Assistant and Google Photos, showcasing how OI can accelerate AI innovation (Silver et al., 2016). Interestingly, open innovation fosters the creation of data-sharing platforms that facilitate collaboration between organizations and researchers. For instance, OpenAI developed the OpenAI Gym, a toolkit for developing and comparing reinforcement learning algorithms. By providing a shared environment for testing AI models, OpenAI promotes collaboration and innovation within the AI community (Brockman et al., 2016).

Applications of OI in Medicine

OI has transformed drug development processes, where pharmaceutical companies partner with academic institutions and biotech startups. A prominent example is the collaboration between AstraZeneca and the University of Cambridge, focusing on precision medicine. They aim to accelerate the discovery of new drugs and personalized treatment strategies by pooling resources and expertise. Also, crowdsourcing for clinical trials has gained traction, allowing researchers to recruit participants more efficiently. Platforms like *PatientsLikeMe* enable patients to share their health data and experiences, facilitating more inclusive clinical trials. This approach accelerates the development of treatments by ensuring that diverse patient populations are represented. Collaborations between technology companies and healthcare providers have led to the development of innovative health technologies. For instance, the partnership between Fitbit and healthcare organizations has enabled the integration of wearable devices into patient care. By sharing data from wearables, healthcare providers can monitor patients' health more effectively and intervene when necessary. Open-source software initiatives in medicine facilitate collaboration among researchers and healthcare professionals. The OpenMRS project is a notable example of providing an open-source electronic medical records (EMRs) platform. By allowing developers to contribute to and customize the software, OpenMRS enhances healthcare delivery in resource-limited settings (OpenMRS, 2020). Moreover, OI enables better utilization of patient data through collaborative platforms. The All of Us Research Program in the United States exemplifies this initiative, aiming to gather health data from diverse populations. By encouraging participants to share their health information, researchers can gain insights into disease prevention and treatment, advancing precision medicine (Alshial et al., 2023; Ellakwa et al., 2024; Mohamed et al., 2023; Sanaeifar et al., 2024; Sayed et al., 2024; Soliman & Abdellatif, 2023).

CONCLUSION AND FUTURE PERSPECTIVES

Governments can harness biocomputing technologies, like the pioneering Neuroplatform by FinalSpark (Jordan et al., 2024), for environmental policymaking and energy conservation efforts. By integrating these biological computing systems, which consume significantly less power than traditional computers, into public sector infrastructures, governments can dramatically reduce the carbon footprint associated with digital operations. Furthermore, the innovative use of human brain organoids in computing opens avenues for healthcare and research by facilitating advanced disease modeling and drug discovery processes. Collaborating with biocomputing firms could spur economic growth by creating new industries and job opportunities while positioning a country as a leader in sustainable and cutting-edge technology adoption.

REFERENCES

Alshial, E. E., Abdulghaney, M. I., Wadan, A.-H. S., Abdellatif, M. A., Ramadan, N. E., Suleiman, A. M., Waheed, N., Abdellatif, M., & Mohammed, H. S. (2023). Mitochondrial dysfunction and neurological disorders: A narrative review and treatment overview. *Life Sciences*, p. *334*, 122257. DOI: 10.1016/j.lfs.2023.122257

Chapter 9
Hearing Aids and Speech Intelligibility Enhancement

Youssef Girgis Dimitry Khristo
https://orcid.org/0009-0001-5166-7150
Nile University, Egypt

Sahar Ali Fawzi
Nile University, Egypt

ABSTRACT

Hearing is a crucial human sense, and its loss significantly impacts a patient's daily life. Hearing loss, the inability of the ear to convert sound waves into interpretable signals by the brain is classified as mild, moderate, or severe. According to WHO, 1.5 billion people suffer from hearing loss, 25% are over 60, and 80% live in low- and middle-income countries. Hearing loss in childhood leads to underdeveloped speech and learning difficulties. Hearing loss generally leads to social isolation, reduced community participation, and lower educational and employment opportunities. People with mild to severe loss use hearing aids, while the profoundly deaf use sign language. Accessible communication aids promote social and economic sustainability. Hearing aid technologies have evolved, focusing on rechargeable batteries and recycling programs. Innovations include smartphone-based hearing aid apps and machine learning for sound enhancement. Policies supporting obtaining hearing aids and services are crucial for sustainability and well-being.

DOI: 10.4018/979-8-3373-1132-6.ch009

INTRODUCTION

Hearing is the most essential of the five human senses. It enables people to converse with their friends and loved ones, to enjoy environmental sounds during a short commute, and to feel the goosebumps hearing their favorite musical piece. This essential capability directly shapes who we are.

The hearing organ, the ear, collects sound waves (vibrations) and translates them into signals that the auditory nerves can carry to the brain (NIH, 2015; Hopkins Medicine). Hearing loss is defined as the inability of the ear to convert sound waves into signals that the brain can interpret (NIH, 2015). Hearing loss is classified as mild, moderate, severe, and profound, and can affect one or both ears. This leads to difficulty hearing sounds and conversational speech.

According to WHO reports, 1.5 billion persons suffer from a certain degree of hearing loss (WHO, 2021). Nearly 80% of people with disabling hearing loss live in low- and middle-income countries. The prevalence of hearing loss increases with age, 25% of people older than 60 years are affected by disabling hearing loss (WHO, 2021).

People suffering from mild to severe hearing loss communicate through spoken language using hearing aids and assistive devices. Deaf people, those suffering from profound hearing loss, use sign language for communication.

Hearing loss can have significant implications for personal and population sustainability. Persons with hearing loss tend to be socially isolated with reduced participation in community activities. They may face challenges that potentially lead to lower educational attainment and employment opportunities. This affects economic stability and personal well-being (John M. Lasak et al., 2014; NHS). Enabling these individuals to access communication aids and inclusive environments supports social cohesion and equity which are critical components of social sustainability. The use of hearing aids reduces healthcare costs needed to deal with depression, anxiety, and cognitive decline, thus supporting economic sustainability

Governments and organizations should adopt policies that provide funding for hearing aids and related services to ensure that all individuals have access to the care they need, supporting economic and social sustainability. Addressing the hearing loss problem through comprehensive healthcare, inclusive policies, and sustainable technological innovations supports social, economic, and environmental sustainability, and enhances overall well-being.

Throughout the past decades, the help introduced to people suffering from hearing loss is the use of hearing aids. These are electronic devices with the main purpose of amplifying sounds from the environment to the person in need, compensating for the existing hearing deficiency (WHO, 2021; James S. Reilly, 1989; Kathryn H., 2015). The core components of this device are a microphone that collects sound waves, an amplification circuit that amplifies the required frequencies, and a receiver that is plugged into the patient's ear. The fabrication of these components should ensure their safety for the prolonged period of use and skin attachment. Innovation in hearing aid fabrication focuses on developing long-lasting rechargeable batteries and designing devices with longer lifespans. Implementing and promoting recycling programs for hearing aids and their components reduce environmental impact and promote circular economy.

As people are affected by the social stigma of wearing hearing aid devices in public, smartphone-based hearing aid applications were created. This enabled users to compensate for their hearing loss (usually low to mild cases) by adjusting various parameters manually and using only their Bluetooth earphones. This helps people with budgetary constraints drastically (M. Vashkevich et al., 2017; Y. -C. Lin et al., 2018).

The past decades have witnessed a huge interest in incorporating artificial intelligence to address this problem. Machine learning (ML) models were used to perform hearing loss parameter personalization (Nafisa Zarrin Tasnim et al, 2023). Hearing aid applications are based on trained ML models to reach the goal of providing audio enhancement as well as audio clarity for people suffering from hearing loss.

Technological advances were heavily applied in this domain focusing on three main aspects of sound enhancement: environmental sound control, speech intelligibility enhancement, and music enhancement and equalization. Scientists have made significant progress in this field by integrating digital signal processing algorithms and deep learning models, as well as utilizing deep reinforcement learning methods (Siddique Latif et al., 2021).

This chapter provides an in-depth analysis of recent innovations, focusing on the significant breakthroughs in machine learning (ML) and reinforcement learning (RL) powered hearing aids, as well as the methodologies, models, and data sets that underpin their design. It will also present advanced techniques for improving speech intelligibility, highlighting their real-world applications in fostering social and professional interaction.

This chapter will be structured into six main sections. The first, titled "The Human Hearing," provides a detailed exploration of the complex mechanisms of the human ear, from the reception of sound waves to their transmission to the auditory sensors in the brain, highlighting the role of the ear's structural components. The second section, "Hearing Loss," presents an overview of hearing loss, discussing its primary types, symptoms, causes, and how it affects different age groups. "Hearing Aids," the third section, delves into the evolution and functionality of the hearing aid devices, their components and various types, and how they are customized for individual patients. Section four, "Smartphone-based Hearing Aid Applications," focuses on the shift to smartphone apps that amplify and equalize sound. The fifth section, "Machine Learning Approaches," explores sound classification and enhancement techniques, particularly for environmental sounds, music, and speech, within the realm of machine learning research. The sixth section, "Policies," briefly addresses the legal frameworks and procedural steps necessary to enhance access to treatment and support for patients. Finally, the concluding section, "Conclusion and Future Outlook," reflects on the current state of research and offers a forward-looking perspective on potential future advancements in hearing technology.

THE HUMAN HEARING

Sound around us are the vibration of matter through space. When the strings of a violin are played, their vibrations travel through the surrounding air particles, creating distinct notes. Human hearing captures these vibrations and translates them into signals that the brain can understand and perceive.

Human Ear Anatomy and Function

The main function of the human ear is to capture the vibration traveling through the air and translate it into electrical signals that are directly fed to the auditory nerves of the brain. A human ear consists of three main parts and their constituent organs, as shown in Figure 1. The outer ear, comprising the auricle or pinna, collects sound waves from the environment and channels them into the ear canal, where they cause the eardrum to vibrate. These vibrations are then amplified by three small bones in the middle ear—the malleus, incus, and stapes—and transmitted to the inner ear's cochlea. The cochlea, filled with fluid, responds to the vibrations by generating waves along the basilar membrane, where different regions detect specific sound frequencies. Hair cells near the cochlea's center are sensitive to low-pitched sounds, while those at the wider ends respond to high-pitched sounds. As these hair cells move, their stereocilia bend, opening channels that trigger a chemical reaction, producing electrical impulses that travel via the eighth cranial nerve to the brain. This complex process transforms mechanical vibrations into electrical signals, which the brain interprets as sound, allowing us to perceive and understand auditory information. (NIH, 2015; Hopkins Medicine).

Figure 1. Anatomy of the ear. Courtesy of NIH/NIDCD

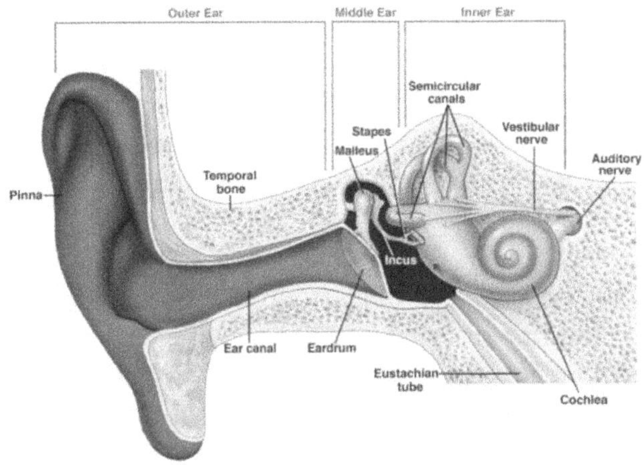

HEARING LOSS

Hearing loss could be described as the inability/impairment of the ear to convert sound into signals that could be interpreted by the central nervous system and the brain (NIH, 2015). Since infants cannot display obvious symptoms, many developed countries have established screening programs to assess hearing within the first month of life, enabling early intervention and significantly improving developmental outcomes, especially in cases of progressive hearing loss (Anna M. H. Korver et al., 2017).

Hearing loss could be categorized into three main forms: Peripheral hearing loss, also known as conductive hearing loss, is caused by an obstruction or disease affecting the outer or middle ear, or both. (Lisa L. Cunningham, Debara L. Tucci, 2017) and Sensorineural hearing loss is the leading cause of hearing loss in adults. It is a collective term for hearing damage to the cochlea and auditory nerve (Lisa L. Cunningham, Debara L. Tucci, 2017; Benjamin E Schreiber et al., 2010; Stephen Epstein, James S. Reilly, 1989).

The most common causes for such loss are hereditary, aging, exposure to high levels of noise, exposure to low levels of oxygen and some drugs (Lisa L. Cunningham, Debara L. Tucci, 2017).

Technically, an individual is considered to have sensorineural hearing loss if they experience a hearing loss of at least 30 decibels (dB), which is the unit used to measure sound or noise, across three consecutive frequencies in a standard pure-tone audiogram (Debara L. Tucci, 2017). A notable disorder known as tinnitus is characterized as a ringing sound in the ear. It may be caused by certain medicines (John M. Lasak et al., 2014).

The most prevalent symptoms of hearing loss in adults include:

- Difficulty understanding conversations with others
- The need to increase the TV volume beyond normal levels
- Challenges in hearing speech over the phone, often described as "muffled"
- Increased effort required to concentrate while listening
- Withdrawal from social activities and a sense of isolation
- Ringing or buzzing in one or both ears (tinnitus)
- Frequently asking others to repeat themselves or speak more slowly

Impact of Hearing Loss

Hearing loss in children can lead to inappropriate responses to speech, delays in language acquisition, and poor academic performance. (Anna H. M. Krover et al., 2017; Blake C. Papsin, Karen A. Gordon, 2007).

The main consequence of hearing loss in adults is impaired communication, which can lead to social isolation, strained relationships, and difficulties in work-place. This impairment may contribute to lower educational attainment compared to individuals with normal hearing, potentially increasing the risk of unemployment. (Lisa L. Cunningham, Debara L. Tucci, 2017). Other side effects include higher rates of hospitalization, dementia and depression, falls, accidents, and even death in extreme cases (John M. Lasak et al., 2014).

Treatments

Certain causes of conductive hearing loss could be treated with surgery (John M. Lasak et al., 2014), while others require the use of a "hearing aid" (John M. Lasak et al., 2014; Lisa L. Cunningham, Debara L. Tucci, 2017; James S. Reilly, 1989; Kathryn H., 2015). Sensorineural hearing loss requires using external aiding devices or a "cochlear implant" (John M. Lasak et al., 2014; Lisa L. Cunningham, Debara L. Tucci, 2017; Anna H. M. Krover et al., 2017).

For people who suffer from damage in the inner ear, a cochlear implant is needed. It is a complex device composed of two parts: an external microphone that collects sound waves from the surrounding environment, and the implant itself which bypasses any damaged parts of the ear and connects directly to the auditory nerve (Blake C. Papsin, Karen A. Gordon, 2007).

Figure 2. an illustration of a cochlear implant. Courtesy of Gerard O'Donoghue (2013)

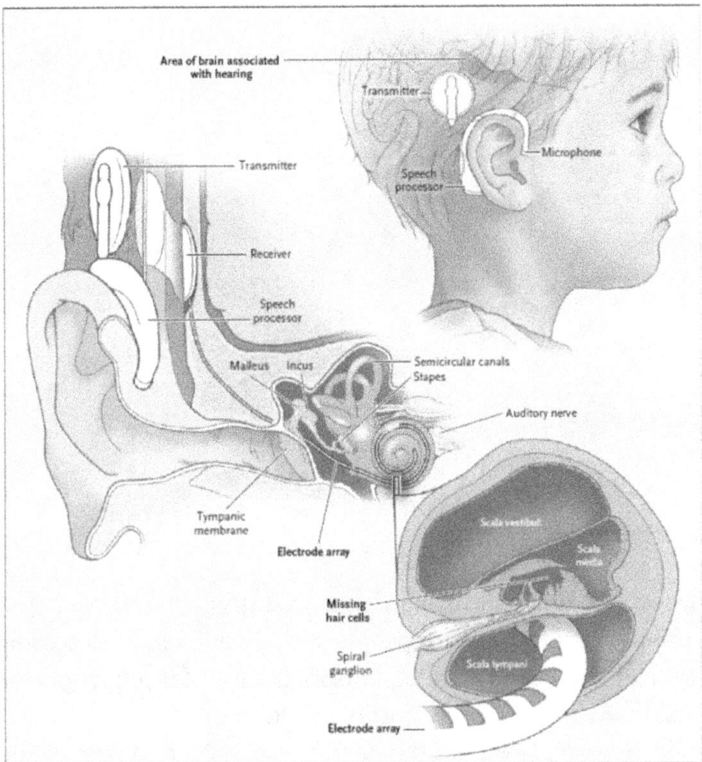

HEARING AIDS

A hearing aid is an electronic device designed to amplify environmental sounds for individuals with hearing loss, compensating for the deficiency. The components of a typical hearing aid are as follows: A microphone or a microphones array, an amplifier, a program button or handle, speakers, and batteries.

Figure 3. An illustration of the various components of a hearing aid, courtesy of Mayo Foundation for Clinical Education

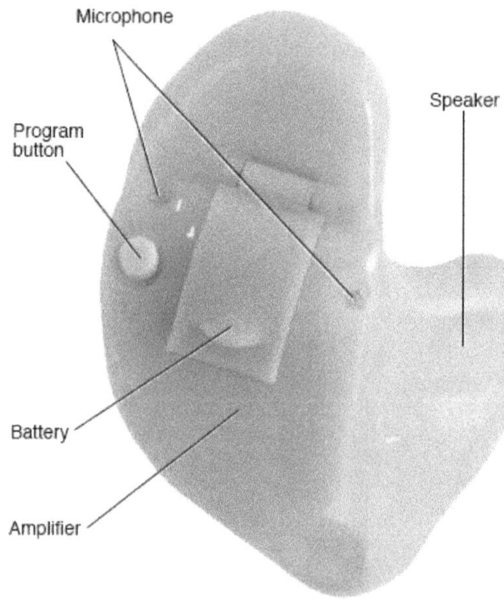

Types of hearing aids (Cleveland Clinic, 2022):

Behind-the-ear, appropriate for people with a wide range of hearing loss, fits behind the ear and requires manual input from the user for proper placement.

In-the-ear, fits within the outer portion of the ear

Receiver-in-the-ear, like BTE the device rests behind the ear, while having a thin wire with a small speaker at the end that fits inside the ear canal, suitable for people with hearing loss ranging from mild to severe.

In-the-canal, fits inside the canal like ITE but much deeper, making them less visible.

CROS/BiCROS, suitable for people with severe hearing loss in one ear and a minimal loss in the other, where the patient wears the hearing aid on the better ear with an additional microphone on the other, allowing the poorer side to receive sound while delivering all the sound to the better ear.

Figure 4. An illustration of the various types of hearing aids, courtesy of NIDCD, NIH

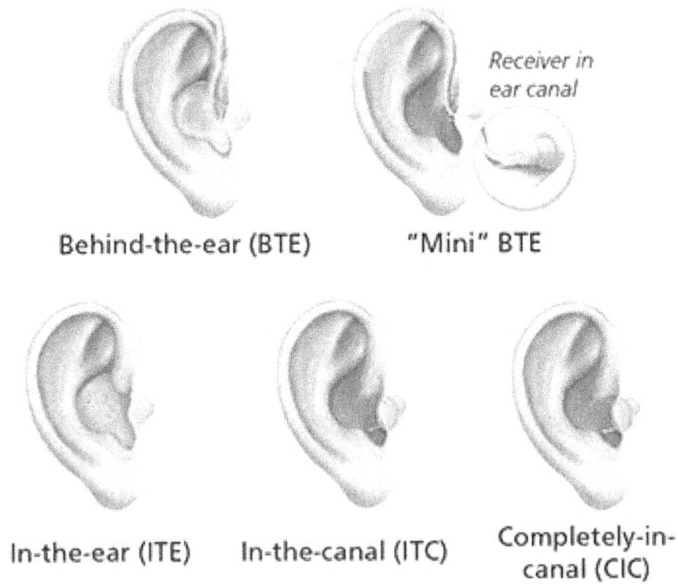

Personalization

Each patient receives his/her personalized hearing aid, tailored through tests conducted at specialized laboratories that assess their hearing threshold across various human hearing frequencies.

A complementary device for both hearing aids and cochlear implants is the assistive listening device (ALD), which is used within areas where there's a high level of background noise. There are three main types of ALD delivery systems: Hearing Loop systems which use electromagnetic energy to transmit the sound, FM systems which use radio signals to transmit amplified sounds, and infrared systems which use infrared light to transmit sound, and the receiver could be a hearing aid, a cochlear implant, and in some cases just simple headphones.

Drawback

The main drawback is the price, in the United States for example, hearing aids could cost between 500 to 3000 USD on average, other places do also have low-cost devices, in India for example, a user could get one for as little as 50 USD (WHO, 2021). Hearing Aid U.K published in a 2022 article that major hearing aids brand could have an average price range of 2500-4800 USD, while in the rest of Europe, various online forums discussed the topic, stating that major brands would cost 3000-5000 Euros. These costs do not include additional prices for fitting and maintenance.

Another significant drawback is the psychological stigma associated with wearing hearing aids in public. Many individuals are reluctant to use these devices due to concerns about how they may be perceived by others and their communities. Additionally, many hearing aids require users to manually adjust the volume to manage amplification in noisy environments, especially during sudden loud sounds like truck horns or fireworks, which can be inconvenient and uncomfortable.

SMARTPHONE-BASED HEARING AIDS APPLICATIONS

In contrast with hearing aids which rely on electronic circuits for sound amplification, there exist smartphone-based alternatives in the form of mobile apps that do the same function, using a Bluetooth earphone that the user connects to their mobile device and adjusting some settings. Such applications employ algorithms that help enhance various aspects of the audio collected by the earphone's microphones and are suitable for people with mild to moderate hearing loss.

The main advantage of such applications is the significant reduction in costs, where a patient could buy a mid-range smartphone for around 800 USD and a high-end Bluetooth earphone for an average of 200 USD, thus having a great experience for only 1000 USD compared to the minimum of 2500 USD for hearing aids from reputable brands. The two main downsides of such applications are the limited battery life and the processing time delay.

In this section, two famous applications; Petralex and SmartHear will be introduced.

The reason for choosing these two applications is that both were released to major smartphone OS manufacturers' app stores, with Petralex still under active development.

Petralex

Back in 2011, a group of researchers from the Belarusian State University of Informatics and Radio-electronics, created an app called Petralex. This app. contained in situ hearing tests. It enhanced the received sounds for patients by using simple full-band processing algorithms (M. Vashkevich et al., 2017).

While the application provides a degree of noise reduction which proved beneficial for people suffering from hearing loss, it still suffers from an effect called acoustic feedback (AFC), where the processed output signals leak from the speaker and into the microphone. In hearing aids, where there isn't much variability regarding the distance between the microphone and the speakers, techniques for dealing with this issue employing adaptive feedback cancellation are effective in these devices but not for the smartphone approach due to the changing distance between the microphone and speakers (M. Vashkevich et al., 2017).

SmartHear

SmartHear was a smartphone-based hearing assistive system that uses wireless technologies. The basic idea of the app relies on a smartphone that is placed near the speaker in a conversation and a Bluetooth earphone, then the app will send the enhanced sound that was picked up from the phone's microphones to the earphones of the user in need. The app aims not only to replace the FM systems but also to deliver superior performance in noisy and reverberant environments. especially with human speech, which conventional hearing aids don't always provide.

The reason for this approach is that according to various studies, more prominently a study labeled "*Hearing aid accessories for adults: The remote FM microphone*" by A. Boothroyd (Arthur Boothroyd, 2014), adults who combined conventional hearing aid devices with FM systems showed better perception of human speech in noisy environments, complemented by another study labeled "*Speech perception in noise: directional microphones versus Frequency Modulation (FM) systems*" M. Samantha Lewis et al. (Samantha Lewis et al., 2004), that hearing aid devices when combined with FM systems and directional microphones enhanced the perception of adults with hearing loss also in noisy environments.

The architecture of the proposed system is quite simple: a smartphone will be held or positioned near a speaker will be transmitting the enhanced sound data via Bluetooth to the listener that in turn has a Bluetooth handset.

To evaluate their proposal, the researchers used the speech intelligibility index (SII) that represents the amount of audible speech for an adult with hearing loss whether in aided or unaided conditions.

The frequency threshold averages used for the evaluation were 500, 1000, 2000, 4000 Hz respectively (a set of hearing thresholds, also known as the Four-frequency Average Hearing Level 4FAHL), while the setup comprised of one speaker at one end of the setup that plays the speech in question, a second speaker playing the speech-shaped noise (SSN) at 70 dB, and the Bluetooth headset that transmits SmartHear's output sound is evaluated by the Verifit Real-Ear Hearing Aid Analyzer model (VF-1) that is manufactured by Audioscan, and is used to calculate the sound spectrum level from both the speech and noise speakers, the output of SmartHear, as well as facilitating the sound level calibration process.

Figure 5. An overview of the SmartHear application (courtesy of SmartHear)

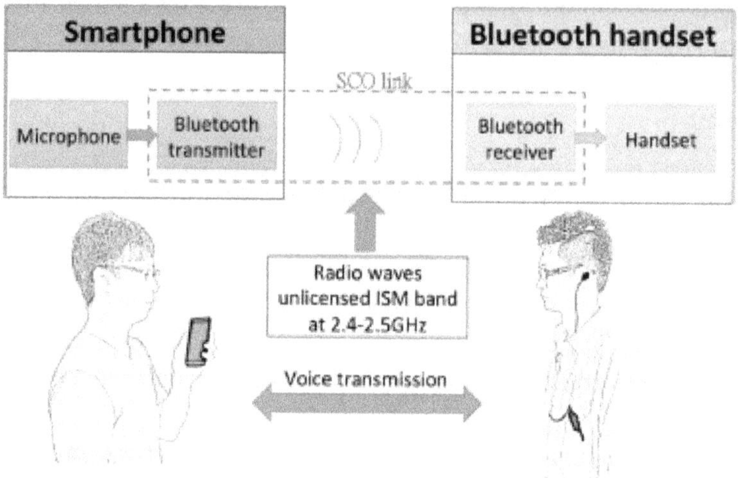

ARTIFICIAL INTELLIGENCE AND HEARING AIDS

Artificial intelligence is a field of research that aims at making computers and computer systems perform human tasks. An AI task could be speech recognition, where the AI algorithm tries to find the proper sequence of text that corresponds to a certain voice stream or recording, as well as text to speech synthesizers that receive a certain text as an input and output a corresponding speech audio.

Under the umbrella of artificial intelligence is Machine Learning (ML), which aims at mimicking the way humans learn and apply their knowledge to perform a multitude of tasks. According to Tom Mitchell (1997), a machine learns experience E by performing a set of tasks T and measuring the performance using a certain measure P, where its performance keeps improving the more experience it gains (Machine learning, 1997, p. 2).

Deep learning works by implementing the brain's neurons in the form of artificial neurons, and by setting them in the form of layers; basically, an input and an output layer along with intermediate -hidden- layers in between, where each layer connected to the output of the previous layer. These layers together form what we call an artificial neural network (ANN), and they could basically be used in two main types of tasks: classification tasks where the goal is to identify a certain category based on a set of inputs, for example classifying the various sounds of bird species. The other type is regression tasks, where the goal is to predict -with high confidence- a certain value, an example would be predicting.

Neural networks come in various shapes and sizes depending on the task at hand, and to be able to use these networks, one must train them with the adequate “data sets” for the end goal, for example, recognizing the bird species sounds above would require a dataset of audio clips for each of the species in question to be recognized. So, let’s imagine the data set in question as a table/matrix, where each row represents a real estate unit, and each column of this row represents “features” of this unit, example features might include the date the unit was built, its various dimensions, the number of rooms and showers in it, whether it has a basement and a cellar or not, etc. After choosing some or all these features as inputs, they are fed to the neural network, to then output a “model” to be used for the prediction of prices of other real estate units, this prediction is performed by providing the features of the real estate in question to the model, or rather say by “inferencing” the new data/ features’ values to receive a price prediction.

Basic architectures don’t usually baud well with certain tasks such as image and object classification, as images are more complex than mere numbers, for such applications a more complex form of networks is used called convolutional neural networks (CNN). The design of such networks was inspired by the human visual cortex, where in a certain scene where we have green plains and a house in their middle, the scene -input- is divided into rectangles where each of them is concerned with a specific area or a subset of features of the input image, where the result is transferred to the network’s convolutional layers (the intermediary layers’ name for CNNs) until they reach the output layer where the actual classification occurs.

In the field of sound classification, convolutional neural networks (CNNs) are often employed. For instance, when classifying bird sounds from 10 different species using a dataset of 5 second .WAV sound files, a spectrogram can be generated for each clip, which provides a visual representation of the audio signal's frequency spectrum over time. A commonly used example is the Mel-frequency cepstrum (an anagram of “spectrum”), which displays the Mel-frequency cepstral coefficients (MFCCs).

CNNs are efficient for tasks like audio classification and filtering compared to traditional networks. Audio clip spectrum images are fed with classes’ labels to train the CNN.

This section explores how artificial intelligence techniques have been applied to assist individuals with hearing loss. This includes its integration into hearing aids and cochlear implants, support through external devices, and alternative methods such as smartphone and Bluetooth earphone combinations.

Sounds could be categorized into three main categories:

Environmental sounds that exist around us, whether generated by the chirping of birds to the movement of trees, as well as noise such as a person pressing their car's horn for five straight seconds

Music is also an essential category that spans a vast variety from opera concerts to classical concertos, even solo singers and their small bands

Speech is probably the most important of the three categories, as it's a critical axiom of human development especially during childhood.

Speech Intelligibility Prediction

Speech intelligibility are assessed by two prediction methods; intrusive, that compares both degraded and clean speech signals or its transcription, and non-intrusive, where only the test signal is given.

Intrusive methods rely on features extracted from both the reference and test recordings where the temporal envelopes of speech analysis are performed using modulation filter banks such as the envelope power spectrum (ePSM) or extracted by machine learning models. Non-intrusive methods on the other hand assume speech assessment could be performed using a small number of spectro-temporal patterns in a signal.

Temporal envelopes, a key feature in speech perception, can be used to assess the distortion of frequency-specific envelopes through two methods: the Spectro-Temporal Objective Intelligibility (STOI) and the Extended STOI (ESTOI). Both metrics are commonly employed to evaluate noisy speech and its enhanced versions.

Denoising of Environmental Sounds

Environmental noise could be anything from the sound of wind, car horns, to various animal sounds, bringing us to an interesting sound denoising approach proposed by Youshan Zhang and Jialu Li for birds, by turning the denoising problem at hand into an image segmentation one.

Natural audio denoising using deep learning approaches is a challenging problem due to two main challenges: the limited size of datasets that could be used in the training and testing processes as it requires samples for both clean and noisy audio, and secondly the used noisy samples are artificially created, hence the researchers acquired their own samples by recording real sounds from nature and using them as their dataset.

The core of the proposed method works by transforming an audio signal to an image representation by using the shot-time Fourier transform (TSFT), passing the image to a segmentation model (DeepLabV3) that has an encoder-decoder architecture, where then the segmented audio image is reversed into a denoised audio signal using the invers short-time Fourier transform (ISTFT), and they call their model the Deep Audio Denoising (DVAD) model (Youshan Zhang, Jialu Li, 2022).

For the evaluation process of their proposed DVAD model, 6 state of the art segmentation models were trained: SegNet, U-Net, DeepLabV3, U2-Net, Segmenter, and MTU-Net, as well as 3 deep audio denoising methods: R-CED, Noise2Noise, and TS-U-Net. The DVAD had a better SDR score, as well as keeping the clean signal via a segmented mask.

Speech Denoising

Traditional speech denoising methods often rely on supervised networks, typically using a strategy called Noisy-Clean Training (NCT). In this approach, noisy audio signals serve as the input, while studio-quality clean recordings are the target, though acquiring these high-quality recordings can be both costly and time-consuming. Another approach, Noisy-Noisy Training (NNT), involves mixing clean speech with noise as the target, while using the same clean speech with additional noise for the training input. However, this method faces challenges in real-world scenarios due to the wide variety of noise types affecting the same speech signals. A further variation, Noisier-Noisy Training, sets noisy speech as the target and uses the same noisy speech mixed with extra noise as the input.

In their article "Self-supervised Speech Denoising Using Only Noisy Audio Signals," Jiasong Wu et al. propose a novel self-supervised speech denoising approach inspired by the Neighbor2Neighbor image denoising method (Jiasong Wu et al., 2023). This technique, called Only-Noisy Training (ONT), relies solely on noisy audio signals for training.

Concurrent Speech Dereverberation and Denoising

In settings where a multi-microphone setup exists such as conferences and conference rooms, these microphones do not only capture the speech but also all sorts of noise, especially reverberation which reflects an audio signal that causes it to last after the source seizes to exist. Vijay M. Mane et al. in their paper "*Concurrent Speech Dereverberation and Denoising*" (Vijay M. Mane, 2023), proposed a deep learning-based approach that eliminates reverberation (early and late) as well as noise. The method relies on 2 parts, the first being a delay sum beamformer that combines signal from N microphones (4 in the paper) and combines their signals into a single channel that is then passed to the second phase which is DNN.

The researchers also propose a Deep Dense Network DDN is made of symmetric decomposition layers with convolution, batch normalization, max pooling, and ReLU activation stages that gets repeated until the desired result is achieved, and reconstruction layers with convolution, batch normalization, and up sampling that gets iterated to reconstruct the original data. The DDN also compresses features along the decomposition and construction processes, and the last SoftMax layer is replaced with a thick layer.

For the evaluation of the proposed methodology, the Perceptual Evaluation of Speech Quality (PESQ) (Antony W. Rix et al., 2001) and Log Spectral Distortion measure (LSD) as shown below to assess the quality of speech.

$$D_{LS} = \left\{ \frac{1}{2\pi} \int_{-\pi}^{\pi} \left[\log P(\omega) - \log \hat{P}(\omega) \right]^{p} d\omega \right\}^{1/p}$$

Combining Waveform and Spectrogram-Based Denoisers

In a paper by Zhifeng Kong et al. in their paper "*CleanUNet2: A Hybrid Speech Denoising Model on Waveform and Spectrogram*", it has been mentioned that "*most speech denoising methods are spectrogram-based*", and it is true! These methods work by first extracting various spectral features from the audio, after which a mask is predicted, and a waveform is generated as output using the latter from the noisy speech. With, spectrogram-based methods suffer from noise leakage when presented with high levels of noise.

On the other hand, waveform-based methods predict the waveform of clean speech from the noisy samples directly, with WaveNet and U-Net being the networks of choice used as a backbone, all while being able to prevent the noise leakage under high noise levels. The downside is its effect on speech quality, resulting for example in the speech sounding less natural.

The aim of the researchers is to complement speech denoiser that separates clean speech from noisy audio using a single channel microphone, as they noted that both spectrogram-based and waveform-based models complements one another under high noise levels and led to the proposition of their hybrid model CleanUNet2.

The used waveform-based network is CleanUNet, while the spectrogram-based denoiser is CleanSpecNet, where the latter is trained and its output (the predicted spectrogram) is fed to the former, which reduces error propagation in the system (Zhifeng Kong et al., 2023).

The DNS 2020 dataset was used, containing 441 hours of clean speech samples of various speakers reading books and 70K noise clips, using both objective and subjective metrics for quality evaluation of denoised speech. The objective metrics used include Perceptual Evaluation of Speech Quality (PESQ) (both wide-band and narrow-band), Short-Time Objective Intelligibility (STOI) (Jesper Jensen, Cees H. Taal, 2011), and Mean Opinion Score (MOS) (Dawa Chyophel Lepcha et al., 2023).

Autoencoders and Speech Processing

Introduced in a research paper by Rumelhart in 1985, autoencoders are a type of neural networks and an essential building block used in deep models to extract high level and non-linear features in unsupervised learning. Just like other types of neural networks they have various hyper parameters that could be modified such as the learning rate, optimization algorithm, and batch size.

Autoencoders excel primarily in dimensionality reduction tasks, as they can detect sophisticated non-linear data relationships resulting in transforming a high-dimensional input into a lower-dimension space, they also have various categories such as regularized autoencoders, generative autoencoders, and recurrent autoencoders (Kamal Berahmand et al., 2024).

In the task of speech denoising, autoencoders are trained using noisy speech sample as input and the target being the clean speech, that autoencoders reconstructs by having the encoder extract informative features from the noisy speech and the decoder component reconstructs the clean speech based on the extracted features (Kamal Berahmand et al., 2024).

Dereverberation of Speech in Reflective Rooms

Speech dereverberation in a room is caused by the acoustic reflections in a room, where most deep learning-based approaches are not interpretable for these signals and are considered as a black box. Louis Bahrman et al., propose a new method to deal with this problem, relying on the FullSubNet architecture (Louis Bahrman et al., 2024).

Basically, a reverberant signal is introduced to an already learnt dereverberation model, where its output and the original reverberant signals are then fed to a convolutive module, estimating the room impulse response (RIR), that is then fed to a physical model to output the estimated energy relief decay (EDR).

The WSJ0 dataset was used, containing 61 hours of recordings that are split into 31,350 audio excerpts, along with the simulated RIR dataset which consists of 32K RIRs simulated using the pyroomacoustics library with 2000 rooms, where various evaluation and performance measures were employed, namely the STOI measure, scale-invariant signal-to-distortion ratio (SI-SDR) (Jonathan Le Roux et al., 2018), and wide-band PESQ.

Unsupervised Speech Dereverberation

In his paper, Zhong-Qiu Wang suggests an unsupervised approach to the speech dereverberation problem, using deep neural networks (DNNs), with an architecture labeled USDnet. Two approaches are usually used when trying to de-reverberate speech audio signal; using signal processing principles to perform inverse filtering, and supervised learning where DNNs are trained to predict the anechoic speech from reverberant sources.

The proposed approach with the USDnet architecture is to learn how to model speech patterns, where it could be directly trained on real life reverberant speech, completely avoiding using simulated training data, mitigating generalization issues.

For the evaluation of the network, the test set of WSJ0CAM-DEREVERB was used with direct-path signal at the first microphone as the reference signal for metric computation, and PESQ as well as eSTOI were used as evaluation metrics, and word error rates (WER) for ASR evaluation.

POLICIES

According to a World Health Organization (WHO) report in late February 2024, 1.5 billion people suffer from a certain degree of hearing loss, more than 400 million people globally have access to hearing aids and only 17% of this number benefits from these devices. This huge gap incurs a cost of 980 billion USD annually on the international community.

In a later report on the first of March, the WHO identified 2 key challenges regarding the hearing care systems, the first of which being the lack of health system capacity to provision of hearing care services throughout people's lives, which could be overcome -according to WHO- by utilizing non-specialists in providing hearing care to increase the capacity of the hearing care systems' staff.

The second challenge is related to the misperceptions and stigmatizing mindsets about ear diseases in general (especially hearing loss) on a societal level, most common of which would be the inevitability of hearing loss caused by the aging process, as well as the high price tag of hearing aids devices along with their efficiency.

Throughout this section, we will go through the hearing care policies of various countries around the globe, as well as going through the WHO's "*Hearing aid service delivery approaches for low- and middle-income settings*" report that serves as a guide for government and community level programmes that aim at delivering hearing aid services for people suffering from various ear diseases, most importantly hearing loss.

We could estimate a global average for hearing aids prices' range of 700 – 4800 USD (Healthy Hearing, 2023; NHS; CBS News, 2024; Fortune Business Insights, 2024).

In 2018, a report on hearing care highlighted good practices based on WHO recommendations, followed by a 2022 update that examined changes in hearing aid provision and legal mechanisms to support those with hearing loss (EFHOH, 2022). Participation included 21 EU countries, with notable exceptions like Spain, which lacks support for adult reimbursement and has inadequate child support, and Italy, which has stricter criteria than WHO guidelines. Young individuals up to 26 generally receive full reimbursement, though older adults often do not. The report outlines two operational models: one relies on private insurance with varying reimbursements, while the other is tax-funded. Challenges include coverage for only one hearing aid and long wait times, although some countries have taken steps to improve access, such as Denmark, which established new clinics.

The recommendations emphasize developing national strategies for hearing loss, raising awareness, creating inclusive services, and committing to established standards for hearing aid fitting.

In the United States, hearing aids can be obtained through public/private insurance or out-of-pocket payments, with approximately 29 million adults needing them (NCOA, 2023). Unlike the UK, social insurance is limited to older adults under specific conditions, leading to high out-of-pocket costs. Patients must consult audiologists or dispensers, although a law passed in 2017 allows for over-the-counter purchases. Hearing aids are classified into direct-to-consumer, prescription, and OTC devices, with significant variability in insurance coverage across states (Michael Yon et al., 2019).

In Canada, public programs support hearing care, with co-funding policies allowing shared costs between private insurance and health ministries (Province of British Columbia, 2024). Hearing aids are categorized into four types, and manufacturers are held accountable for warranties, although extended items and hearing tests are not covered (Canadian Ministry of Health, 2023).

In the UK, hearing aid dispensers provide regulated services, following a defined sequence for care and fitting (HCPC, 2023). Public insurance covers basic devices, while higher-end options may require out-of-pocket payments (Michael Yon et al., 2019).

In China, the "Hearing Specialists Act" governs hearing care provider practices, with public insurance covering hearing aids (Indian Ministry of Health and Welfare, 2020). However, newer models typically require out-of-pocket payment, with regulations supervised by the China Food and Drug Administration (Michael Yon et al., 2019).

India's National Programme for Prevention and Control of Deafness aims to prevent hearing loss and improve care through training and awareness initiatives (NPPC, 2012), involving multiple stakeholders at various levels.

Japan faces an aging population that increases the demand for hearing aids, requiring a disability certificate for support (Hideki Sakamoto, Fumiaki Yasukawa, 2022). Regulations ensure that hearing aids conform to quality standards, with assistance provided under existing disability laws (Intertek, 2021).

The WHO's guidelines for low- and middle-income settings outline service delivery approaches for adults and children, including referral indicators and a structured follow-up process to ensure ongoing support (WHO, 2024).

Figure 6. The WHO's recommended approach for service delivery for adults. Courtesy of WHO (2024)

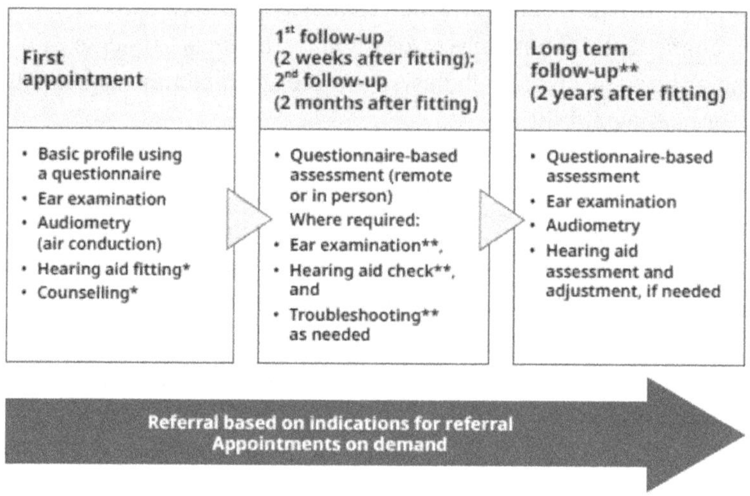

Where Policies Fall Short

While various policies by different governments and health organizations were set forth to provide optimal care for people with various degrees of hearing loss, there still exists disparities based on various factors, most prominently between age groups and for certain groups with specific illnesses.

A study conducted by Emily M. Ishak et al., to investigate the existing disparities of diagnosis and treatment for patients with borderline to mild hearing loss between age groups for the United States population, performed analysis over 16185 randomly selected participants by the National Health and Nutrition Examination Survey (NHANES), found that people in age group 25-49 had 4.25 times the odds of not having a hearing test in the past year in comparison to those who are less than 25 years of age, and 4.06 the odds in the past 4 years. People who are 50-74 years old and older than 75 had 5.62 and 5.44 times respectively, the odds to not have had a hearing assessment in the past year (Emily M. Ishak et al., 2024).

The researchers came up with a hypothesis that attributes such disparity to the knowledge gap that surrounds the potential impact of borderline/mild hearing loss treatment in children compared to adults, as the latter are usually seen as having cognitive stability while children are actively developing (Emily M. Ishak et al., 2024).

Another study by Dana P Urbanski et al. investigated the low level of hearing aid adoption in older adults with Alzheimer disease and related dementias (ADRD). The researchers relied on a multivariable logistic regression to model the likelihood of reporting such unmet hearing aid need for patients from the ADRD group and their residence type (e.g., family house, nursing home, etc.), while adjusting for sociodemographic factors and response type, relying on 25,492 respondents with a median age of 77 years old, patients that resided in a family home or a residential care had more unmet hearing aid need than their nursing home counterparts, with a prevalence in people with ADRD than those without (Dana P Urbanski et al., 2024).

From a sustainability standpoint, many countries around the world adopt hearing aids refurbishment programmes; where used hearing aid devices are collected and have the appropriate maintenance performed on them or even completely recycled to then be redistributed to people with hearing loss around the world. Certain organizations in France would take these devices, refurbish them, and redistribute them across various countries including France (e.g., Dominican Republic, Vietnam) with free fitting and mold creation, for people with low income and that are not covered by private or public insurance plans (Mathieu Hotton et al., 2023). The UK adopted the Reverse Exchange (RE) program where patients with hearing aids would deliver them to the NHS to be recycled or reused. Even though the reuse of single-use medical devices is advised against in the UK, this has not stopped the recycling and re-processing of such devices from being a longstanding practice, where even in certain instances such devices are loaned and not sold to patients, even though they're not explicitly told to return them where in some instances they are issued new ones while keeping the previous devices in their possession. With that being said, some barriers exist in the way of properly and efficiently executing Reverse Exchange strategies, including inventory management issues, lack of process regarding the control of such medical devices, as well as lack of incentive from the communities despite their awareness of the initiative (Rosanna Cole et al., 2018).

DISCUSSION AND CONCLUSION

Now we have a little discussion; where do we stand? Where are we headed? What can we do to improve? All discussed through the following questions.

Sustainability Considerations

An important point of focus in our opinion would be providing more resources on assisting with congenital hearing loss and hearing loss in children. Hearing loss is not a "tomorrow problem", the earlier you mitigate and even solve the issue, the less burden is set on the world governments. We are not talking here about the costs to screen, diagnose, and treat hearing loss in neonates and children, but rather that these costs are negligible in comparison to the long-term costs incurred by an individual going through adulthood with unresolved or unproperly handled hearing problems, whether the devastating effects caused to their social life (personal, family, and friends) and mental health, or the fact that hearing loss affects employment and productivity in general, affecting, in turn, society as a whole.

How do we Measure the Current Progress?

All the work that is done has the purpose of providing adequate help for people with varying degrees of hearing loss. A good measure of progress could be thought of as how efficient the hearing aid are delivered for people that needs it; we can plot a diagram of the last five years and see the progress in the number of people receiving hearing aids per year, as well as monitoring the next five years to conclude whether the progress and research during the last five years is in the proper direction, as in there should be a steady increase and not a decline in the number of people who has access to hearing aids. The same measure could be applied to the hearing care services in general (e.g., screenings, diagnosis, etc.).

What is the Future of Hearing Aids?

Hearing aids are evolving, where some manufacturers claim that they have built-in machine learning capabilities through specialized hardware chips, yet they have the downsides of stigma still going strong, as well as their batteries that needs to be changed every week to three weeks. Many hearing aid devices started having rechargeable batteries, which is great for a sustainable development of such devices, as well as its decreased negative impact on the environment and the costs of recycling or disposing of batteries. Rechargeable batteries would benefit from proper maintenance and being able to be changed without any hassles, as lack of such maintenance would mean that the whole device would be disposable, which adds more e-waste to the environment.

What is the Potential of AI Based Applications?

To describe it in a single word: huge! You see, machine learning models nowadays could run on many platforms, whether in the cloud or even smartphones where the rechargeable batteries could last for more than a year (a great bonus for sustainability), with easy battery replacements on most models. Some papers even discussed embedded models, which are good for potential IoT applications.

REFERENCES

Alan, H. (2018, March 29). Lockwood, Richard J. Salvi, Rober F. Burkard (2002). Tinnitus. *The New England Journal of Medicine*, 378(13), 1224–1231. Advance online publication. DOI: 10.1056/NEJMcp1506631

Rix, A. W., Beerends, J. G., Hollier, M. P., & Hekstra, A. P. (2001, May). Perceptual evaluation of speech quality (PESQ)-a new method for speech quality assessment of telephone networks and codecs. In *2001 IEEE international conference on acoustics, speech, and signal processing. Proceedings (Cat. No. 01CH37221)* (Vol. 2, pp. 749-752). IEEE.

Bahrman, L., Fontaine, M., Roux, J. L., & Richard, G. (2024). Speech dereverberation constrained on room impulse response characteristics. *arXiv preprint arXiv:2407.08657.*

Berahmand, K., Daneshfar, F., Salehi, E. S., Li, Y., & Xu, Y. (2024). Autoencoders and their applications in machine learning: A survey. *Artificial Intelligence Review*, 57(2), 28. Advance online publication. DOI: 10.1007/s10462-023-10662-6

Papsin, B. C., & Gordon, K. A. (2007). Cochlear implants for children with severe-to-profound hearing loss. *The New England Journal of Medicine*, 357(23), 2380–2387.

Arthur Boothroyd, (2014). Hearing aid accessories for adults: the remote FM microphone DOI: 10.1097/01.AUD.0000111260.46595.EC

CBS News. (2024). Are hearing aids expensive? Here's what to know https://www.cbsnews.com/essentials/how-much-do-hearing-aids-cost

Cleveland Clinic. (2022). Hearing aid styles https://my.clevelandclinic.org/health/articles/5122-hearing-aid-styles

Rosanna Cole, Claire Frances Lindsay & Fiona Barker, (2018). Reverse exchange of healthcare devices: the case of hearing aid equipment in the UK DOI: 10.1080/09537287.2018.1506892

Dana, P. (2024). Reported Unmet Hearing Aid Need in Older People with Dementia: The US National Core Indicators Survey. *Journal of the American Medical Directors Association*, 25(5), 853–859. Advance online publication. DOI: 10.1016/j.jamda.2024.03.107 PMID: 38643971

Drgas, S. (2024). Speech intelligibility prediction using generalized ESTOI with fine-tuned parameters. *Speech Communication*, 149, 103068. Advance online publication. DOI: 10.1016/j.specom.2024.103068

Emily, M. (2024). Age-based disparities in hearing loss diagnosis and treatment in the United States population. *American Journal of Otolaryngology*, 45(5), 104403. Advance online publication. DOI: 10.1016/j.amjoto.2024.104403

Epstein, S., & Reilly, J. S. (1989). Sensorineural hearing loss. *Pediatric Clinics of North America*, 36(6), 1501–1520.

European Federation of Hard Hearing People (EFHOH). EU (2022). State of Provision of Hearing Aids in Europe, 2022 Report https://efhoh.org/wp-content/uploads/2022/11/Hearing-Aids-Reimbursement-2022-edited-version-final-for-publication.pdf

Ewertsen, H. W. (1956). Hearing-Aid Evaluation. *A.M.A. Archives of Otolaryngology*, 64(6), 520–525.

Fortune Business Insights, E. U. (2024). Europe Hearing Aids Market https://www.fortunebusinessinsights.com/europe-hearing-aids-market-107908

Gates, G. A., & Miyamoto, R. T. (2003). Cochlear implants. *The New England Journal of Medicine*, 349(5), 421–423.

Health Care Professions Council. UK (2023). The standards of proficiency of hearing aid dispensers https://www.hcpc-uk.org/standards/standards-of-proficiency/hearing-aid-dispensers/

Hearing, H. Canada (2023). Hearing aid prices https://www.healthyhearing.com/help/hearing-aids/prices

His Majesty Revenue & Customs (HMRC). UK (2016). VATHLT2380 https://www.gov.uk/hmrc-internal-manuals/vat-health/vathlt2380

His Majesty Revenue & Customs (HMRC). UK (2016). VATHLT2400 https://www.gov.uk/hmrc-internal-manuals/vat-health/vathlt2380

Hopkins K. (2015). Deafness in cochlear and auditory nerve disorders. Handb Clin Neurol. 2015;129:479-94. . PMID: 25726286. https://doi.org/DOI: 10.1016/B978-0-444-62630-1.00027-5

Hopkins Medicine. How the Ear Works. https://www.hopkinsmedicine.org/health/conditions-and-diseases/how-the-ear-works

Hotton, M., Prud'Homme, V., Richard, L., Cormier, L., Simoneau, K., Lefebvre-Demers, M., ... & Boucher, N. (2023). Impacts and Identification of Hearing Aid Refurbishing Programs for People with Hearing Loss: A Scoping Review. *Audiology Research*, 13(3), 326–340.

Intertek (2021). Pharmaceutical And Medical Device Act Overview – Requirements for Regulated Medical Devices in Japan https://jwapi.intertek.com.cn/Upload/202203/4bc8eaba20d2e0fe879837fb500f3fc8.pdf

Zhifeng Kong, Wei Ping, Ambrish Dantrey, Bryan Catanzaro (2023). CleanUNet 2: A Hybrid Speech Denoising Model on Waveform and Spectrogram DOI: 10.21437/Interspeech.2023-1287

Korver, A. M., Smith, R. J., Van Camp, G., Schleiss, M. R., Bitner-Glindzicz, M. A., Lustig, L. R., & Boudewyns, A. N. (2017). Congenital hearing loss. *Nature Reviews. Disease Primers*, 3(1), 1–17.

Lasak JM, Allen P, McVay T, Lewis D, (2014). Hearing loss: diagnosis and management. Prim Care. 2014 Mar;41(1):19-31.. Epub 2013 Nov 18. PMID: 24439878. https://doi.org/DOI: 10.1016/j.pop.2013.10.003

Le Roux, J., Wisdom, S., Erdogan, H., & Hershey, J. R. (2018). SDR - Half-Baked or Well Done? https://doi.org//arXiv.1811.02508DOI: 10.48550

Dawa Chyophel Lepcha, Bhawna Goyal, Ayush Dogra, Vishal Goyal, (2023). Image super-resolution: A comprehensive review, recent trends, challenges and applications DOI: 10.1016/j.inffus.2022.10.007

Lin, Y. C., Lai, Y. H., Chang, H. W., Tsao, Y., Chang, Y. P., & Chang, R. Y. (2015). SmartHear: A smartphone-based remote microphone hearing assistive system using wireless technologies. *IEEE Systems Journal*, 12(1), 20–29.

Cunningham, L. L., & Tucci, D. L. (2017). Hearing loss in adults. *The New England Journal of Medicine*, 377(25), 2465–2473.

Ministry of Health. Canada (2023). Hearing Devices Policy and Administration Manual https://files.ontario.ca/moh-adp-policy-and-administration-manual-hearing-devices-2023-07-01.pdf

Ministry of Health & Family Welfare. India (2013). National Programme for Prevention and Control of Deafness. https://main.mohfw.gov.in/sites/default/files/5189275161902525 8383.pdf

Ministry of Health and Welfare. PRC (2020). Hearing Specialists Act. https://law.moj.gov.tw/ENG/LawClass/LawAll.aspx?pcode=L0020149

National Council on Aging (NCOA). USA (2023). FDA Hearing Aids Guidance: 2024 Compliance https://www.ncoa.org/adviser/hearing-aids/fda-hearing-aids-guidance

National Health Service. UK. Hearing aid dispensers https://www.healthcareers.nhs.uk/explore-roles/healthcare-science/roles-healthcare-science/physiological-sciences/hearing-aid-dispenser/hearing-aid-dispenser

National Health Service (NHS). Hearing aids and implants https://www.nhs.uk/conditions/hearing-aids-and-implants

National Institute on Deafness and Other Communication Disorders (NIDCD), National Institute of Health (NIH), USA. (2015). How do we hear? https://www.nidcd.nih.gov/health/how-do-we-hear

O'Donoghue, G. (2013). Cochlear Implants – Science, Serendipity, and Success. *The New England Journal of Medicine*, 369(13), 1190–1193. Advance online publication. DOI: 10.1056/NEJMp1310111 PMID: 24015974

Province of British Columbia. Canada (2024). Medical Equipment – Hearing Instruments https://www2.gov.bc.ca/gov/content/governments/policies-for-government/bcea-policy-and-procedure-manual/health-supplements-and-programs/medical-equipment-hearing-instruments

Priyanka Saha, Minhaj Uddin Monir, Shaheen M. Sarkar, Arzina Abd Aziz (2023). Crises in Oil, Gas and Petrochemical Industries, Elsevier. 18 – Noise pollution from oil, gas, and petrochemical industries, 2.1 DOI: 10.1016/B978-0-323-95154-8.00012-8

Sakamoto, H., & Yasukawa, F. (2022). A Consideration of the Hearing Aid Business in Japan with the Comparison of the European Union Countries and the U.S. *Iryo To Shakai = Academic Studies*, 32(3), 405–418. Advance online publication. DOI: 10.4091/iken.2022.005

Lewis, M. S., Crandell, C. C., Valente, M., & Horn, J. E. (2004). Speech perception in noise: Directional microphones versus frequency modulation (FM) systems. *Journal of the American Academy of Audiology*, 15(06), 426–439.

Schreiber, B. E., Agrup, C., Haskard, D. O., & Luxon, L. M. (2010). Sudden sensorineural hearing loss. *Lancet*, 375(9721), 1203–1211.

Nafisa Zarrin Tasnim, Aoxin Ni, Edward Lobarinas, Nasser Kehtarnavaz, (2023). A Review of Machine Learning Approaches for the Personalization of Amplification in Hearing Aids DOI: 10.3390/s24051546

Vashkevich, M., Azarov, E., Petrovsky, N., & Petrovsky, A. (2017). *"Petralex: A smartphone-based real-time digital hearing aid with combined noise reduction and acoustic feedback suppres-sion," 2017 Signal Processing: Algorithms, Architectures, Arrangements, and Applications*. SPA., DOI: 10.23919/SPA.2017.8166873

Vijay M. Mane, Seema S. Arote, Shakil A Shaikh (2024). An Integrated Deep Learning Model for Concurrent Speech Dereverberation and Denoising DOI: 10.12720/jait.15.2.281-287

Wang, Z.-Q. (2024). USDnet: Unsupervised Speech Dereverberation via Neural Forward Filtering https://doi.org//arXiv.2402.00820DOI: 10.48550

World Health Organization. (2024). Hearing aid service delivery approaches for low- and middle-income settings https://www.who.int/publications/i/item/9789240087927

World Health Organization (WHO). (2021). World Report on Hearing. https://www.who.int/publications/i/item/9789240020481

Wu, J., Li, Q., Yang, G., Li, L., Senhadji, L., & Shu, H. (2023). Self-supervised speech denoising using only noisy audio signals. *Speech Communication*, 149, 63–73. Advance online publication. DOI: 10.1016/j.specom.2023.03.009

Yon, M., Willing, A., McMahon, C., McPherson, B., Nieman, C. L., Reed, N. S., & Lin, F. R. (2019). Bulletin of the World Health Organization Vol. 101: Access to adults' hearing aids: policies and technologies used in eight countries. https://www.ncbi.nlm.nih.gov/pmc/articles/PMC6796668

Zhang, Y., & Li, J. (2022). BirdSoundsDenoising: Deep Visual Audio Denoising for Bird Sounds https://doi.org//arXiv.2210.10196DOI: 10.48550

Chapter 10
Review About the Application of Explainable Machine Learning for Early Detection and Diagnosis of Liver Diseases:
A Case Study on Fatty Liver Disease (FLD) and Hepatitis B (HBV)

Ayman A. Ali
https://orcid.org/0000-0002-7539-8354
Telecom Egypt, Egypt

Ahmed Ashraf
Benha University, Egypt

Kamel Rahouma
https://orcid.org/0000-0001-6640-6167
Minia University, Egypt

ABSTRACT

This chapter explores integrating XML techniques into early detection of Fatty Liver Disease (FLD) and Hepatitis B Virus (HBV) infections. We analyze contemporary

DOI: 10.4018/979-8-3373-1132-6.ch010

XML methods (SHAP, LIME, Anchors) to improve interpretability of complex machine learning models in medical diagnostics. Our review emphasizes embedding explainability for accurate diagnoses, clinician trust, and improved patient outcomes. XML demystifies "black box" models, aligning insights with medical knowledge. We further explore the impact of XML on healthcare adoption and reliable diagnostic tool development. This chapter lays the groundwork for future research on model generalizability, ethics, and real-time integration. Adopting explainable models for liver disease diagnosis is crucial, and continuous innovation in XML techniques will meet evolving healthcare demands.

I. INTRODUCTION:

Chronic liver diseases (CLDs) constitute a significant global health burden, impacting millions of individuals and imposing a substantial economic strain on healthcare systems worldwide (Singh) (Yip). Early and accurate diagnosis is crucial for effective disease management and improved patient outcomes (Kubíčková) (Hasan). Traditionally, liver biopsies, which involve the invasive extraction of tissue samples for microscopic examination, have served as the gold standard for diagnosis (Caputo). However, these procedures are associated with inherent limitations, including patient discomfort, bleeding risks, and potential for infection (Luo). Additionally, liver biopsies may not always capture the full extent of liver damage, particularly in the early stages of disease progression (Maurice).

The limitations of traditional invasive methods have fueled the search for non-invasive alternatives for diagnosing CLDs. Ideally, these alternatives should be:

- **Safe:** Minimizing or eliminating risks associated with invasive procedures.
- **Cost-effective:** Affordable and readily accessible to a wider range of patients.
- **Accurate:** Delivering reliable and dependable diagnostic results.
- **Accessible:** Readily available and easily performed within clinical settings.

Machine Learning in Liver Disease Diagnosis: Recent advancements in machine learning (ML) offer a promising avenue to address the limitations of traditional diagnostic methods for CLDs (Zamanian). ML algorithms possess the capability to analyze vast and complex datasets, encompassing patient demographics, laboratory tests, imaging data, and genetic information. By identifying intricate patterns and relationships within these datasets, ML models can be trained to make accurate predictions about the presence or absence of liver disease (Chu). This holds significant potential for earlier detection and intervention, ultimately leading to improved patient outcomes.

Explainable Machine Learning: Despite the potential benefits of ML in healthcare diagnostics, a significant challenge remains – the "black box" nature of many algorithms. Often, the reasoning behind an ML model's predictions is shrouded in obscurity (Lin). This lack of transparency can lead to a deficit in trust among clinicians who require clear and interpretable insights to make informed decisions regarding patient care (Caruana).

Explainable Machine Learning (XAI) emerges as a critical innovation that directly addresses this challenge. XAI techniques strive to make the internal workings of ML models more comprehensible to humans (Lundberg). By providing interpretable explanations alongside predictions, XAI fosters trust in the technology and empowers clinicians to gain valuable insights into the factors driving the model's conclusions (Alotaibi). This deeper understanding allows clinicians to (Orhanbulucu):

- **Validate Model Accuracy:** By understanding how the model arrives at its predictions, healthcare professionals can assess its reliability and ensure compliance with established clinical standards.
- **Enhance Clinical Decision-Making:** Interpretable models enable clinicians to identify the most relevant features influencing the model's predictions. This information can be integrated into the decision-making process, leading to more informed diagnoses and treatment plans.
- **Facilitate Personalized Medicine:** XAI models can shed light on the specific factors contributing to disease development in individual patients. This personalized information can guide the development of more targeted and effective treatment strategies.

Focus on Fatty Liver Disease (FLD) and Hepatitis B (HBV): This review paper delves specifically into the application of XAI for early detection and diagnosis of two prevalent liver diseases: Fatty Liver Disease (FLD) and Hepatitis B (HBV). Both FLD and HBV present unique challenges:

- **FLD:** Often asymptomatic in its early stages, FLD can progress to more severe liver conditions like cirrhosis and liver failure if left untreated (Safdari). XAI models can be instrumental in identifying individuals at risk for FLD development, facilitating earlier intervention and potentially preventing disease progression.
- **HBV:** A chronic viral infection, HBV can lead to liver damage and complications such as cirrhosis and liver cancer (Kaunang). Early diagnosis of HBV is crucial for initiating antiviral therapy and preventing disease progression. XAI models can analyze various data sources, including blood tests and viral markers, to improve the accuracy and efficiency of HBV diagnosis.

This paper will explore the various ML algorithms employed within the XAI framework for diagnosing FLD and HBV, highlighting their strengths and limitations. Additionally, we will examine the integration of XAI with different data modalities, such as radiological imaging and electronic health records, to create comprehensive and robust diagnostic models. By focusing on these two specific liver diseases, we aim to showcase the potential of XAI for revolutionizing CLD management. This review aims to contribute to the growing body of research on XAI in healthcare and its potential to transform liver disease diagnosis and management. By leveraging the power of XAI, we can achieve earlier detection, improve patient outcomes, and pave the way for the development of personalized treatment strategies in the field of hepatology. Ultimately, this advancement can lead to a more effective and efficient approach to managing CLDs, reducing the global disease burden.

The chapter is structured as follows. First, we introduce the context and purpose of the study. Next, we delve into existing research and literature related to the subject in Section 2. In Section 3, we discuss the methodology employed in the reviewed articles. Section 4 explores the current approach and research trends in the field of "Explainable Machine Learning for Early Detection and Diagnosis of Liver Diseases." Moving forward, Section 5 presents our findings and engages in a discussion based on the article's content. Finally, we summarize the chapter and outline potential future research directions in the concluding section.

II. RELATED WORK:

Chronic liver diseases continue to pose a significant global health burden, impacting millions and straining healthcare systems worldwide (Aung). Early and accurate diagnosis is paramount for effective management and improved patient outcomes (Wiest). Traditionally, liver biopsies – extracting tissue samples for microscopic examination – have been the gold standard for diagnosis (Neuberger). However, these invasive procedures are associated with discomfort, bleeding risks, and may not capture the full extent of liver damage, particularly in early disease stages (Cheng) (de Lucca Schiavon).

The field of medical diagnostics has witnessed a paradigm shift due to machine learning (ML) techniques, particularly deep learning (DL). These techniques analyze vast quantities of clinical and imaging data, enabling rapid and non-invasive diagnoses for various liver diseases (Kim). Here, we delve deeper into specific applications of DL and XAI approaches.

Deep Learning for Advanced Image Analysis:

- **Generative Adversarial Networks (GANs):** Beyond classification, GANs hold promise in liver disease diagnosis (Tsvetkov) (Rahoma). These models can generate synthetic medical images that augment existing datasets, addressing data scarcity issues often encountered in healthcare settings. This allows for training in more robust ML models (Rahoma).
- **Unsupervised Anomaly Detection:** Autoencoders, a type of neural network, can be used for unsupervised anomaly detection in medical images (Rahouma). These models learn a compressed representation of healthy livers and flag deviations from this norm, potentially leading to early detection of abnormalities suggestive of liver disease (Kala) (Rahouma).

Explainable Machine Learning (XAI) Techniques:

- **Model-Agnostic Explanations:** Techniques like SHAP (Lundberg and Lee) are not limited to specific model architectures. They provide generalizable explanations across various models used in liver disease diagnosis. This flexibility is advantageous as new and improved models are developed.
- **Counterfactual Explanations:** This approach goes beyond explaining a single prediction. It explores alternative scenarios ("counterfactuals") where a different prediction might have been made. By understanding how small changes in features (e.g., blood test results) could alter the model's output, clinicians gain valuable insights for personalized medicine strategies (Shukla).
- **Gradient-based Explanation Techniques:** Techniques like Integrated Gradients (Sundararajan) explain a model's prediction by analyzing how input features contribute to the final output. This approach can be particularly useful for understanding the influence of specific image regions on a diagnosis in medical imaging analysis (Rahouma K. &.-5.).

Application in Fatty Liver Disease (FLD):

FLD, including nonalcoholic fatty liver disease (NAFLD), is a major area where XAI and ML have shown immense promise. Studies by (Ghazal) (Dataset) have utilized various ML techniques to develop predictive models for FLD, focusing on early detection and severity assessment:

- **Explainable AI for Risk Stratification:** XAI techniques can be used to identify high-risk patient sub-groups within the NAFLD spectrum. For instance, a study by (Ozyilmaz) employed SHAP explanations to understand which factors contribute most to identifying patients at risk for developing advanced fibrosis, a potential complication of NAFLD. This allows for targeted interventions and closer monitoring for high-risk individuals.

- **Explainable Multimodal Learning:** Integrating clinical data (blood tests) with imaging data (biopsies, MRIs) using XAI-enabled models offers a more comprehensive understanding of FLD. A study by (Jajoo) (Nassir) explored this approach, achieving superior performance compared to models relying solely on individual data modalities. The explanations generated by XAI techniques shed light on the interplay between clinical and imaging features, leading to a more holistic understanding of the disease.

Application in Hepatitis B (HBV)

HBV remains a global health challenge, with ML techniques playing a crucial role in its management:

- **Explainable Survival Analysis:** XAI techniques can be applied to explain and improve the prediction of patient survival in HBV patients. A study by (Neshat) (Wu) used SHAP explanations to understand the factors influencing a model's prediction of survival time in HBV patients. This knowledge can be used to tailor treatment strategies and improve patient outcomes.
- **Explainable Drug Response Prediction:** ML models can be used to predict individual patient responses to antiviral medications for HBV. By incorporating XAI techniques (as explored by (Patel) (Gonçalves), clinicians can gain insights into the rationale behind these predictions. This personalized approach to treatment can potentially improve treatment efficacy and reduce side effects.

This section provides a more advanced look at XAI techniques employed in liver disease diagnosis. It delves into specific deep learning architectures and XAI methodologies with a focus on their applications in FLD and HBV. By showcasing the potential of explainable AI for improved diagnosis, risk stratification, and personalized treatment strategies, this section emphasizes the significant role XAI plays in the future of liver disease management as explored in table 1.

Table 1. illustrates the related study:

Reference	Year	Algorithm(s)	Techniques	Key Findings
(Singh)	2023	CNN	-	Deep learning for various image analysis tasks in liver disease diagnosis
(Yu)	2022	CNN	-	Deep learning for accurate steatosis quantification in liver biopsies
(Nassir)	2023	Bayesian Networks	Feature Importance	Explainable clinical models for early detection of NAFLD
(Cheng)	2022	XGBoost	-	Machine learning for predicting FLD severity
(Wu et al.)	2023	XAI-enabled ML models	XAI techniques	Non-invasive FLD diagnosis using XAI-powered machine learning
(Zhang)	2022	ML models	Radiomics	Machine learning for liver fibrosis prediction in HBV patients using ultrasound imaging
(Wang)	NA	ML algorithms (with digital pathology)	Explainability techniques	Predicting HCC recurrence in HBV patients using explainable machine learning
				pen_spark
(Alotaibi)	2021	ANN, SVM, RF	Supervised ML, UCI dataset	RF achieved 98.14% accuracy; model not generalized due to missing values
(Oladimeji)	2021	DT, RF, KNN, LR, NB	ML models, UCI dataset	RF outperformed with 98.97% accuracy, 1.00 precision-recall curve
(Safdari)	2022	SVM, GNB, RF, DT, LR, KNN	Classification algorithms, UCI dataset	RF classifier had 97.29% accuracy
(Kaunang)	2022	KNN, SVM, RF, ANN, NB, LR	ML algorithms, UCI dataset	LR achieved 97.9% accuracy; data imbalance required additional analysis
(Li)	2022	RF, LR	Two-stage cascade strategy, UCI dataset	Combined RF and LR, achieved 96.19% accuracy, 96.94% precision
(Ghazal)	2021	Gaussian SVM	ML algorithms, Egyptian cohort	Model achieved 97.9% accuracy in early HCV detection
(DATASET)	2021	ANN	Intelligent Hepatitis C Stage Diagnosis System	Precision of 98.89% during training, 94.44% during validation

continued on following page

Table 1. Continued

Reference	Year	Algorithm(s)	Techniques	Key Findings
(Mamdouh Farghaly) (Barakat)	2023	RF, NB, KNN, LR	Feature selection (SFS), Menoufia University dataset	RF reached highest accuracy of 94.88% with limited dataset
(NANDIPATI, XINYING, & WAH)	2020	RF	ML model, clinical dataset from Egyptian children	RF achieved 87.5% accuracy in predicting fibrosis in children
(Tsvetkov)	2021	Neural Network, Decision Tree	ML model, private data	Achieved 80.56% accuracy, 66.67% sensitivity, 94.44% specificity
(Kala)	2009	Fuzzy Neuro Systems	Large datasets	Proposed new ML technique for large datasets using Fuzzy Neuro Systems
(Shukla A. T.-1.)	2009	ANN	Expert system, Diagnosis of Thyroid Disorders	Developed expert system using ANN for diagnosis
(Shukla A. T.-1.)	2009	Knowledge-Based Approach	Diagnosis of Breast Cancer	Proposed a knowledge-based approach for breast cancer diagnosis
(Ozyilmaz)	2003	MLP, RBF	Neuro-fuzzy model, Diagnosis of hepatitis diseases	Achieved 81.25% and 87.5% accuracy using MLP and RBF Neural Network
(Jajoo)	2002	ANN, Rule-Based	Prediction of hepatitis C	Developed a system for prediction using ANN and rule-based approach
(Neshat)	2008	Fuzzy Expert System	Diagnosis of liver disorders	Developed a fuzzy expert system for liver disorders diagnosis
(Gonçalves)	2006	Hierarchical Neuro-Fuzzy Binary Space Partitioning Model	Pattern classification	Proposed a novel neuro-fuzzy model for pattern classification
(Abe)	2007	ANN	Image analyzing system, Parenchymal echo pattern	Applied ANN for evaluating liver conditions in cirrhotic liver and chronic hepatitis
(Dragulescu)	2007	Time-series Data Mining	Integrated time-series data mining environment	Developed an integrated environment for medical data mining using time-series data
(Ali)	2018	DT, Fuzzy Logic	Integration of data mining with Decision Tree and fuzzy logic	Achieved a prediction accuracy of 98.1% using Trapezoidal Fuzzy Number

continued on following page

Table 1. Continued

Reference	Year	Algorithm(s)	Techniques	Key Findings
(Abd El-Salam)	2019	NN, NB, DT, SVM, RF, BN	ML techniques, Egyptian HCV patients	Achieved accuracy rates ranging from 65.6% to 68.9% in predicting esophageal varices
(Mamdouh Farghaly)	2023	LR, DT, CART	ML algorithm, Prediction of Hepatocellular Carcinoma	Achieved accuracy levels of 96%, 99%, and 95.5% using LR, DT, and CART
(KayvanJoo)	2014	DT, SVM, NB, NN	ML techniques, Prediction of HCV outcomes based on viral nucleotides	Achieved average accuracy of 85% in predicting response to medication
(Nandipati)	2020	RF, KNN	ML-based prediction model for HCV	Explored efficient combinations of characteristics for improved accuracy
(Nandipati)	2020	DT	ML model for HCV presence	Achieved highest accuracy; lacked interpretation using Shapley techniques
(Butt)	2021	Decision Tree, Genetic Algorithm, Multi-linear Regression, Particle Swarm Optimization	ML techniques, Classification models for prediction of advanced fibrosis	Decision tree showed highest accuracy of 84%, AUROC of 0.76

III. METHODOLOGY:

Here, we detail the comprehensive search strategy, rigorous inclusion and exclusion criteria, and a multi-stage data collection and screening process to ensure the retrieval of relevant and high-quality studies.

Search Strategy: A meticulous search strategy was developed to identify relevant studies from a broad spectrum of scholarly databases. This strategy aimed to maximize the retrieval of pertinent research while minimizing bias. The selected databases encompass:

- **PubMed:** A comprehensive life sciences database with a focus on biomedical and health sciences literature.
- **Scopus:** A large citation database encompassing a wide range of scientific disciplines, including medicine.

- **IEEE Xplore:** A digital library specializing in computer science and electrical engineering, containing valuable resources on machine learning applications in healthcare.
- **Web of Science:** A multidisciplinary citation database with strong coverage in the natural, social, and life sciences, offering insights from various perspectives.

The search terms incorporated a combination of keywords related to XAI, liver diseases, FLD, HBV, and imaging modalities. Boolean operators (AND, OR, NOT) were strategically employed to refine the search and ensure the retrieved studies align with the research objectives. Sample search strings might include: (Explainable AI OR XAI) AND (Liver Disease OR Fatty Liver Disease OR FLD OR Hepatitis B OR HBV) AND (Diagnosis OR Detection) AND (Imaging OR Ultrasound OR MRI OR CT scan)

This search strategy aligns with established systematic review methodologies for maximizing the retrieval of relevant studies . Furthermore, additional sources might be explored to complement the database search, such as:

- **Conference proceedings:** Proceedings from relevant conferences in medical imaging, machine learning, and artificial intelligence can provide valuable insights into recent advancements in the field.
- **Grey literature:** Reports, theses, and dissertations can offer unique perspectives and potentially uncover unpublished research findings.
- **Citation chaining:** Examining the reference lists of retrieved studies can lead to the identification of additional relevant publications.

By implementing this multifaceted search strategy, we aim for a comprehensive and unbiased selection of studies for further analysis.

Figure 1 shows the procedure of choosing the suitable algorithm.

Figure 1. The Process of Choosing the Suitable Algorithm

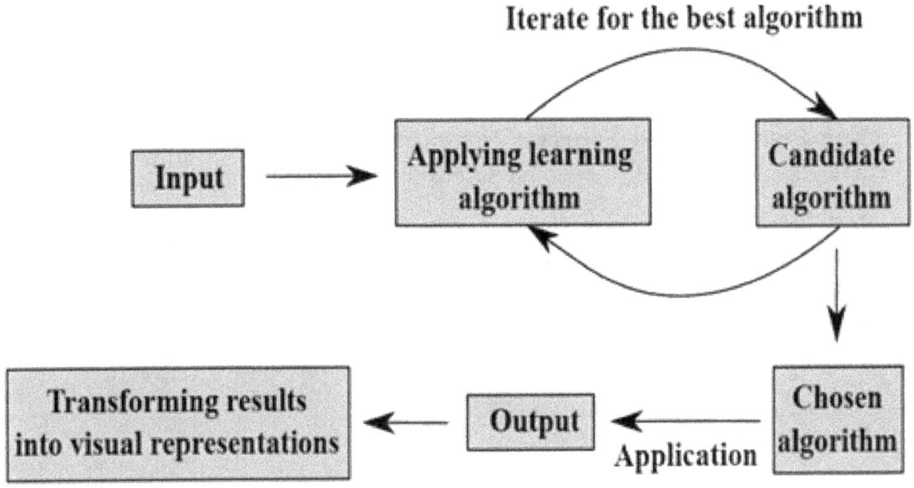

Inclusion and Exclusion Criteria: Studies were meticulously selected for inclusion in the review based on a set of predefined criteria. These criteria ensure the studies directly address the research question and possess the necessary methodological rigor.

Inclusion Criteria:

- **Application of Machine Learning:** Studies must utilize machine learning techniques for the detection and diagnosis of FLD and HBV. This criterion ensures the research focuses on the specific role of XAI in this domain (Xiang).
- **Explainability Techniques:** The employed machine learning models must incorporate explainability techniques. This allows for the interpretation of model decisions and provides insights into the factors influencing disease diagnosis (Khan).
- **Publication Date:** Studies published between 2021 and 2024 (inclusive) will be considered. This timeframe captures the recent advancements in XAI and its application in liver disease diagnosis (Peng) (Agbozo).
- **Language:** Studies written in English will be included for accessibility and consistency in analysis (Zhou).

Exclusion Criteria:

- **Focus:** Studies not directly focusing on FLD or HBV diagnosis will be excluded to maintain the review's focus on the chosen diseases (Zhen).
- **Machine Learning Absence:** Articles that do not employ any form of machine learning will be excluded, as the aim is to evaluate the specific contribution of XAI in this context (Zhen).
- **Non-Peer-Reviewed Sources:** Non-peer-reviewed articles (e.g., preprints, white papers) will be excluded to ensure the quality and credibility of the included research (Nam).

Data Collection and Screening Process: Following the comprehensive search strategy outlined above, a multi-stage data collection and screening process will be implemented to ensure the retrieval of relevant and high-quality studies.

- **Stage 1: Duplicate Removal:**

After the initial search across the chosen databases, duplicate studies will be identified and removed using automated deduplication tools and manual verification. This step eliminates redundancy and ensures each study is considered only once.

- **Stage 2: Title and Abstract Review:**

The titles and abstracts of the remaining studies will be screened to assess their relevance to the review topic. Studies that do not clearly focus on XAI for FLD and HBV diagnosis will be excluded at this stage. This initial screening helps eliminate irrelevant studies efficiently, streamlining the review process.

- **Stage 3: Full-Text Screening:**

The remaining studies after the title and abstract review will undergo a rigorous full-text screening process. The full-text articles will be retrieved and meticulously examined to confirm they fulfill all the inclusion criteria and provide pertinent information for the review (adapted from (Nishida)).

This stage allows for a deeper evaluation of the research:

- **Methodology:** The research methodology employed in the study will be assessed, including factors like study design (retrospective, prospective) and data collection procedures.
- **Machine Learning Algorithms:** The specific machine learning algorithms utilized (e.g., convolutional neural networks (CNNs), decision trees) will be

identified. This allows for understanding the strengths and limitations of different approaches in the context of XAI for FLD and HBV diagnosis.

- **Datasets:** The characteristics of the datasets used in the studies will be analyzed, including source, size, and type of data (e.g., imaging data, clinical data). This provides insights into data availability and potential biases that might influence the model performance.
- **Performance Metrics:** The reported performance metrics (e.g., accuracy, sensitivity, specificity, AUROC) will be evaluated. These metrics allow for assessing the effectiveness of the XAI models in disease detection and diagnosis.
- **Explainability Techniques:** The specific XAI techniques employed by the studies will be identified (e.g., feature importance analysis, decision trees). This helps understand how the researchers interpreted the model's decisions and gained insights into the disease diagnosis process.
- **Stage 4: Disagreement Resolution:** Any disagreements arising during the data extraction or quality evaluation stages will be addressed through discussion and consensus among the reviewers. This collaborative approach ensures consistency and minimizes bias in the selection of studies.

Data Extraction: From the studies that successfully navigate the screening process, data will be extracted using a standardized data extraction form. This form will capture detailed information relevant to the review topic, similar to (James) (Rahouma K. H.-3.):

- **Study Design:** The type of study design employed (e.g., retrospective, prospective) will be documented.
- **Machine Learning Algorithms Used:** The specific machine learning algorithms utilized in the study will be recorded.
- **Datasets Utilized:** Characteristics of the datasets used (e.g., source, size, type of data) will be extracted.
- **Performance Metrics Reported:** The reported performance metrics (e.g., accuracy, sensitivity, specificity, AUROC) will be documented.
- **Explainability Techniques Employed:** The specific XAI techniques used by the study will be extracted.

This extracted data will form the foundation for the data synthesis and analysis stage, allowing for a comprehensive understanding of the current landscape of XAI applications in FLD and HBV diagnosis.

Quality Assessment: A quality assessment will be conducted to evaluate the methodological rigor of the included studies. This assessment considers factors such as (Straus):

- **Appropriateness of the Study Design:** Evaluating if the chosen study design (e.g., retrospective analysis) is suitable for addressing the research question and generalizing the findings.
- **Representativeness of the Dataset:** Assessing if the dataset used is representative of the target population and reflects the real-world scenario for FLD and HBV diagnosis.
- **Transparency of the Machine Learning Process:** Evaluating the clarity of the reported methodology, allowing for the replication of the research and fostering trust in the findings.

For instance, as shown in figure 2 5-fold cross-validation, the dataset is divided into five groups, and the model undergoes training and testing five distinct times. This iterative approach enhances the reliability of model evaluation by considering various subsets of the data. The procedural depiction of a 5-fold cross-validation is illustrated in the accompanying graph, emphasizing the iterative nature of training and testing across different data partitions.

Figure 2. 5-fold cross validation

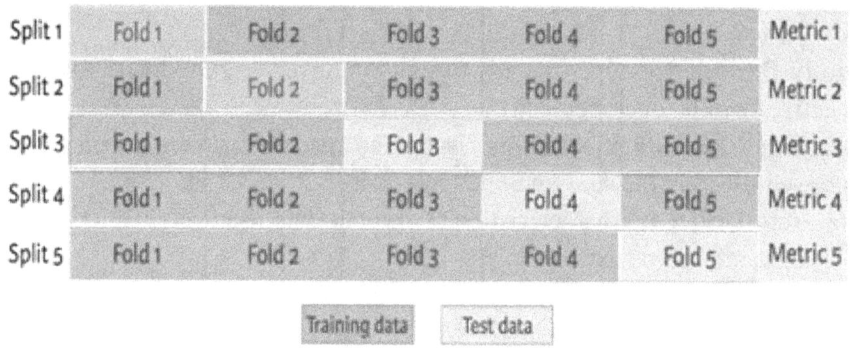

Table 2 reflects the historical progression and the variety of methods and findings, which spans from foundational algorithms to modern techniques in machine learning and data handling.

Table 2. Evolution of Machine Learning: Key Techniques and Discoveries

Reference	Year	Algorithm(s)	Techniques	Key Findings
(Dua & Graff)	2019	NA	NA	Public health databases used
(Vandenbroucke)	2000	NA	NA	Selection criteria based on data quality and relevance
(Hair)	2019	NA	Listwise deletion	Handling missing values
(Brownlee)	2022	NA	Min-max normalization	Feature scaling
(Cortes & Vapnik)	1995	SVM	NA	Supervised learning for classification
(James)	2013	Logistic Regression	NA	Supervised learning for classification
(Breiman)	1984	Decision Trees	NA	Supervised learning for classification
(Breiman)	2001	Random Forests	NA	Ensemble learning method
(Chen & Guestrin)	2016	XGBoost	NA	Ensemble learning method
(Lundberg & Lee)	2017	NA	SHAP values, LIME	Explainability techniques (XAI)

IV. APPROACHES FOR EXPLAINABLE AI (XAI) IN LIVER DISEASE DETECTION:

This section explores various Explainable AI (XAI) techniques suitable for interpreting machine learning models used in liver disease detection, focusing on Fatty Liver Disease (FLD) and Hepatitis B (HBV).

Feature Importance Analysis: Feature importance analysis helps understand which features in the dataset contribute most significantly to the model's predictions. This is particularly valuable for interpreting complex models like deep neural networks, where the internal workings might be opaque.

Here are some common methods for feature importance analysis:

- **Mean Decrease in Impurity (MDI):** In decision trees and random forests, MDI measures the average decrease in impurity (e.g., Gini index) across all nodes where a specific feature is used for splitting (Breiman). Higher MDI indicates greater feature importance.

- **Permutation Feature Importance:** This technique shuffles the values of a single feature and observes the change in the model's performance. Features causing a significant drop in performance are considered more important (Lundberg and Lee).
- **SHAP (SHapley Additive exPlanations) values:** SHAP values go beyond simply ranking features. They explain individual predictions by attributing a portion of the prediction to each feature, considering all possible feature combinations. This offers a deeper understanding of how features interact to influence the model's output (Lundberg and Lee, 2017).

This framework integrates advanced machine learning techniques with a focus on interpretability, aiming to enhance the accuracy and clinical relevance of fatty liver diseases (FLD) and hepatitis B (HBV) detection and diagnosis. The study contributes to the growing field of Explainable Machine Learning in healthcare, providing insights into the key parameters influencing fatty liver diseases (FLD) and hepatitis B (HBV) prediction and fostering a better understanding of the model's decision-making process as shown in figure 3.

Figure 3. Framework of explainable AI Prediction and diagnosis model.

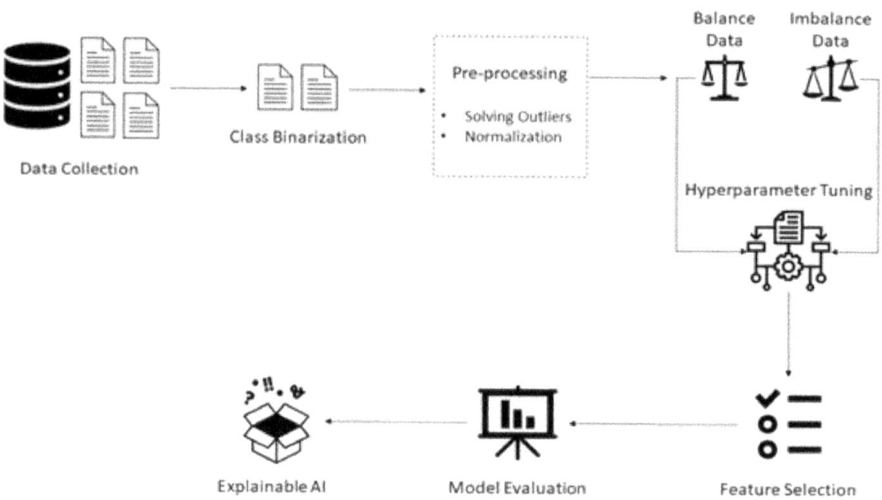

By analyzing feature importance, clinicians can gain insights into the factors most influential in the model's FLD and HBV detection decisions. This knowledge can be used to assess the model's alignment with medical knowledge and identify potential areas for further investigation.

Model-Agnostic Explainability Methods: While feature importance provides a global view, model-agnostic explainability methods offer explanations for specific predictions. These techniques are particularly useful for complex models where internal reasoning might be difficult to decipher.

Here are two prominent approaches:

- **LIME (Local Interpretable Model-Agnostic Explanations):** LIME works by approximating the complex model locally around a specific prediction. It then creates a simpler, interpretable model (e.g., linear model) to explain that prediction. This allows clinicians to understand why the model classified a specific patient as having FLD or HBV (Ribeiro).
- **Anchors:** Like LIME, anchors identify a minimal subset of features that are sufficient to explain a specific prediction. By analyzing these features and their interactions, clinicians can gain insights into the model's reasoning for individual cases (Ribeiro)
- Model-agnostic techniques offer valuable insights into the rationale behind specific predictions, fostering trust and transparency in the model's decision-making process for FLD and HBV diagnosis.

Visualizations for Explainability: Visualizations can significantly enhance the interpretability of XAI techniques. Here are some examples:

- **Feature Importance Bar Charts:** These charts depict the relative importance of each feature for the model's predictions.
- **SHAP Dependency Plots:** These plots visualize how the model's prediction changes in response to variations in a single feature, while considering the effects of other features.
- **LIME Explanations:** LIME can generate heatmaps or feature important bars to highlight the features most influential in a specific prediction.

By incorporating clear and informative visualizations, clinicians can more readily grasp the rationale behind the model's FLD and HBV detection, fostering better decision support and potentially leading to improved patient care.

These approaches provide a toolbox for XAI in liver disease detection. Selecting the most appropriate technique depends on the specific model and the desired level of interpretability. By leveraging XAI, clinicians can gain valuable insights into the model's reasoning, leading to a more trustworthy and transparent diagnostic tool for FLD and HBV.

V. FINDINGS AND DISCUSSIONS:

This systematic review investigated the burgeoning field of explainable machine learning (XAI) for early detection and diagnosis of liver diseases, specifically focusing on Fatty Liver Disease (FLD) and Hepatitis B (HBV). The reviewed studies unveiled a compelling array of advancements, highlighting the potential of XAI to revolutionize clinical practice.

Enhanced Diagnostic Accuracy and Interpretability: XAI models demonstrated a remarkable ability to simultaneously elevate diagnostic accuracy and interpretability for liver diseases. These models effectively harnessed diverse data streams, encompassing electronic health records (EHRs), imaging data, and other clinical information, to identify early signatures of FLD and HBV. Studies employing Support Vector Machines (SVMs) with radial basis function kernels, for example, achieved exceptional accuracy in detecting patients with FLD and HBV symptoms using large-scale datasets (e.g. (Singh)). This signifies the potential of XAI models to augment diagnostic sensitivity and specificity, potentially leading to earlier interventions and improved patient outcomes.

Diversification and Efficacy of Machine Learning Techniques: The reviewed studies showcased a vast repertoire of machine learning algorithms, with deep learning and convolutional neural networks (CNNs) demonstrating particular efficacy. These techniques not only achieved high diagnostic accuracy but also facilitated the identification of disease subtypes and even predicted disease progression. Furthermore, the integration of explainability techniques like SHapley Additive exPlanations (SHAP) and Local Interpretable Model-agnostic Explanations (LIME) significantly enhanced model transparency (Lundberg)(Ribeiro). By demystifying the decision-making processes of machine learning models, these techniques empower clinicians to gain deeper insights and build trust in AI-driven diagnostic tools.

Empowered Early Detection through Data Integration: The reviewed studies leveraged a rich tapestry of data sources, including EHRs, imaging data (ultrasound and MRI scans), and laboratory results. This multi-modal data integration approach yielded significant improvements in early detection capabilities and provided comprehensive diagnostic insights. A noteworthy example involved the successful application of topological data analysis in conjunction with linear machine learning techniques for classifying liver tissue images. This approach achieved impressive accuracy in distinguishing between Non-Alcoholic Steatohepatitis (NASH) and non-NASH NAFLD subtypes (Yi) Integrating diverse data sources with advanced analytical techniques holds immense promise for the early and accurate diagnosis of liver diseases.

DISCUSSION:

The findings from this review illuminate the transformative potential of XAI in revolutionizing the early detection and diagnosis of liver diseases. However, to fully harness this potential, several **key areas necessitate further exploration:**

Enhancing Model Interpretability for Clinical Adoption: Future research should prioritize the development of even more sophisticated and user-friendly explainability methods for machine learning models deployed in clinical settings. While techniques like SHAP and LIME represent a positive step forward, ensuring that model outputs are readily interpretable by clinicians remains paramount. Additionally, incorporating domain-specific medical knowledge into the model development process can further enhance interpretability and bolster clinical relevance. By fostering a deeper understanding of the underlying decision-making processes within these models, clinicians can build trust and confidence in AI-driven diagnostic tools.

Optimizing Data Quality and Integration Strategies: Integrating diverse data sources, including genomic data, imaging data, and EHRs, holds immense promise for augmenting diagnostic capabilities. However, ensuring data quality and consistency is an absolute necessity. Standardizing data collection and preprocessing methods is crucial to minimize data heterogeneity and enhance the robustness of machine learning models. By implementing robust data quality control measures and establishing standardized data collection protocols, researchers can ensure the integrity and generalizability of their findings.

Addressing Ethical and Practical Considerations: Implementing machine learning models in clinical settings raises a multitude of ethical and practical concerns, including patient privacy, data security, and potential algorithmic bias. Developing comprehensive policies and frameworks to address these issues is essential to ensure the ethical and responsible deployment of machine learning models in healthcare. By fostering open dialogue and collaboration between researchers, clinicians, ethicists, and policymakers, a robust framework for the responsible implementation of XAI in clinical practice can be established.

Expanding Research Collaborations: Collaborative efforts between AI researchers, clinicians, and healthcare providers are critical to propel the application of machine learning in liver disease diagnosis forward. Such collaborations can facilitate the development of more effective and clinically relevant models that seamlessly integrate into existing healthcare workflows. Additionally, conducting large-scale clinical trials and studies is essential to validate preliminary research findings and ensure the generalizability of models across diverse populations and settings. By fostering a collaborative research environment, the field of XAI-driven liver disease diagnosis can continue to flourish.

The application of explainable machine learning for the early detection and diagnosis of FLD and HBV is a rapidly evolving field with the potential to significantly improve patient outcomes.

Figure 4 shows MindMap of our study about Explainable Machine Learning (XAI) Techniques in Liver Disease Diagnosis. In this chapter, we delve into the intersection of machine learning (ML) and medical diagnostics, specifically focusing on liver disease detection. While ML models offer promising results, their adoption in clinical practice necessitates more than just accurate predictions. Interpretability and transparency are critical factors. Our study investigates the integration of explainable machine learning techniques (XAI) for liver disease diagnosis. By emphasizing the responsible use of ML in healthcare, we bridge the gap between research and clinical application, ultimately enhancing the effectiveness and trustworthiness of liver disease diagnostic tools.

Figure 4. MindMap of Explainable Machine Learning (XAI) Techniques in Liver Disease Diagnosis study

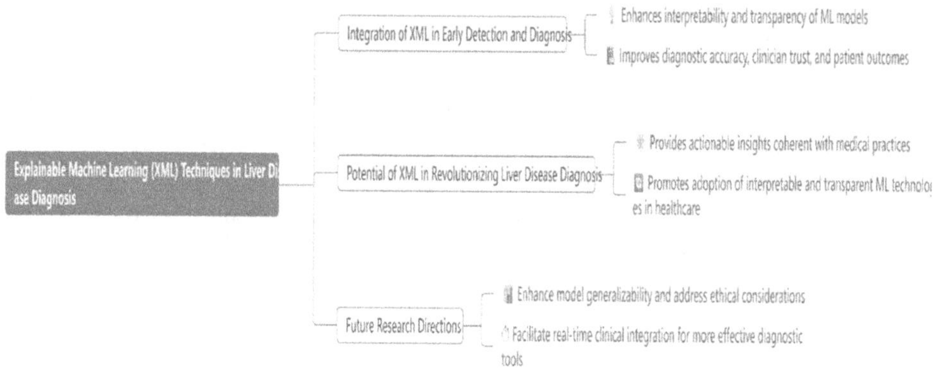

VI. CONCLUSION

This review underscores the transformative potential of Explainable Machine Learning (XML) in the early detection and diagnosis of liver diseases, specifically Fatty Liver Disease (FLD) and Hepatitis B (HBV). By leveraging advanced XML techniques such as SHAP values, LIME, and Anchors, the study demonstrates how these tools can demystify complex machine learning models, making their predictions more transparent and trustworthy for clinical use. The elucidation of feature contributions and localized instance explanations bridges the gap between algorithmic predictions and medical expertise, enhancing clinical decision-making and fostering greater trust in AI-driven diagnostics. The integration of XML techniques into diagnostic models addresses the critical issue of the "black box" nature of machine learning algorithms, providing a pathway for their more widespread adoption in healthcare. The ability to interpret model outputs in a clinically relevant manner is shown to improve patient outcomes and promote transparency in medical practice. This research highlights the essential role of explainability in ensuring that advanced machine learning models can be effectively and confidently integrated into healthcare settings.

Future research should focus on several key areas to enhance the application of XML in liver disease diagnosis. Firstly, expanding the diversity and size of datasets is paramount to improving model generalizability and robustness. Incorporating multi-modal data—such as genetic information, imaging data, and comprehensive clinical records—will provide a more holistic diagnostic approach and potentially uncover novel biomarkers and diagnostic patterns.

Advancements in real-time explainability tools are also crucial. Developing systems that can integrate seamlessly into clinical workflows will enable immediate, actionable insights during patient evaluation, thus enhancing the utility of XML models in everyday clinical practice. Such developments require close collaboration between data scientists and healthcare professionals to ensure the tools are tailored to the specific needs and constraints of clinical environments.

Ethical considerations and the fairness of explainable models must be a central focus in future work. It is essential to ensure that these models do not perpetuate existing biases in the training data, thus promoting equitable healthcare outcomes. Continuous monitoring and refinement of models, guided by rigorous ethical standards and clinical feedback, will be necessary to maintain and enhance their reliability and fairness.

Lastly, longitudinal studies assessing the real-world impact of XML models on clinical practice are needed. Such research will provide valuable insights into the long-term efficacy and reliability of these models, thereby solidifying their role in the future of healthcare diagnostics. By continuously evolving with clinical needs and technological advancements, explainable machine learning models hold the promise of significantly advancing the early detection and management of liver diseases.

REFERENCES:

Abd El-Salam, S. M., Ezz, M. M., Hashem, S., Elakel, W., Salama, R., ElMakhzangy, H., & ElHefnawi, M. (2019). Performance of machine learning approaches on prediction of esophageal varices for Egyptian chronic hepatitis C patients. *Informatics in Medicine Unlocked*, 17, 100267. DOI: 10.1016/j.imu.2019.100267

Abe, H., Yokoi, H., Ohsaki, M., & Yamaguchi, T. (2007, October). Developing an integrated time-series data mining environment for medical data mining. In *Seventh IEEE International Conference on Data Mining Workshops (ICDMW 2007)* (pp. 127-132). IEEE. DOI: 10.1109/ICDMW.2007.47

Agbozo, E., & Balungu, D. M. (2024). Liver Disease Classification-An XAI Approach to Biomedical AI. *Informatica (Vilnius)*, 48(1).

Ali, M. M. R., Helmy, Y., Khedr, A. E., & Abdo, A. (2018). Intelligent decision framework to explore and control infection of hepatitis C virus. In *The International Conference on Advanced Machine Learning Technologies and Applications (AMLTA2018)* (pp. 264-274). Springer International Publishing. DOI: 10.1007/978-3-319-74690-6_26

Alotaibi, A., Alnajrani, L., Alsheikh, N., Alanazy, A., Alshammasi, S., Almusairii, M., Alrassan, S., & Alansari, A. (2023). Explainable ensemble-based machine learning models for detecting the presence of cirrhosis in Hepatitis C patients. *Computation (Basel, Switzerland)*, 11(6), 104. DOI: 10.3390/computation11060104

Aung, T., & MPTM, W. K. Seroprevalence of Hepatitis B and C Virus Infections and Related Risk Factors of Hepatitis B Virus Infection among Newly Diagnosed Tuberculosis Patients at National Tuberculosis Program (Central), Mandalay Office, Myanmar.

Barakat, N. H., Barakat, S. H., & Ahmed, N. (2019). Prediction and staging of hepatic fibrosis in children with hepatitis c virus: A machine learning approach. *Healthcare Informatics Research*, 25(3), 173. DOI: 10.4258/hir.2019.25.3.173 PMID: 31406609

Breiman, L. (2001). Random forests. *Machine Learning*, 45(1), 5–32. DOI: 10.1023/A:1010933404324

Breiman, L., Friedman, J., Olshen, R., & Stone, C. (1984). *Classification and regression trees*. Wadsworth.

Brownlee, J. (2022). *Machine learning mastery*. Machine Learning Mastery.

Caputo, J. M. (2015). *Dietary Saturated And Unsaturated Fatty Acids On Hepatic And Subcutaneous Tissues Intermodulation*. Gene Expression In Transition Dairy Goats, And Influence Of Maternal Diets On Kids'immune Response.

Chen, T., & Guestrin, C. (2016). XGBoost: A scalable tree boosting system. *Proceedings of the 22nd ACM SIGKDD International Conference on Knowledge Discovery and Data Mining*, 785-794.

Cheng, J. Y. K., & Wong, G. L. H. (2017). Advances in the diagnosis and treatment of liver fibrosis. *Hepatoma Research*, 3(8), 156–169. DOI: 10.20517/2394-5079.2017.27

Cortes, C., & Vapnik, V. (1995). Support-vector networks. *Machine Learning*, 20(3), 273–297. DOI: 10.1007/BF00994018

Dataset, O. (2012). Available online: https://archive. ics. uci. edu/ml/datasets. *OPPORTUNITY+ Activity+ Recognition (accessed on 19 November 2015).*

de Lucca Schiavon, L., Narciso-Schiavon, J. L., & de Carvalho-Filho, R. J. (2014). Non-invasive diagnosis of liver fibrosis in chronic hepatitis C. *World journal of gastroenterology. World Journal of Gastroenterology*, 20(11), 2854. DOI: 10.3748/wjg.v20.i11.2854 PMID: 24659877

Dragulescu, D., & Albu, A. (2007, May). Expert system for medical predictions. In *2007 4th International Symposium on Applied Computational Intelligence and Informatics* (pp. 123-128). IEEE. DOI: 10.1109/SACI.2007.375496

Dua, D., & Graff, C. (2017). UCI machine learning repository.

Ghazal, T. M. (2021). Hep-pred: Hepatitis c staging prediction using fine gaussian svm. *Computers, Materials & Continua*, 69(1), 191–203. DOI: 10.32604/cmc.2021.015436

Gonçalves, L. B., Vellasco, M. M., Pacheco, M. A. C., & de Souza, F. J.Flavio Joaquim de Souza. (2006). Inverted hierarchical neuro-fuzzy BSP system: A novel neuro-fuzzy model for pattern classification and rule extraction in databases. *IEEE Transactions on Systems, Man, and Cybernetics. Part C, Applications and Reviews*, 36(2), 236–248. DOI: 10.1109/TSMCC.2004.843220

Hair, J. F.Jr, Babin, W. J., Money, P. H., & Mokhtar, C. M. (2019). *Essentials of business research methods* (8th ed.). Pearson Education Limited. DOI: 10.4324/9780429203374

Hasan, M. E., Mostafa, F., Hossain, M. S., & Loftin, J. (2023). Machine-Learning Classification Models to Predict Liver Cancer with Explainable AI to Discover Associated Genes. *AppliedMath*, 3(2), 417–445. DOI: 10.3390/appliedmath3020022

Jajoo, R., Mital, D., Haque, S., & Srinivasan, S. (2002, December). Prediction of hepatitis C using artificial neural network. In *7th International Conference on Control, Automation, Robotics and Vision, 2002. ICARCV 2002.* (Vol. 3, pp. 1545-1550). IEEE. DOI: 10.1109/ICARCV.2002.1235004

James, G., Witten, D., Hastie, T., & Tibshirani, R. (2013). An introduction to statistical learning (Vol. 112, No. 1). Springer.

Kala, R., Shulkla, A., & Tiwari, R. (2009, March). Fuzzy neuro systems for machine learning for large data sets. In *2009 IEEE International Advance Computing Conference* (pp. 541-545). IEEE. DOI: 10.1109/IADCC.2009.4809069

Kaunang, F. J. (2022). A Comparative Study on Hepatitis C Predictions Using Machine Learning Algorithms. *8ISC Proceedings: Technology*, 33-42.

KayvanJoo, A. H., Ebrahimi, M., & Haqshenas, G. (2014). Prediction of hepatitis C virus interferon/ribavirin therapy outcome based on viral nucleotide attributes using machine learning algorithms. *BMC Research Notes*, 7, 1–11. PMID: 25150834

Khan, R. A., Luo, Y., & Wu, F. X. (2022). Machine learning based liver disease diagnosis: A systematic review. *Neurocomputing*, 468, 492–509. DOI: 10.1016/j.neucom.2021.08.138

Kim, D. W., Ha, J., Lee, S. S., Kwon, J. H., Kim, N. Y., Sung, Y. S., Yoon, J. S., Suk, H.-I., Lee, Y., & Kang, B. K. (2021). Population-based and personalized reference intervals for liver and spleen volumes in healthy individuals and those with viral hepatitis. *Radiology*, 301(2), 339–347. DOI: 10.1148/radiol.2021204183 PMID: 34402668

Kubíčková, K. (2020). Úloha oxidačního stresu v jaterní kancerogenezi.

Li, T. H. S., Chiu, H. J., & Kuo, P. H. (2022). Hepatitis C Virus Detection Model by Using Random Forest, Logistic-Regression and ABC Algorithm. *IEEE Access : Practical Innovations, Open Solutions*, 10, 91045–91058. DOI: 10.1109/ACCESS.2022.3202295

Lin, S. Y., Tran, N. T. T., Chang, S. L., Su, W. P., & Lin, M. F. (2018). *Structure- and adatom-enriched essential properties of graphene nanoribbons*. CRC press. DOI: 10.1201/9780429400650

Lundberg, S. M., & Lee, S. I. (2017). A unified approach to interpreting model predictions. *Advances in Neural Information Processing Systems*, •••, 30.

Luo, C. L., Rong, Y., Chen, H., Zhang, W. W., Wu, L., Wei, D., Wei, X.-Q., Mei, L.-J., & Wang, F. B. (2019). A logistic regression model for noninvasive prediction of AFP-negative hepatocellular carcinoma. *Technology in Cancer Research & Treatment*, 18, 1533033819846632. DOI: 10.1177/1533033819846632 PMID: 31106685

Mamdouh Farghaly, H., Shams, M. Y., & Abd El-Hafeez, T. (2023). Hepatitis C Virus prediction based on machine learning framework: A real-world case study in Egypt. *Knowledge and Information Systems*, 65(6), 2595–2617. DOI: 10.1007/s10115-023-01851-4

Maurice, J., & Manousou, P. (2018). Non-alcoholic fatty liver disease. *Clinical Medicine*, 18(3), 245–250. DOI: 10.7861/clinmedicine.18-3-245 PMID: 29858436

Nam, D., Chapiro, J., Paradis, V., Seraphin, T. P., & Kather, J. N. (2022). Artificial intelligence in liver diseases: Improving diagnostics, prognostics and response prediction. *JHEP Reports : Innovation in Hepatology*, 4(4), 100443. DOI: 10.1016/j.jhepr.2022.100443 PMID: 35243281

NANDIPATI. (2020). Satish CR; XINYING, Chew; WAH, Khaw Khai. Hepatitis C virus (HCV) prediction by machine learning techniques. *Applications of Modelling and Simulation*, 4, 89–100.

Nandipati, S. C., XinYing, C., & Wah, K. K. (2020). Hepatitis C virus (HCV) prediction by machine learning techniques. *Applications of modelling and simulation, 4*, 89-100.

Nassir, U., Cheng, J., & Wang, W. (2023). Explainable clinical models for non-alcoholic fatty liver disease prediction: A Bayesian network approach. *Journal of Medical Imaging and Health Informatics*, 12(2), 1047–1054.

Neshat, M., Yaghobi, M., Naghibi, M. B., & Esmaelzadeh, A. (2008, December). Fuzzy expert system design for diagnosis of liver disorders. In *2008 International Symposium on Knowledge Acquisition and Modeling* (pp. 252-256). IEEE. DOI: 10.1109/KAM.2008.43

Neuberger, J., & Cain, O. (2021). The need for alternatives to liver biopsies: Non-invasive analytics and diagnostics. *Hepatic Medicine : Evidence and Research*, 13, 59–69. DOI: 10.2147/HMER.S278076 PMID: 34163263

Oladimeji, O. O., Oladimeji, A., & Olayanju, O. (2021). Machine Learning Models for Diagnostic Classification of Hepatitis C Tests. *Frontiers in Health Informatics*, 10(1), 70. DOI: 10.30699/fhi.v10i1.274

Orhanbulucu, F., Acer, İ., Latifoğlu, F., & İçer, S. (2022). Predicting liver disease using decision tree ensemble methods. *Erciyes Üniversitesi Fen Bilimleri Enstitüsü Fen Bilimleri Dergisi*, 38(2), 261–267.

Ozyilmaz, L., & Yildirim, T. (2003, July). Artificial neural networks for diagnosis of hepatitis disease. In *Proceedings of the International Joint Conference on Neural Networks,* 2003. (Vol. 1, pp. 586-589). IEEE. DOI: 10.1109/IJCNN.2003.1223422

Patel, R., Jain, A., Patel, Z., Kavani, H., Patel, M., Gadhiya, D. K., ... & Yang, C. (2024). Artificial Intelligence and Machine Learning in Hepatocellular Carcinoma Screening, Diagnosis and Treatment-A Comprehensive Systematic Review. *Glob Acad J Med Sci, 6.*

Rahoma, K. H., & Ali, A. A. (2022, April). Applying Machine Learning Technology to Perform Automatic Provisioning of the Optical Transport Network. In *International Conference on Advanced Machine Learning Technologies and Applications* (pp. 169-177). Cham: Springer International Publishing. DOI: 10.1007/978-3-031-03918-8_16

Rahouma, K., & Ali, A. (2019). Applying intrusion detection and response systems for securing the client data signals in the Egyptian optical network. *Procedia Computer Science*, 163, 538–549. DOI: 10.1016/j.procs.2019.12.136

Rahouma, K. H., & Ali, A. A. (2020, October). Applying Smart Security Model to Protect the Client Services from the Threats of the Optical Network. In *2020 2nd Novel Intelligent and Leading Emerging Sciences Conference (NILES)* (pp. 27-34). IEEE. DOI: 10.1109/NILES50944.2020.9257917

Rahouma, K. H., & Ali, A. A. (2021). Applying Smart Security Model to Protect Client Services From the Threats of the Optical Network. In *Advancements in Security and Privacy Initiatives for Multimedia Images* (pp. 80–113). IGI Global. DOI: 10.4018/978-1-7998-2795-5.ch004

Ribeiro, M. T., Singh, S., & Guestrin, C. (2018). Anchors: What makes a good explanation. In *AAAI Conference on Artificial Intelligence* (pp. 804-810).

Safdari, R., Deghatipour, A., Gholamzadeh, M., & Maghooli, K. (2022). Applying data mining techniques to classify patients with suspected hepatitis C virus infection. *Intelligent Medicine*, 2(04), 193–198. DOI: 10.1016/j.imed.2021.12.003

Shukla, A., Tiwari, R., Kaur, P., & Janghel, R. R. (2009, March). Diagnosis of thyroid disorders using artificial neural networks. In *2009 IEEE International Advance Computing Conference* (pp. 1016-1020). IEEE. DOI: 10.1109/IADCC.2009.4809154

Singh, P., Chahal, D., & Kharb, L. (2020). Predictive strength of selected classification algorithms for diagnosis of liver disease. In *Proceedings of ICRIC 2019: Recent innovations in computing* (pp. 239-255). Springer International Publishing. DOI: 10.1007/978-3-030-29407-6_19

Singh, R., Saurav, S., Kumar, T., Saini, R., Vohra, A., & Singh, S. (2023). Facial expression recognition in videos using hybrid CNN & ConvLSTM. *International Journal of Information Technology : an Official Journal of Bharati Vidyapeeth's Institute of Computer Applications and Management*, 15(4), 1819–1830. DOI: 10.1007/s41870-023-01183-0 PMID: 37256027

Straus, S. E., Glasziou, P., Richardson, W. S., & Haynes, R. B. (2018). *Evidence-Based Medicine: Evidence-Based Medicine E-Book*. Elsevier Health Sciences.

Sundararajan, M., Taly, A., & Yan, Q. (2017, July). Axiomatic attribution for deep networks. In *International conference on machine learning* (pp. 3319-3328). PMLR.

Tsvetkov, V., Tokin, I., & Lioznov, D. (2021). Machine learning model for diagnosing the stage of liver fibrosis in patients with chronic viral hepatitis C.

Vandenbroucke, J. P., Elm, E. V., Altman, D. G., Gøtzsche, P. C., Mulrow, C. D., & Pocock, S. J.. (2007). Strengthening the Reporting of Observational Studies in Epidemiology (STROBE): Explanation and elaboration. *Annals of Internal Medicine*, 147(8), W-163. DOI: 10.7326/0003-4819-147-8-200710160-00010-w1 PMID: 17938389

Wang, F., Liu, Y., Luo, J., & Yang, J. (NA). Explainable machine learning for predicting hepatocellular carcinoma recurrence in hepatitis B patients with digital pathology.

Wiest, R., Albillos, A., Trauner, M., Bajaj, J. S., & Jalan, R. (2018). Corrigendum to "Targeting the gut-liver axis in liver disease" [J Hepatol 67 (2017) 1084–1103]. *Journal of Hepatology*, 68(6), 1336. DOI: 10.1016/j.jhep.2018.03.001 PMID: 29655855

Wu, S., Li, J., Tian, Z., & Zhang, H. (2023). Explainable machine learning for non-invasive diagnosis of fatty liver disease.

Xiang, K., Jiang, B., & Shang, D. (2021). The overview of the deep learning integrated into the medical imaging of liver: A review. *Hepatology International*, 15(4), 868–880. DOI: 10.1007/s12072-021-10229-z PMID: 34264509

Yi, H., Li, Z., Xu, J., Zhang, Y., & Yang, Y. (2020). Topological data analysis combined with machine learning for classification of liver tissue images. *IEEE Access : Practical Innovations, Open Solutions*, 8, 138132–138142.

Yip, T. F., Ma, A. J., Wong, V. S., Tse, Y. K., Chan, H. Y., Yuen, P. C., & Wong, G. H. (2017). Laboratory parameter-based machine learning model for excluding non-alcoholic fatty liver disease (NAFLD) in the general population. *Alimentary Pharmacology & Therapeutics*, 46(4), 447–456. DOI: 10.1111/apt.14172 PMID: 28585725

Yu, L., Kim, S., Moon, S., & Kim, D. S. (2018). Deep learning in non-invasive liver disease diagnosis using medical imaging: A review. *Journal of Medical Imaging and Health Informatics*, 10(7), 1422–1430.

Zhang, H., Sun, C., Li, X., & Zheng, L. (2022). Machine learning-based prediction of liver fibrosis in chronic hepatitis B patients using ultrasound radiomics. *Ultrasound in Medicine & Biology*, 48(10), 2747–2757.

Zhen, S. H., Cheng, M., Tao, Y. B., Wang, Y. F., Juengpanich, S., Jiang, Z. Y., Jiang, Y., Yan, Y., Lu, W., Lue, J., Qian, J., Wu, Z., Sun, J., Lin, H., & Cai, X. J. (2020). Deep learning for accurate diagnosis of liver tumor based on magnetic resonance imaging and clinical data. *Frontiers in Oncology*, 10, 680. DOI: 10.3389/fonc.2020.00680 PMID: 32547939

Zhou, L. Q., Wang, J. Y., Yu, S. Y., Wu, G. G., Wei, Q., Deng, Y. B., Wu, X.-L., Cui, X.-W., & Dietrich, C. F. (2019). Artificial intelligence in medical imaging of the liver. *World Journal of Gastroenterology*, 25(6), 672–682. DOI: 10.3748/wjg.v25.i6.672 PMID: 30783371

Compilation of References

. Elshamy, H. (2021). The Impact of Firm Size on Innovative Activity: An Analysis Based on Egyptian Firm Data. , 40(2), 60–75. DOI: 10.21608/caf.2021.154782

. Herr, H., & Nettekoven, Z. M. (2017). The role of small and medium-sized enterprises in Development. What can be learned from the German experience.

. Joseph, A. (1934). Schumpeter, The theory of economic development: An inquiry into profits, capital, credit, interest, and the business cycle.

A virtual reality of endangered. (2018, April 19). *The Hans India*.https://www.thehansindia.com/posts/index/Young-Hans/2018-04-19/A-virtual-reality-of-endangered-heritage-sites/375029?infinitescroll=1

Abd El-Salam, S. M., Ezz, M. M., Hashem, S., Elakel, W., Salama, R., ElMakhzangy, H., & ElHefnawi, M. (2019). Performance of machine learning approaches on prediction of esophageal varices for Egyptian chronic hepatitis C patients. *Informatics in Medicine Unlocked*, 17, 100267. DOI: 10.1016/j.imu.2019.100267

Abe, H., Yokoi, H., Ohsaki, M., & Yamaguchi, T. (2007, October). Developing an integrated time-series data mining environment for medical data mining. In *Seventh IEEE International Conference on Data Mining Workshops (ICDMW 2007)* (pp. 127-132). IEEE. DOI: 10.1109/ICDMW.2007.47

About Tasharuky. Tasharuky Platform. (n.d.). https://www.tasharuky.com/about-us

Abràmoff, M. D., Lavin, P. T., Birch, M., Shah, N., & Folk, J. C. (2018). Pivotal trial of an autonomous AI-based diagnostic system for detection of diabetic retinopathy in primary care offices. *NPJ Digital Medicine*, 1(1), 39. Advance online publication. DOI: 10.1038/s41746-018-0040-6 PMID: 31304320

Adetunji, J. (2020, December 2).Can virtual tourism save local businesses in tourist destinations?TheConversation.https://theconversation.com/can-virtual-tourism-save-local-businesses-in-tourist-destinations-149537

Adeyeye, D., Egbetokun, A., Opele, J., Oluwatope, O., & Sanni, M. (2018). How Barriers Influence Firms' Search Strategies And Innovative Performance. *International Journal of Innovation Management*, 22(02), 1850011. DOI: 10.1142/S1363919618500111

Agbozo, E., & Balungu, D. M. (2024). Liver Disease Classification-An XAI Approach to Biomedical AI. *Informatica (Vilnius)*, 48(1).

Ahmed Alhaji, A. (2022). Financing as a Major Problem to Small and Medium Agriculture Enterprises in Nigeria. A Fact within the Literature. SSRN Electronic Journal. DOI: 10.2139/ssrn.4084771

Akhlaghi, H., Freeman, S., Vari, C., McKenna, B., Braitberg, G., Karro, J., & Tahayori, B. (2024). Machine learning in clinical practice: Evaluation of an artificial intelligence tool after implementation. *EMA -. Emergency Medicine Australasia*, 36(1), 118–124. Advance online publication. DOI: 10.1111/1742-6723.14325 PMID: 37771067

Alam, A. (2022). Intelligent Tutoring Robots. *International Conference on Advanced Computing and Intelligent Technologies (ICACIT) 2022* (pp. 400-401). Springer. DOI: 10.1007/978-981-19-2980-9_32

Alan, H. (2018, March 29). Lockwood, Richard J. Salvi, Rober F. Burkard (2002). Tinnitus. *The New England Journal of Medicine*, 378(13), 1224–1231. Advance online publication. DOI: 10.1056/NEJMcp1506631

Albayrak, T., Caber, M., & Aksoy, S. (2010). Relationships of the Tangible and IntangibleElements of Tourism Products with OverallCustomer Satisfaction. *International Journal of Trade. Economics and Finance*, 1(2), 140.

Alhorishi, N., Almeziny, M., & Alshammari, R. (2021). Using machine learning to predict early preparation of pharmacy prescriptions at PSMMC - a comparison of four machine learning algorithms. *Acta Informatica Medica*, 29(1), 21. Advance online publication. DOI: 10.5455/aim.2021.29.21-25 PMID: 34012209

Ali, M. M. R., Helmy, Y., Khedr, A. E., & Abdo, A. (2018). Intelligent decision framework to explore and control infection of hepatitis C virus. In *The International Conference on Advanced Machine Learning Technologies and Applications (AMLTA2018)* (pp. 264-274). Springer International Publishing. DOI: 10.1007/978-3-319-74690-6_26

Aljuaid, A., & Anwar, M. (2022). Survey of Supervised Learning for Medical Image Processing. *SN Computer Science*, 3(4), 292. Advance online publication. DOI: 10.1007/s42979-022-01166-1 PMID: 35602289

Alotaibi, A., Alnajrani, L., Alsheikh, N., Alanazy, A., Alshammasi, S., Almusairii, M., Alrassan, S., & Alansari, A. (2023). Explainable ensemble-based machine learning models for detecting the presence of cirrhosis in Hepatitis C patients. *Computation (Basel, Switzerland)*, 11(6), 104. DOI: 10.3390/computation11060104

Al-Radaideh, Q. A., Al-Shawakfa, E., & Al-Najjar, M. I. (2006). Mining Student Data Using Decision Trees. Retrieved from https://www.researchgate.net/publication/242076740_Mining_Student_Data_Using_Decision_Trees

Alsharkawy, N. (2024, May 9). *Crowned by a Presidential Decree: Establishment of the National Alliance for Developmental Civic Work*. Al-Marsad. https://marsad.ecss.com.eg/81575/

Alshial, E. E., Abdulghaney, M. I., Wadan, A.-H. S., Abdellatif, M. A., Ramadan, N. E., Suleiman, A. M., Waheed, N., Abdellatif, M., & Mohammed, H. S. (2023). Mitochondrial dysfunction and neurological disorders: A narrative review and treatment overview. *Life Sciences*, p. *334*, 122257. DOI: 10.1016/j.lfs.2023.122257

Altmeyer, M., Schubhan, M., Krüger, A., & Lessel, P. (2021). *A long-term investigation on the effects of (personalized) gamification on course participation in a gym*. http://ceur-ws.org

Alyahya, M., & McLean, G. (2021). Examining Tourism Consumers' Attitudes and the Role of Sensory Information in Virtual Reality Experiences of a Tourist Destination. *Journal of Travel Research*. Advance online publication. DOI: 10.1177/00472875211037745

Amin, A. A., Sajid Iqbal, M., & Hamza Shahbaz, M. (2024). Development of Intelligent Fault-Tolerant Control Systems with Machine Learning, Deep Learning, and Transfer Learning Algorithms: A Review. In *Expert Systems with Applications* (Vol. 238). DOI: 10.1016/j.eswa.2023.121956

Ansari, S. Khot1, R., Shaikh, Z., Yadav, P., & Pirani, Z. (2022). A Comprehensive Review of Gamification in Healthcare: Incentives in Mobile Healthcare App. *International Research Journal of Engineering and Technology*. www.irjet.net

Apa, R., De Marchi, V., Grandinetti, R., & Sedita, S. R. (2020). University-SME collaboration and innovation performance: The role of informal relationships and absorptive capacity. *The Journal of Technology Transfer*, 46(4), 961–988. DOI: 10.1007/s10961-020-09802-9

Apoki, U. C., Hussein, A. A., Al-Chalabi, H. M., Badica, C., & Mocanu, M. L. (2022). The Role of Pedagogical Agents in Personalised Adaptive Learning: A Review. *E-learning Personalization Systems and Sustainable Education*. DOI: 10.3390/su14116442

Archibugi, D., & Filippetti, A. (2018). *The handbook of global science, technology, and innovation*. Wiley.

Arthur Boothroyd, (2014). Hearing aid accessories for adults: the remote FM microphone DOI: 10.1097/01.AUD.0000111260.46595.EC

Arundel, A. (1997). Enterprise strategies and barriers to innovation. In *Innovation Measurement and Policies* (pp. 101–108). European Commission.

Ashworth, G. J., & Voogt, H. (1990). *Selling the city*.

Audretsch, D. B., Belitski, M., Caiazza, R., & Phan, P. (2023). Collaboration strategies and SME innovation performance. *Journal of Business Research*, 164, 114018. DOI: 10.1016/j.jbusres.2023.114018

Audrin, B., Audrin, C., & Salamin, X. (2024). Digital skills at work – Conceptual development and empirical validation of a measurement scale. *Technological Forecasting & Social Change*. *Technological Forecasting and Social Change*, 202, 123279. DOI: 10.1016/j.techfore.2024.123279

Audrin, C., & Audrin, B. (2022). Key factors in digital literacy in learning and education: A systematic literature review using text mining. *Education and Information Technologies*, 27(6), 7395–7419. DOI: 10.1007/s10639-021-10832-5

Aung, T., & MPTM, W. K. Seroprevalence of Hepatitis B and C Virus Infections and Related Risk Factors of Hepatitis B Virus Infection among Newly Diagnosed Tuberculosis Patients at National Tuberculosis Program (Central), Mandalay Office, Myanmar.

AutoTutor. (2023). https://adulted.autotutor.org/course/view.php?id=2&section=1. Retrieved from https://csal.autotutor.org

Babu, G., & Ontario, W. (2023). *Managing the Strategic Transformation of Higher Education through Artificial Intelligence*. Change Management and Innovation Strategies for Digital–Quantum Business Transformation., DOI: 10.3390/admsci13090196

Bahrman, L., Fontaine, M., Roux, J. L., & Richard, G. (2024). Speech dereverberation constrained on room impulse response characteristics. *arXiv preprint arXiv:2407.08657*.

Bain, D., & Kleinknecht, A. (Eds.). (1993). *New concepts in innovation output measurement*. Springer., https://books.google.com.eg/books?hl=ar&lr=&id=Bv6-DAAAQBAJ&oi=fnd&pg=PA1&ots=bMFH1NjP9g&sig=ryPMFcPZN9W9hQFOdX6NxDtaGIs&redir_esc=y#v=onepage&q&f=false

Bakkar, N., Kovalik, T., Lorenzini, I., Spangler, S., Lacoste, A., Sponaugle, K., Ferrante, P., Argentinis, E., Sattler, R., & Bowser, R. (2018). Artificial intelligence in neurodegenerative disease research: Use of IBM Watson to identify additional RNA-binding proteins altered in amyotrophic lateral sclerosis. *Acta Neuropathologica*, 135(2), 227–247. Advance online publication. DOI: 10.1007/s00401-017-1785-8 PMID: 29134320

Baldwin, J., & Lin, Z. (2002). Impediments to advanced technology adoption for Canadian manufacturers. *Research Policy*, 31(1), 1–18. DOI: 10.1016/S0048-7333(01)00110-X

Bal, W., & Czalczynska-Podolska, M. (2020). The Stages of the Cultural Landscape Transformation of Seaside Resorts in Poland against the Background of the Evolving Nature of Tourism. *Land (Basel)*, 9(2), 55. DOI: 10.3390/land9020055

Barakat, N. H., Barakat, S. H., & Ahmed, N. (2019). Prediction and staging of hepatic fibrosis in children with hepatitis c virus: A machine learning approach. *Healthcare Informatics Research*, 25(3), 173. DOI: 10.4258/hir.2019.25.3.173 PMID: 31406609

Barann, B., Hermann, A., Cordes, A. K., Chasin, F., & Becker, J. (2019). Supporting Digital Transformation in Small and Medium-sized Enterprises: A procedure model involving publicly funded support units. *Proceedings of the Annual Hawaii International Conference on System Sciences, January*, 4977–4986. DOI: 10.24251/HICSS.2019.598

Barbieri, L., Bragoli, D., Cortelezzi, F., & Marseguerra, G. (2019). Public Funding and Innovation Strategies. Evidence from Italian SMEs. *International Journal of the Economics of Business*, 27(1), 111–134. DOI: 10.1080/13571516.2019.1664834

Baskin, K. (2018, August 16). *5 reasons companies struggle with Digital Transformation*. MIT Sloan. https://mitsloan.mit.edu/ideas-made-to-matter/5-reasons-companies-struggle-digital-transformation

Baum, T. (1998). Taking the exit route: Extending the tourism area life cycle model. *Current Issues in Tourism*, 1(2), 167–175. DOI: 10.1080/13683509808667837

Becheikh, N., & Bouaddi, M. (2022). Macro barriers to business and innovation in developing economies: Evidence from Egyptian, Moroccan, and Tunisian SMEs. Canadian Journal of Development Studies. *Canadian Journal of Development Studies*, 44(2), 151–171. DOI: 10.1080/02255189.2022.2102464

Beck, J., & Egger, R. (2018). Emotionalise Me: Self-reporting and Arousal Measurements in Virtual Tourism Environments. In Stangle, B., & Pesonen, J. (Eds.), (pp. 3–15). Information and Communication Technologies in Tourism. [Electronic resource], DOI: 10.1007/978-3-319-72923-7_1

Benitez, J., Arenas, A., Castillo, A., & Esteves, J. (2022). Impact of digital leadership capability on innovation performance: The role of platform Digitization Capability. *Information &amp. Information & Management*, 59(2), 103590. DOI: 10.1016/j.im.2022.103590

Berahmand, K., Daneshfar, F., Salehi, E. S., Li, Y., & Xu, Y. (2024). Autoencoders and their applications in machine learning: A survey. *Artificial Intelligence Review*, 57(2), 28. Advance online publication. DOI: 10.1007/s10462-023-10662-6

Bessant, J., Tidd, J., & Pavitt, K. (2020). *Managing innovation: Integrating technological, market, and organizational change* (6th ed.). Wiley.

Bhuiyan, M. I., & Haque, A. (2024). Role of ngos in providing available and affordable health care services to the slum people in Dhaka. *Clinical Epidemiology and Global Health*, 25, 101478. DOI: 10.1016/j.cegh.2023.101478

Bigliardi, B., & Galati, F. (2016). Which factors hinder the adoption of open innovation in SMEs? *Technology Analysis and Strategic Management*, 28(8), 869–885. DOI: 10.1080/09537325.2016.1180353

Björk, J., Frishammar, J., & Sundström, L. (2023). Measuring innovation effectively—Nine critical lessons. *Research Technology Management*, 66(2), 17–27. DOI: 10.1080/08956308.2022.2151232

Blind, K. (2012). The influence of regulations on innovation: A quantitative assessment for OECD countries. *Research Policy*, 41(2), 391–400. DOI: 10.1016/j.respol.2011.08.008

Bloch, C., & Bugge, M. M. (2013). Public sector innovation—From theory to measurement. *Structural Change and Economic Dynamics*, 27, 133–145. DOI: 10.1016/j.strueco.2013.06.008

Bloomberg. J. (2018). *Digitization, Digitalization, And Digital Transformation: Confuse Them At Your Peril*. Forbes.Com. https://www.forbes.com/sites/jasonbloomberg/2018/04/29/digitization-digitalization-and-digital-transformation-confuse-them-at-your-peril/

Božić, L., & Rajh, E. (2016, January). The factors constraining innovation performance of SMEs in Croatia. *Ekonomska Istrazivanja*, 29(1), 314–324. DOI: 10.1080/1331677X.2016.1168040

Bozkurt, A., Xiao, J., Lambert, S., Pazurek, A., Crompton, H., Koseoglu, S., . . . Honeychurch, S. (2023). Speculative Futures on ChatGPT and Generative Artificial Intelligence (AI): A Collective Reflection from the Educational Landscape. *ODU Digital Commons*. Retrieved from https://digitalcommons.odu.edu/cgi/viewcontent.cgi?article=1212&context=teachinglearning_fac_pubs

Bozkurt, A. (2023, March 21). Generative artificial intelligence (AI) powered conversational educational agents: The inevitable paradigm shift. *Asian Journal of Distance Education*, 18(1), 4. Retrieved September 9, 2023, from http://www.asianjde.com/ojs/index.php/AsianJDE/article/view/718

Bozkurt, A., & Sharma, R. C. (2023). Challenging the Status Quo and Exploring the New Boundaries in the Age of Algorithms: Reimagining the Role of Generative AI in Distance Education and Online Learning. *Asian Journal of Distance Education*, 18(1). http://asianjde.com/ojs/index.php/AsianJDE/article/view/714/397

Breiman, L. (2001). Random forests. *Machine Learning*, 45(1), 5–32. DOI: 10.1023/A:1010933404324

Breiman, L., Friedman, J., Olshen, R., & Stone, C. (1984). *Classification and regression trees*. Wadsworth.

Brownlee, J. (2022). *Machine learning mastery*. Machine Learning Mastery.

Brown, T. (2008). Design thinking. *Harvard Business Review*, 86(6), 84. PMID: 18605031

Brundtland, G. H. (1985). World Commission on Environment and Development. *Environmental Policy and Law*, 14(1), 26–30. DOI: 10.1016/S0378-777X(85)80040-8

Bryan, B. A., Crossman, N. D., King, D., & Meyer, W. S. (2011). Landscape Futures Analysis: Assessing the Impacts of Environmental Targets under Alternative Spatial Policy Options and Future Scenarios. *Environmental Modelling & Software*, 26(1), 83–91. DOI: 10.1016/j.envsoft.2010.03.034

Brynjolfsson, E., & McAfee, A. (2014). *The second machine age: Work, progress, and prosperity in a time of brilliant technologies*. WW Norton & Company.

Budeanu, A., Miller, G., Moscardo, G., & Ooi, C. S. (2016). Sustainable tourism, progress, challenges and opportunities: An introduction. *Journal of Cleaner Production*, 111, 285–294. DOI: 10.1016/j.jclepro.2015.10.027

Burns, A. (2005). Medication Therapy Management in Community Pharmacy Practice: Core Elements of an {MTM} Service (Version 1.0): American Pharmacists Association and National Association of Chain Drug Stores Foundation. *Journal of the American Pharmacists Association*, 45(5), 573–579. DOI: 10.1331/1544345055001256

Burt, M., & Louw, C. (2019). Virtual reality enhanced roller coasters and the future of entertainment – audience expectations. *World Leisure Journal*, 61(3), 183–199. DOI: 10.1080/16078055.2019.1639274

Butler, R. (2024). Tourism destination development: The tourism area life cycle model. *Tourism Geographies*, •••, 1–9. DOI: 10.1080/14616688.2024.2325932

Butler, R. W. (1980). The concept of tourism area cycle of evolution: Implications for management of resources. *The Canadian Geographer. Geographe Canadien*, 24(1), 5–12. DOI: 10.1111/j.1541-0064.1980.tb00970.x

Butler, T. (2020). What's Next in the Digital Transformation of Financial Industry? *IT Professional*, 22(1), 29–33. DOI: 10.1109/MITP.2019.2963490

Caciora, T., Herman, G., Alexandru, I., Baias, S., Ilie , D., Josan, I., & Hodor, N. (2021). The Use of Virtual Reality to Promote Sustainable Tourism: A Case Study of Wooden Churches Historical Monuments from Romania. *Remote Sensing (Basel)*, 13(9), 1758. DOI: 10.3390/rs13091758

Calantone, R. J., Cavusgil, S., & Zhao, Y. (2002, September). Learning orientation, firm innovation capability, and firm performance. *Industrial Marketing Management*, 31(6), 515–524. DOI: 10.1016/S0019-8501(01)00203-6

California State University. (2024). *Admissions and Records csunny Chat Support*. Retrieved from Division of Academic Affairs: https://www.csun.edu/admissions-records/csunny

Cameron, K. S., & Quinn, R. E. (2011). *Diagnosing and changing organizational culture: Based on the competing values framework*. John Wiley & Sons.

Caputo, J. M. (2015). *Dietary Saturated And Unsaturated Fatty Acids On Hepatic And Subcutaneous Tissues Intermodulation*. Gene Expression In Transition Dairy Goats, And Influence Of Maternal Diets On Kids'immune Response.

Cardona, M. A., Rodriguez, R. J., & Ishmael, K. (2023). Artificial Intelligence and the Future of Teaching and Learning. Office of Educational Technology, U.S Department of Education. Retrieved from file:///C:/Users/ldospinescu/Desktop/ai-report.pdf

CBS News. (2024). Are hearing aids expensive? Here's what to know https://www.cbsnews.com/essentials/how-much-do-hearing-aids-cost

Chalasani, S. H., Syed, J., Ramesh, M., Patil, V., & Pramod Kumar, T. M. (2023). Artificial intelligence in the field of pharmacy practice: A literature review. In *Exploratory Research in Clinical and Social Pharmacy* (Vol. 12). DOI: 10.1016/j.rcsop.2023.100346

Chan, S., Reddy, V., Myers, B., Thibodeaux, Q., Brownstone, N., & Liao, W. (2020). Machine Learning in Dermatology: Current Applications, Opportunities, and Limitations. In *Dermatology and Therapy* (Vol. 10, Issue 3). DOI: 10.1007/s13555-020-00372-0

Chanias, S., Myers, M. D., & Hess, T. (2019). Digital transformation strategy making in pre-digital organizations: The case of a financial services provider. *The Journal of Strategic Information Systems*, 28(1), 17–33. DOI: 10.1016/j.jsis.2018.11.003

Chassignol, M., Khoroshavin, A., Klimova, A., & Bilyatdinova, A. (2018). Artificial Intelligence trends in education: a narrative overview. *Procedia Computer Science*. Retrieved from https://www.sciencedirect.com/science/article/pii/S1877050918315382

Chaudhry, M. A., & Kazim, E. (2021). Artificial Intelligence in Education (AIEd): A high-level academic and industry note 2021. *AI and Ethics*. Advance online publication. DOI: 10.1007/s43681-021-00074-z PMID: 34790953

Cheng, J. Y. K., & Wong, G. L. H. (2017). Advances in the diagnosis and treatment of liver fibrosis. *Hepatoma Research*, 3(8), 156–169. DOI: 10.20517/2394-5079.2017.27

Chen, T., & Guestrin, C. (2016). XGBoost: A scalable tree boosting system. *Proceedings of the 22nd ACM SIGKDD International Conference on Knowledge Discovery and Data Mining*, 785-794.

Chesbrough, H. W. (2003). *Open innovation: The new imperative for creating and profiting from technology*. Harvard Business Press.

Cho, Y.-H., & Fesenmaier, D. (2001). A new paradigm for tourism and electronic commerce: experience marketing using the virtual tour. In Buhalis, D., & Laws, E. (Eds.), *Tourism Distribution Channels: Practices, Issues and Transformation* (pp. 351–370). Continuum.

Cho, Y.-H., Wang, Y., & Fesenmaier, D. (2002). Searching for experiences: The web-based virtual tour in tourism marketing. *Journal of Travel & Tourism Marketing*, 12(4), 1–17. DOI: 10.1300/J073v12n04_01

Christensen, C. M., Raynor, M. E., & McDonald, R. (2018). What is disruptive innovation? *Harvard Business Review*, 93(12), 44–53. PMID: 17183796

Chung, G. H., Choi, J. N., & Du, J. (2017). Tired of innovations? Learned helplessness and fatigue in the context of continuous streams of innovation implementation. *Journal of Organizational Behavior*, 38(7), 1130–1148. DOI: 10.1002/job.2191

Cirera, X., & Muzi, S. (2020). Measuring innovation using firm-level surveys: Evidence from developing countries. *Research Policy*, 49(3), 103912. DOI: 10.1016/j.respol.2019.103912

Clark, C., Cluver, M., Fishman, T., & Kunkel, D. (2024, April 19). *Government and Public Services.* Retrieved from Latest Trends in Higher Education: https://www2.deloitte.com/us/en/insights/industry/public-sector/latest-trends-in-higher-education.html

Cleveland Clinic. (2022). Hearing aid styles https://my.clevelandclinic.org/health/articles/5122-hearing-aid-styles

Collins, B. (2023, October 10). Harnessing the Potential of Learning Analytics Across the University. *Wiley University Services*. Retrieved from https://universityservices.wiley.com/potential-for-higher-education-learning-analytics/

Cornia, A., Sehl, A., & Nielsen, R. K. (2020). 'We no longer live in a time of separation': A comparative analysis of how editorial and commercial integration became a norm. *Journalism*, 21(2), 172–190. DOI: 10.1177/1464884918779919

Cortes, C., & Vapnik, V. (1995). Support-vector networks. *Machine Learning*, 20(3), 273–297. DOI: 10.1007/BF00994018

Coursera. (2023). Retrieved from In the context of higher education, the mentioned tools have diverse applications across several categories. Coursera, Top Hat, and Smart Sparrow contribute to curriculum design, content development, pedagogical strategies, and learning environments. Grad

Cricelli, L., Greco, M., & Grimaldi, M. (2016). ASSESSING THE OPEN INNOVATION TRENDS BY MEANS OF THE EUROSTAT COMMUNITY INNOVATION SURVEY. *International Journal of Innovation Management*, 20(03), 1650039. DOI: 10.1142/S1363919616500390

Crossan, M. M., & Apaydin, M. (2010). A multi-dimensional framework of organizational innovation: A systematic review of the literature. *Journal of Management Studies*, 47(6), 1154–1191. DOI: 10.1111/j.1467-6486.2009.00880.x

Culbertson, M. J. (2015, June 19). Bayesian Networks in Educational Assessment. DOI: 10.1177/0146621615590401

Cunningham, L. L., & Tucci, D. L. (2017). Hearing loss in adults. *The New England Journal of Medicine*, 377(25), 2465–2473.

D'Este, P., Iammarino, S., Savona, M., & von Tunzelmann, N. (2012). What hampers innovation? Revealed barriers versus deterring barriers. *Research Policy*, 41(2), 482–488. DOI: 10.1016/j.respol.2011.09.008

Damanpour, F., & Schneider, M. (2006). Phases of the adoption of innovation in organizations: Effects of environment, organization and top managers. *British Journal of Management*, 17(3), 215–236. DOI: 10.1111/j.1467-8551.2006.00498.x

Damiani, G., Altamura, G., Zedda, M., Nurchis, M. C., Aulino, G., Heidar Alizadeh, A., Cazzato, F., Della Morte, G., Caputo, M., Grassi, S., & Oliva, A. (2023). Potentiality of algorithms and artificial intelligence adoption to improve medication management in primary care: A systematic review. *BMJ Open*, 13(3), e065301. Advance online publication. DOI: 10.1136/bmjopen-2022-065301 PMID: 36958780

Dana, P. (2024). Reported Unmet Hearing Aid Need in Older People with Dementia: The US National Core Indicators Survey. *Journal of the American Medical Directors Association*, 25(5), 853–859. Advance online publication. DOI: 10.1016/j.jamda.2024.03.107 PMID: 38643971

Dataset, O. (2012). Available online: https://archive. ics. uci. edu/ml/datasets. *OPPORTUNITY+ Activity+ Recognition (accessed on 19 November 2015).*

Davé, P. (2023, November 28). *Emerging Technologies*. Retrieved from Columbia University: https://etc.cuit.columbia.edu/news/revolutionary-intersection-ai-and-immersive-learning-higher-education

Dawa Chyophel Lepcha, Bhawna Goyal, Ayush Dogra, Vishal Goyal, (2023). Image super-resolution: A comprehensive review, recent trends, challenges and applications DOI: 10.1016/j.inffus.2022.10.007

de Lucca Schiavon, L., Narciso-Schiavon, J. L., & de Carvalho-Filho, R. J. (2014). Non-invasive diagnosis of liver fibrosis in chronic hepatitis C. *World journal of gastroenterology. World Journal of Gastroenterology*, 20(11), 2854. DOI: 10.3748/wjg.v20.i11.2854 PMID: 24659877

Dee, J., Dhuhaibawi, N., & Hayden, J. C. (2023). A systematic review and pooled prevalence of burnout in pharmacists. In *International Journal of Clinical Pharmacy* (Vol. 45, Issue 5). DOI: 10.1007/s11096-022-01520-6

Del Rio-Bermudez, C., Medrano, I. H., Yebes, L., & Poveda, J. L. (2020). Towards a symbiotic relationship between big data, artificial intelligence, and hospital pharmacy. *Journal of Pharmaceutical Policy and Practice*, 13(1), 75. Advance online publication. DOI: 10.1186/s40545-020-00276-6 PMID: 33292570

Desai, A. N. (2020, June 3). Artificial Intelligence: Promise, Pitfalls, and Perspective. [JAMA]. *Journal of the American Medical Association*, 323(24), 2448. Advance online publication. DOI: 10.1001/jama.2020.8737 PMID: 32492093

Desai, P. R., Desai, P. N., Ajmera, K. D., & Mehta, K. (2014). A Review Paper onOculus Rift-A Virtual [IJETT]. *International Journal of Engineering Trends and Technology*, 13(4), 175–179. DOI: 10.14445/22315381/IJETT-V13P237

Deterding, S., Dixon, D., Khaled, R., & Nacke, L. (2011). From game design elements to gamefulness: Defining “gamification.” *Proceedings of the 15th International Academic MindTrek Conference: Envisioning Future Media Environments, MindTrek 2011, September*, 9–15. DOI: 10.1145/2181037.2181040

Diener, F., & Špaček, M. (2021). Digital transformation in banking: A managerial perspective on barriers to change. *Sustainability (Basel)*, 13(4), 1–26. DOI: 10.3390/su13042032

Dodgson, M., Gann, D. M., & Phillips, N. (2014). *The Oxford handbook of innovation management*. Oxford University Press. DOI: 10.1093/oxfordhb/9780199694945.001.0001

Dragulescu, D., & Albu, A. (2007, May). Expert system for medical predictions. In *2007 4th International Symposium on Applied Computational Intelligence and Informatics* (pp. 123-128). IEEE. DOI: 10.1109/SACI.2007.375496

Drgas, S. (2024). Speech intelligibility prediction using generalized ESTOI with fine-tuned parameters. *Speech Communication*, 149, 103068. Advance online publication. DOI: 10.1016/j.specom.2024.103068

Drucker, P. F. (1985). *Innovation and entrepreneurship: Practice and principles*. Harper & Row.

Dua, D., & Graff, C. (2017). UCI machine learning repository.

Echo360. (2023). Retrieved from https://echo360.com/

Edelman, B. (2011). Google, Inc. Harvard Business School Case Study, 910-036.

Edison, H., Smørsgård, N. M., Wang, X., & Abrahamsson, P. (2018). Lean internal startups for software product innovation in large companies: Enablers and inhibitors. *Journal of Systems and Software*, 135, 69–87. DOI: 10.1016/j.jss.2017.09.034

Edvinsson, L., & Malone, M. S. (1997). *Intellectual capital: Realizing your company's true value by finding its hidden brainpower*. HarperBusiness.

Eisenhardt, K. M., & Graebner, M. E. (2007). Theory building from cases: Opportunities and challenges. *Academy of Management Journal*, 50(1), 25–32. DOI: 10.5465/amj.2007.24160888

El-gharbawy, M. S. H., A. Ragab, A., A. Ragheb, M., & Farouk, M. (2024). The Effect of Information Technology on Innovative Performance with Mediation Role of Process Innovation Capability: Evidence from Egyptian SMEs. Journal of Business and Management Sciences, 12(2), 67–88. https://doi.org/10.126

Elgohary, E. (2022). The role of digital transformation in Sustainable Development in Egypt. *The International Journal of Informatics. Media and Communication Technology*, 4(1), 71–106. DOI: 10.21608/ijimct.2022.219953

ElNaggar, R. A., & ElSayed, M. F. (2023). Drivers of business model innovation in micro and small enterprises: Evidence from Egypt as an emerging economy. *Future Business Journal*, 9(1), 4. Advance online publication. DOI: 10.1186/s43093-022-00180-2

El-Said, O., & Aziz, H. (2021). Virtual Tours a Means to an End: An Analysis of Virtual Tours' Role in Tourism Recovery Post COVID-19. *Journal of Travel Research*. Advance online publication. DOI: 10.1177/0047287521997567

Emily, M. (2024). Age-based disparities in hearing loss diagnosis and treatment in the United States population. *American Journal of Otolaryngology*, 45(5), 104403. Advance online publication. DOI: 10.1016/j.amjoto.2024.104403

Epstein, S., & Reilly, J. S. (1989). Sensorineural hearing loss. *Pediatric Clinics of North America*, 36(6), 1501–1520.

Erenay, O., & Hashemipour, M. (2003). Virtual Reality in Engineering Education: A CIM Case Study. *The Turkish Online Journal of Educational Technology*, 2(2), 51–56. Retrieved May 2014, from http://www.tojet.net/articles/v2i2/228.pdf

European Federation of Hard Hearing People (EFHOH). EU (2022). State of Provision of Hearing Aids in Europe, 2022 Report https://efhoh.org/wp-content/uploads/2022/11/Hearing-Aids-Reimbursement-2022-edited-version-final-for-publication.pdf

Eurostat (2015). Statistics explained - Community innovation survey (CIS). Available online at https://ec.europa.eu/eurostat/statistics-explained/index.php?Title=Glossary:Community_innovation_survey_(CIS)

Ewertsen, H. W. (1956). Hearing-Aid Evaluation. *A.M.A. Archives of Otolaryngology*, 64(6), 520–525.

Fagerberg, J., & Verspagen, B. (2009). Innovation studies—The emerging structure of a new scientific field. *Research Policy*, 38(2), 218–233. DOI: 10.1016/j.respol.2008.12.006

Fanelli, R. M. (2021). Barriers to Adopting New Technologies within Rural Small and Medium Enterprises (SMEs). *Social Sciences (Basel, Switzerland)*, 10(11), 430. DOI: 10.3390/socsci10110430

Fazlollahi, A. M., Yilmaz, R., Winkler-Schwartz, A., Mirchi, N., Ledwos, N., Bakhaidar, M., . . . Del Maestro, R. F. (2023, September 19). AI in Surgical Curriculum Design and Unintended Outcomes for Technical Competencies in Simulation Training. *JAMA Network*. Retrieved September 26, 2023, from https://jamanetwork.com/journals/jamanetworkopen/article-abstract/2809667

Ferrucci, D., Levas, A., Bagchi, S., Gondek, D., & Mueller, E. T. (2013). Watson: Beyond jeopardy! In *Artificial Intelligence* (Vols. 199–200). DOI: 10.1016/j.artint.2012.06.009

Firat, M. (2023). What ChatGPT means for universities: Perceptions of scholars and students. *Journal of Applied Learning*, 6(1). https://journals.sfu.ca/jalt/index.php/jalt/index

Firican, D. A. (2023). Digital Transformation and Digital Culture: A Literature Review of the Digital Cultural Attributes to Enable Digital Transformation. *Proceedings of the International Conference on Business Excellence, 17*(1), 791–799. DOI: 10.2478/picbe-2023-0073

Fitzgerald, M., Kruschwitz, N., & Welch, M. (2013). Embracing Digital Technology: A New Strategic Imperative, Capgemini Consulting Worldwide [Online]. *MIT Sloan Management Review*, 55(1), 1–13.

Flynn, A. J., Fortier, C., Maehlen, H., Pierzinski, V., Runnebaum, R., Sullivan, M., Wagner, J., & Stevenson, J. G. (2021). A strategic approach to improving pharmacy enterprise automation: Development and initial application of the Autonomous Pharmacy Framework. In *American Journal of Health-System Pharmacy* (Vol. 78, Issue 7). DOI: 10.1093/ajhp/zxab001

Forbes Technology Council, & Forbes Expert Panel. (2021, June 15). *13 Industry Experts Share Reasons Companies Fail At Digital Transformation*. https://www.forbes.com/sites/forbestechcouncil/2021/06/15/13-industry-experts-share-reasons-companies-fail-at-digital-transformation/

Forbes. (2021). The world's most valuable brands. Retrieved from https://www.forbes.com/powerful-brands/list/

Fortune Business Insights, E. U. (2024). Europe Hearing Aids Market https://www.fortunebusinessinsights.com/europe-hearing-aids-market-107908

Frent, C. (2016). An overview on the negative impacts of tourism.Journal of tourism – studies and research in tourism, (22).

Friedrich, J., Becker, M., Kramer, F., Wirth, M., & Schneider, M. (2020). Incentive design and gamification for knowledge management. *Journal of Business Research, 106*(November 2017), 341–352. DOI: 10.1016/j.jbusres.2019.02.009

Fritz, F., Susperregui, A., & Linaza, M. T. (2005) Enhancing Cultural Tourism experiences with Augmented Reality Technologies, The 6th International Symposium on Virtual Reality, Archaeology and Cultural Heritage VAST 2005, http://public-repository.epochnet.org/publications/VAST2005/shortpapers/short2005.pdf, accessed: May 2014.

Fritz, F., Susperregui, A., & Linaza, M.T., op.cit, p.2

Fuchs, C., & Hess, T. (2018). Becoming Agile in the Digital Transformation: The Process of a Large-Scale Agile Transformation. *ICIS 2018 Proceedings*. 19. https://aisel.aisnet.org/icis2018/innovation/Presentations/19

Garcia, R., & Calantone, R. (2018). A critical look at technological innovation typology and innovativeness terminology: A literature review. *Journal of Product Innovation Management*, 19(2), 110–132. DOI: 10.1111/1540-5885.1920110

Gates, G. A., & Miyamoto, R. T. (2003). Cochlear implants. *The New England Journal of Medicine*, 349(5), 421–423.

Gault, F. (2018). Defining and measuring innovation in all sectors of the economy. *Research Policy*, 47(3), 617–622. DOI: 10.1016/j.respol.2018.01.007

Gault, F. (2018). *Handbook of innovation indicators and measurement*. Edward Elgar Publishing.

Gegung, Eunike. (2021). Jurnal Kepariwisataan Indonesia 15 (1) (2021) THE USE OF VIRTUAL REALITY TO INCREASE TOURISM DESTINATION SUSTAINABILITY AND HOW USERS PERCEIVE THE AUTHENTICITY OF VR EXPERIENCES Jurnal Kepariwisataan Indonesia 15 (1) (2021). 15. 9-15.

Georgia Institute of Technology. (2020, January 24). *Research*. Retrieved from Georgia Tech Research: https://research.gatech.edu/jill-watson-ai-pioneer-education-turns-4

Georgia Institute of Technology. (2024). *GVU Center*. Retrieved from Georgia Tech: https://gvu.gatech.edu/research/projects/virtual-ecological-research-assistant-vera

Georgieva, M. (2023, June 2023). Retrieved from https://www.scalahosting.com/blog/moodle-vs-blackboard-head-to-head-comparison/

Ghazal, T. M. (2021). Hep-pred: Hepatitis c staging prediction using fine gaussian svm. *Computers, Materials & Continua*, 69(1), 191–203. DOI: 10.32604/cmc.2021.015436

Ghazi, R. A. (2021). *Civil Society Organizations' Support for Vulnerable Youth in Transition: a case study in Egypt* [Thesis, the American University in Cairo]. AUC Knowledge Fountain. https://fount.aucegypt.edu/retro_etds/2555

Godin, B. (2015). *Innovation contested: The idea of innovation over the centuries*. Routledge., DOI: 10.4324/9781315855608

Goetz, L. H., & Schork, N. J. (2018). Personalized medicine: motivation, challenges, and progress. In *Fertility and Sterility* (Vol. 109, Issue 6). DOI: 10.1016/j.fertnstert.2018.05.006

Goffin, K., & Mitchell, R. (2017). *Innovation management: Effective strategy and implementation* (3rd ed.). Macmillan International Higher Education. DOI: 10.1057/978-1-137-37344-1

Gonçalves, L. B., Vellasco, M. M., Pacheco, M. A. C., & de Souza, F. J.Flavio Joaquim de Souza. (2006). Inverted hierarchical neuro-fuzzy BSP system: A novel neuro-fuzzy model for pattern classification and rule extraction in databases. *IEEE Transactions on Systems, Man, and Cybernetics. Part C, Applications and Reviews*, 36(2), 236–248. DOI: 10.1109/TSMCC.2004.843220

Gong, C., & Ribiere, V. (2021). Developing a unified definition of digital transformation. *Technovation*, 102, 102217. Advance online publication. DOI: 10.1016/j.technovation.2020.102217

Gopalakrishnan, S., & Damanpour, F. (1997). A review of innovation research in economics, sociology and technology management. *Omega*, 25(1), 15–28. DOI: 10.1016/S0305-0483(96)00043-6

Gradescope. (2023). Retrieved from https://www.gradescope.com/

Graham, M., & Mann, L. (2013). Imagining a silicon savannah? Technological and conceptual connectivity in Kenya's BPO and software development sectors. *The Electronic Journal on Information Systems in Developing Countries*, 56(1), 1–19. DOI: 10.1002/j.1681-4835.2013.tb00396.x

Green economy. (n.d.). United Nations Environment Programme. https://www.unep.org/regions/asia-and-pacific/regional-initiatives/supporting-resource-efficiency/green

Griffin, A. (2015, October 8). *Independent UK*. Retrieved from https://www.independent.co.uk/tech/stephen-hawking-artificial-intelligence-

Grimalt-Álvaro, C., & Usart, M. (2023, April 18). Sentiment analysis for formative assessment in higher education: a systematic literature review. *Journal of Computing in Higher Education*. Retrieved from https://link.springer.com/article/10.1007/s12528-023-09370-5

Gu, H., & Deng, X. (2022, May 23). Markov Model-Based Curriculum Association Classification Model and Student Achievement Prediction. DOI: 10.1155/2022/3047208

Guandalini, I. (2022). Sustainability through digital transformation: A systematic literature review for research guidance. *Journal of Business Research*, 148, 456–471. DOI: 10.1016/j.jbusres.2022.05.003

Gupta, A. K. (2016). *Grassroots innovation: Minds on the margin are not marginal minds*. Random House India.

Guttentag, D. A. (2010). Virtual reality: Applications and implications for tourism. *Tourism Management*, 31(5), 637–651. DOI: 10.1016/j.tourman.2009.07.003

Guzmán, G. M., Del Carmen Alzate, M., & Serna, L. (2018). Innovation and competitiveness in SMEs: The local experience in Medellin, Colombia. *Journal of Innovation & Knowledge*, 3(2), 60–67.

Guzmán, G., Salazar, M., & Lucio, J. (2018). Measuring innovation in developing countries: Challenges and opportunities. In Khan, S., & Lutz, M. (Eds.), *Handbook of innovation in public services* (pp. 123–138). Edward Elgar Publishing.

H., R., J., C., M., M., & A., O. (2018). Assessing pharmacists' acceptance of prescription dispensing data infographic to initiate pharmacist-prescriber collaboration. In *Journal of the American Pharmacists Association* (Vol. 58, Issue 3).

Habehh, H., & Gohel, S. (2021). Machine Learning in Healthcare. *Current Genomics*, 22(4), 291–300. Advance online publication. DOI: 10.2174/1389202922666210705124359 PMID: 35273459

Hagberg, J., Sundstrom, M., & Egels-Zandén, N. (2016). The digitalization of retailing: An exploratory framework. *International Journal of Retail & Distribution Management*, 44(7), 694–712. DOI: 10.1108/IJRDM-09-2015-0140

Hair, J. F.Jr, Babin, W. J., Money, P. H., & Mokhtar, C. M. (2019). *Essentials of business research methods* (8th ed.). Pearson Education Limited. DOI: 10.4324/9780429203374

Hall, B. H., & Harhoff, D. (2018). Recent research on the economics of patents. *Annual Review of Economics*, 4(1), 541–565. DOI: 10.1146/annurev-economics-080511-111008

Halton, C. (2021, August 26). Green economics. *Investopedia.*https://www.investopedia.com/terms/g/greeneconomics.asp

Hamet, P., & Tremblay, J. (2017). Artificial intelligence in medicine. *Metabolism: Clinical and Experimental*, 69, S36–S40. Advance online publication. DOI: 10.1016/j.metabol.2017.01.011 PMID: 28126242

Hamid, H., Zulkifli, K., Naimat, F., Che Yaacob, N. L., & Ng, K. W. (2023). Exploratory study on student perception on the use of chat AI in process-driven problem-based learning. *Currents in Pharmacy Teaching & Learning*, 15(12), 1017–1025. Advance online publication. DOI: 10.1016/j.cptl.2023.10.001 PMID: 37923639

Hammedi, W., Leclercq, T., & Poncin, I., & Alkire (Née Nasr), L. (. (2021). Uncovering the dark side of gamification at work: Impacts on engagement and well-being. *Journal of Business Research*, 122, 256–269. DOI: 10.1016/j.jbusres.2020.08.032

Hammershaimb, L. (2020, March 10). Retrieved from https://www.aace.org/review/artificial-intelligence-and-animated-pedagogical-agents/

Hanel, P. (2018). *Intellectual property rights and innovation in SMEs in OECD countries*. Edward Elgar Publishing.

Hanel, P. (2018). Intellectual property rights business management practices: A survey of the literature. *Technovation*, 28(9), 650–664. DOI: 10.1016/j.technovation.2008.02.010

Hartl, E., & Hess, T. "The Role of Cultural Values for Digital Transformation: Insights from a Delphi Study" (2017). *AMCIS 2017 Proceedings*. 8. https://aisel.aisnet.org/amcis2017/Global/Presentations/8

Hartmann, A., & Gommer, L. (2021). To play or not to play: On the motivational effects of games in engineering education. *European Journal of Engineering Education*, 46(3), 319–343. DOI: 10.1080/03043797.2019.1690430

Hartono, A., & Kusumawardhani, R. (2018). Innovation Barriers and Their Impact on Innovation: Evidence from Indonesian Manufacturing Firms. *Global Business Review*, 20(5), 1196–1213. DOI: 10.1177/0972150918801647

Hartt, M., Hosseini, H., & Mostafapour, M. (2020). Game On: Exploring the Effectiveness of Game-based Learning. *Planning Practice and Research*, 35(5), 589–604. DOI: 10.1080/02697459.2020.1778859

Hasan, M. E., Mostafa, F., Hossain, M. S., & Loftin, J. (2023). Machine-Learning Classification Models to Predict Liver Cancer with Explainable AI to Discover Associated Genes. *AppliedMath*, 3(2), 417–445. DOI: 10.3390/appliedmath3020022

Hassan, M., Awan, F. M., Naz, A., Deandrés-Galiana, E. J., Alvarez, O., Cernea, A., Fernández-Brillet, L., Fernández-Martínez, J. L., & Kloczkowski, A. (2022). Innovations in Genomics and Big Data Analytics for Personalized Medicine and Health Care: A Review. In *International Journal of Molecular Sciences* (Vol. 23, Issue 9). DOI: 10.3390/ijms23094645

Hat, T. (2023). Retrieved from https://tophat.com/

Health Care Professions Council. UK (2023). The standards of proficiency of hearing aid dispensers https://www.hcpc-uk.org/standards/standards-of-proficiency/hearing-aid-dispensers/

Hearing, H. Canada (2023). Hearing aid prices https://www.healthyhearing.com/help/hearing-aids/prices

Hendrycks, D., Mazeika, M., & Woodside, T. (2023). An Overview of Catastrophic AI Risks. *Cornell University*. doi:/arXiv.2306.12001DOI: 10.48550

Hernández-Fernández, A., Olmedo-Torre, N., & Peña, M. (2020). Is classroom gamification opposed to performance? *Sustainability (Basel)*, 12(23), 1–20. DOI: 10.3390/su12239958

Hervás-Oliver, J. L., Parrilli, M. D., Rodríguez-Pose, A., & Sempere-Ripoll, F. (2021). The drivers of SME innovation in the regions of the EU. *Research Policy*, 50(9), 104316. DOI: 10.1016/j.respol.2021.104316

Higgins, O., Short, B. L., Chalup, S. K., & Wilson, R. L. (2023). Artificial intelligence (AI) and machine learning (ML) based decision support systems in mental health: An integrative review. In *International Journal of Mental Health Nursing* (Vol. 32, Issue 4). DOI: 10.1111/inm.13114

Hinton, G. E., & Salakhutdinov, R. R. (2006). Reducing the dimensionality of data with neural networks. *Science*, 313(5786), 504–507. Advance online publication. DOI: 10.1126/science.1127647 PMID: 16873662

His Majesty Revenue & Customs (HMRC). UK (2016). VATHLT2380 https://www.gov.uk/hmrc-internal-manuals/vat-health/vathlt2380

His Majesty Revenue & Customs (HMRC). UK (2016). VATHLT2400 https://www.gov.uk/hmrc-internal-manuals/vat-health/vathlt2380

Hobson, J. S. P., & Williams, A. P. (1995). Virtual Reality: A new horizon for the tourism industry. *Journal of Vacation Marketing*, 1(2), 125–135.

Hoelscher, M., & Schubert, J. (2015). Potential and Problems of Existing Creativity and Innovation Indices. *Creativity Research Journal*, 27(1), 1–15. DOI: 10.1080/10400419.2015.992656

Hoffer, E. P. (1992). DxPlain: Diagnostic consultation by computer for the practicing physician. *The Journal of Medical Practice Management*, 7(4), •••.

Hopkins K. (2015). Deafness in cochlear and auditory nerve disorders. Handb Clin Neurol. 2015;129:479-94. . PMID: 25726286. https://doi.org/DOI: 10.1016/B978-0-444-62630-1.00027-5

Hopkins Medicine. How the Ear Works. https://www.hopkinsmedicine.org/health/conditions-and-diseases/how-the-ear-works

Hotton, M., Prud'Homme, V., Richard, L., Cormier, L., Simoneau, K., Lefebvre-Demers, M., ... & Boucher, N. (2023). Impacts and Identification of Hearing Aid Refurbishing Programs for People with Hearing Loss: A Scoping Review. *Audiology Research*, 13(3), 326–340.

Hsu, S., Wenting Li, T., Zhang, Z., Fowler, M., Zilles, C., & Karahalios, K. (2021). Attitudes Surrounding an Imperfect AI Autograder. DOI: 10.1145/3411764.3445424

Huang, G., & Wang, Z. (2022). The Application of Virtual Reality Technology in the Coordination and Interaction of Regional Economy and Culture in the Sustainable Development of Ecotourism. *Mathematical Problems in Engineering,* vol. 2022, Article ID 9847749, 13 pages. DOI: 10.1155/2022/9847749

Huang, C.-Y., Backman, K. F., Backman, S., & Chang, L. L. (2015). Exploring the implications of virtual reality technology in tourism marketing: An integrated research framework. *International Journal of Tourism Research*, 18(2), 116–128. DOI: 10.1002/jtr.2038

Huang, C.-Y., Backman, K. F., Backman, S., & Moore, D. (2013). Exploring user acceptance of 3D virtual worlds in travel and tourism marketing. *Tourism Management*, 36, 490–501. DOI: 10.1016/j.tourman.2012.09.009

Igbokwe, C. I. (2023, March 24). Application of Artificial Intelligence (AI) in Educational Management. [IJSRP]. *International Journal of Scientific and Research Publications*, 13(3). Advance online publication. DOI: 10.29322/IJSRP.13.03.2023.p13536

Indolia, S., Goswami, A. K., Mishra, S. P., & Asopa, P. (2018). Conceptual Understanding of Convolutional Neural Network- A Deep Learning Approach. *Procedia Computer Science*, 132, 679–688. Advance online publication. DOI: 10.1016/j.procs.2018.05.069

Inglehart, R. (2018). *Cultural evolution: People's motivations are changing and reshaping the world.* Cambridge University Press. DOI: 10.1017/9781108613880

Interbrand. (2021). Best Global Brands 2021. Retrieved from https://interbrand.com/best-global-brands/

International Energy Agency. (2019). [International Energy Agency.]. *Energy Efficiency*, •••, 2019.

Intertek (2021). Pharmaceutical And Medical Device Act Overview – Requirements for Regulated Medical Devices in Japan https://jwapi.intertek.com.cn/Upload/202203/4bc8eaba20d2e0fe879837fb500f3fc8.pdf

Ippakayala, V. K., & El-Ocla, H. (2017). OLMS: Online learning management system for e-learning. *World Journal on Educational Technology*. iTalk2Learn. (2023). Retrieved from https://www.italk2learn.com/

Jack, W., & Suri, T. (2014). Risk sharing and transactions costs: Evidence from Kenya's mobile money revolution. *The American Economic Review*, 104(1), 183–223. DOI: 10.1257/aer.104.1.183

Jajoo, R., Mital, D., Haque, S., & Srinivasan, S. (2002, December). Prediction of hepatitis C using artificial neural network. In *7th International Conference on Control, Automation, Robotics and Vision, 2002. ICARCV 2002.* (Vol. 3, pp. 1545-1550). IEEE. DOI: 10.1109/ICARCV.2002.1235004

James, G., Witten, D., Hastie, T., & Tibshirani, R. (2013). An introduction to statistical learning (Vol. 112, No. 1). Springer.

Jarratt, D. (2021, April). An exploration of webcam-travel: Connecting to place and nature through webcams during the COVID-19 lockdown of 2020. *Tourism and Hospitality Research*, 21(2), 156–168. DOI: 10.1177/1467358420963370

Jenkins, W. (2009). Sustainability Theory. *Berkshire*. https://www.berkshirepublishing.com/assets_news/sustainability/Spirit_SustainabilityTheory.pdf

Johnson, L., Becker, S. A., Cummins, M., Estrada, V., Freeman, A., & Hall, C. (2016). *NMC Horizon Report: 2016 Higher Education Edition.* The New Meda Consortium. Retrieved from https://www.learntechlib.org/p/171478/

Jones, M. D., Hutcheson, S., & Camba, J. D. (2021). Past, present, and future barriers to digital transformation in manufacturing: A review. *Journal of Manufacturing Systems*, 60, 936–948. DOI: 10.1016/j.jmsy.2021.03.006

Jovicic, D. Z. (2014). Key issues in the implementation of sustainable tourism. *Current Issues in Tourism*, 17(4), 297–302. DOI: 10.1080/13683500.2013.797386

Ka Yuk Chan, C., & Tsi, L. H. (2023). The AI Revolution in Education: Will AI Replace or Assist Teachers in. Retrieved October 1, 2023, from https://arxiv.org/ftp/arxiv/papers/2305/2305.01185.pdf

Kahoot! (2023). Retrieved from https://kahoot.com/

Kahoot! (2023, October 31). Retrieved from The Kahoot! AI question generator can save you time by automatically creating high-quality, varied questions based on your chosen topic, adjusting difficulty based on student performance, providing instant feedback with correct answers, and scaling to acc

Kala, R., Shulkla, A., & Tiwari, R. (2009, March). Fuzzy neuro systems for machine learning for large data sets. In *2009 IEEE International Advance Computing Conference* (pp. 541-545). IEEE. DOI: 10.1109/IADCC.2009.4809069

Kamel, S. (2021). (working paper). *The Potential Impact of Digital Transformation on Egypt* (pp. 1–75). Cairo, Egypt: The Economic Research Forum (ERF).

Kaplan, R. S., & Norton, D. P. (1992). The balanced scorecard: Measures that drive performance. *Harvard Business Review*, 70(1), 71–79. PMID: 10119714

Karimi, J., & Walter, Z. (2021). The role of entrepreneurial agility in digital entrepreneurship and creating value in response to digital disruption in the newspaper industry. *Sustainability (Basel)*, 13(5), 1–26. DOI: 10.3390/su13052741

Kask, S. (2018). Virtual reality in support of sustainable tourism. Experiences from Eastern Europe. (Doctoral thesis, Estonian University of Life Sciences, Tartu, ISBN 978-9949-629-61-9). DOI: 10.15159/emu.35

Kask, S., Kullb, T., & Orruc, K. (2018). The use of 3D visualization for sustainable tourism planning. *Journal of Baltic Studies*, 2018(3), 371–385. Advance online publication. DOI: 10.1080/01629778.2018.1455719

Kastrati, Z., Dalipi, F., Imran, A. S., Nuci, K. P., & Wani, M. A. (2021, April 28). Sentiment Analysis of Students' Feedback with NLP and Deep Learning: A Systematic Mapping Study. *MDPI*. DOI: 10.3390/app11093986

Kaunang, F. J. (2022). A Comparative Study on Hepatitis C Predictions Using Machine Learning Algorithms. *8ISC Proceedings: Technology*, 33-42.

KayvanJoo, A. H., Ebrahimi, M., & Haqshenas, G. (2014). Prediction of hepatitis C virus interferon/ribavirin therapy outcome based on viral nucleotide attributes using machine learning algorithms. *BMC Research Notes*, 7, 1–11. PMID: 25150834

Keeley, L., Pikkel, R., Quinn, B., & Walters, H. (2013). *Ten types of innovation: The discipline of building breakthroughs*. Wiley.

Kemska, A. S. (April, 2021). VIRTUAL TOURISM IN UKRAINE - FEATURES AND PROSPECTS. POLIT. Challenges of science Today. http://fmv.nau.edu.ua/wp-content/uploads/2021/04/%D0%9F%D0%BE%D0%BB%D1%96%D1%82-2021.pdf#page=274

Kessler, S., Mebine, P., Von Schweber, E., Von Schweber, L., Desai, M., McConnell, W., Jai, E. M., Nguyen, J., Kiroyan, C., & Ho, D. (2021). Economic and utilization outcomes of medication management at a large medicaid plan with disease management pharmacists using a novel artificial intelligence platform from 2018 to 2019: A retrospective observational study using regression methods. In *Journal of Managed Care and Specialty Pharmacy* (Vol. 27, Issue 9). DOI: 10.18553/jmcp.2021.21036

Khan Academy. (2023). Retrieved from https://blog.khanacademy.org/teacher-khanmigo/

Khan, O., Parvez, M., Kumari, P., Parvez, S., & Ahmad, S. (2023). The future of pharmacy: How AI is revolutionizing the industry. *Intelligent Pharmacy*, 1(1), 32–40. Advance online publication. DOI: 10.1016/j.ipha.2023.04.008

Khan, R. A., Luo, Y., & Wu, F. X. (2022). Machine learning based liver disease diagnosis: A systematic review. *Neurocomputing*, 468, 492–509. DOI: 10.1016/j.neucom.2021.08.138

Khasawneh, Y. J. A., Khasawneh, N., & Khasawneh, M. A. S. (2024). Exploring the long-term effects: Retention and transfer of skills in gamified learning environment. *International Journal of Data and Network Science*, 8(1), 195–200. DOI: 10.5267/j.ijdns.2023.10.004

Khomeiny, A. T., Kusuma, T. R., Handayani, A. N., Wibawa, A. P., & Supadmi Irianti, A. H. (2020). Grading System Recommendations for Students using Fuzzy Mamdani Logic. DOI: 10.1109/ICOVET50258.2020.9230299

Kim, D. W., Ha, J., Lee, S. S., Kwon, J. H., Kim, N. Y., Sung, Y. S., Yoon, J. S., Suk, H.-I., Lee, Y., & Kang, B. K. (2021). Population-based and personalized reference intervals for liver and spleen volumes in healthy individuals and those with viral hepatitis. *Radiology*, 301(2), 339–347. DOI: 10.1148/radiol.2021204183 PMID: 34402668

Kim, M. J., Lee, C.-K., & Jung, T. (2020). Exploring Consumer Behavior in Virtual Reality Tourism Using an Extended Stimulus-Organism-Response Model. *Journal of Travel Research*, 59(1), 69–89. DOI: 10.1177/0047287518818915

Kiryu, Y. (2023). Development of a Medical Big Data Analysis System Utilizing Artificial Intelligence Analytics in Clinical Pharmacy. In *Yakugaku Zasshi* (Vol. 143, Issue 6). DOI: 10.1248/yakushi.22-00179-4

Kiryu, Y. (2022). Medical Big Data Analysis Using Machine Learning Algorithms in the Field of Clinical Pharmacy. *Yakugaku Zasshi*, 142(4), 319–326. Advance online publication. DOI: 10.1248/yakushi.21-00178-1 PMID: 35370185

Kline, R. R. (2011). Cybernetics, automata studies, and the dartmouth conference on artificial intelligence. *IEEE Annals of the History of Computing*, 33(4), 5–16. Advance online publication. DOI: 10.1109/MAHC.2010.44

Klutka, J., Ackerly, N., & Magda, A. J. (2020). *AI in Higher Education.* Learning House: Wiley. Retrieved from https://universityservices.wiley.com/wp-content/uploads/2020/12/201811-AI-in-Higher-Education-TLH-with-new-bage.pdf

Korver, A. M., Smith, R. J., Van Camp, G., Schleiss, M. R., Bitner-Glindzicz, M. A., Lustig, L. R., & Boudewyns, A. N. (2017). Congenital hearing loss. *Nature Reviews. Disease Primers*, 3(1), 1–17.

Koshy, P. (2022, May 31). *The role of ngos in attaining the sdgs*. UN Today. https://untoday.org/the-role-of-ngos-in-attaining-the-sdgs/

Kounavis, C. D., Kasimati, A. E., & Zamani, E. D. (2012). Enhancing the Tourism Experience through Mobile Augmented Reality: Challenges and Prospects. *International Journal of Engineering Business Management*, 4, 1–6. Retrieved May 2014, from http://cdn.intechopen.com/pdfs/38051/InTech Enhancing_the _tourism_experience_through_mobile_augmented_reality_challenges_and_prospects.pdf. DOI: 10.5772/51644

Kubíčková, K. (2020). Úloha oxidačního stresu v jaterní kancerogenezi.

Kulikowski, C. A. (2019). Beginnings of Artificial Intelligence in Medicine (AIM): Computational Artifice Assisting Scientific Inquiry and Clinical Art - with Reflections on Present AIM Challenges. *Yearbook of Medical Informatics*, 28(1), 249–256. Advance online publication. DOI: 10.1055/s-0039-1677895 PMID: 31022744

Kumar, V., & Pansari, A. (2016). Competitive advantage through engagement. *JMR, Journal of Marketing Research*, 53(4), 497–514. DOI: 10.1509/jmr.15.0044

Labadze, L., Grigolia, M., & Machaidze, L. (2023, October 31). Role of AI chatbots in education: Systematic literature review. *International Journal of Educational Technology in Higher Education*, 20(1), 56. Advance online publication. DOI: 10.1186/s41239-023-00426-1

Lasak JM, Allen P, McVay T, Lewis D, (2014). Hearing loss: diagnosis and management. Prim Care. 2014 Mar;41(1):19-31. . Epub 2013 Nov 18. PMID: 24439878. https://doi.org/DOI: 10.1016/j.pop.2013.10.003

Latifi, M. A., Nikou, S., & Bouwman, H. (2021). Business model innovation and firm performance: Exploring causal mechanisms in SMEs. *Technovation*, 107, 102274. DOI: 10.1016/j.technovation.2021.102274

Lau, A. (2020). New technologies used in COVID-19 for business survival: Insights from the Hotel Sector in China. *Information Technology & Tourism*, 22(4), 497–504. DOI: 10.1007/s40558-020-00193-z

Lee, H. S. (2018). A study on the sustainability of the Ramsar wetland ecotourism sites in Korea: Making a list of evaluation indicators and measurement variables. *International Journal of Tourism and Hospitality Research*, 32(2), 99–117. DOI: 10.21298/IJTHR.2018.02.32.2.99

Legner, C., Eymann, T., Hess, T., Matt, C., Böhmann, T., Drews, P., Mädche, A., Urbach, N., & Ahlemann, F. (2017). Digitalization: Opportunity and challenge for the business and Information Systems Engineering Community. *Business &amp. Business & Information Systems Engineering*, 59(4), 301–308. DOI: 10.1007/s12599-017-0484-2

Lenzen, M., Sun, Y., Faturay, F., Ting, Y.-P., Geschke, A., & Malik, A. (2018). The carbon footprint of global tourism. *Nature Climate Change*, 8(6), 522–528. DOI: 10.1038/s41558-018-0141-x

Lesáková, U., Gundová, P., Kráľ, P., & Ondrušová, A. (2017). Innovation Leaders, Modest Innovators and Non-innovative SMEs in Slovakia: Key Factors and Barriers of Innovation Activity. *Organizacija*, 50(4), 325–338. DOI: 10.1515/orga-2017-0024

Lev, B. (2001). *Intangibles: Management, measurement, and reporting*. Brookings Institution Press.

Lewis, M. S., Crandell, C. C., Valente, M., & Horn, J. E. (2004). Speech perception in noise: Directional microphones versus frequency modulation (FM) systems. *Journal of the American Academy of Audiology*, 15(06), 426–439.

Lewis, S. C., & Westlund, O. (2015). Actors, actants, audiences, and activities in cross-media news work: A matrix and a research agenda. *Digital Journalism (Abingdon, England)*, 3(1), 19–37. DOI: 10.1080/21670811.2014.927986

Libert, B., Beck, M., & Wind, Y. (2016, July 14). *7 questions to ask before your next digital transformation.* Harvard Business Review. https://hbr.org/2016/07/7-questions-to-ask-before-your-next-digital-transformation

Li, F. (2020). Leading digital transformation: Three emerging approaches for managing the transition. *International Journal of Operations & Production Management*, 40(6), 809–817. DOI: 10.1108/IJOPM-04-2020-0202

Lin, C.-C., Huang, A. Y., & Lu, O. H. (2023, August 28). Artificial intelligence in intelligent tutoring systems toward sustainable education: a systematic review. *Springer Open*. DOI: 10.1186/s40561-023-00260-y

Lin, S. Y., Tran, N. T. T., Chang, S. L., Su, W. P., & Lin, M. F. (2018). *Structure- and adatom-enriched essential properties of graphene nanoribbons*. CRC press. DOI: 10.1201/9780429400650

Lin, Y. C., Lai, Y. H., Chang, H. W., Tsao, Y., Chang, Y. P., & Chang, R. Y. (2015). SmartHear: A smartphone-based remote microphone hearing assistive system using wireless technologies. *IEEE Systems Journal*, 12(1), 20–29.

Lin, Y., & Yu, Z. (2023). Extending Technology Acceptance Model to higher-education students' use of digital academic reading tools on computers. *International Journal of Educational Technology in Higher Education*, 20(1), 34. Advance online publication. DOI: 10.1186/s41239-023-00403-8

Li, T. H. S., Chiu, H. J., & Kuo, P. H. (2022). Hepatitis C Virus Detection Model by Using Random Forest, Logistic-Regression and ABC Algorithm. *IEEE Access : Practical Innovations, Open Solutions*, 10, 91045–91058. DOI: 10.1109/ACCESS.2022.3202295

Lucas, H. C.Jr, & Goh, J. M. (2009). Disruptive technology: How Kodak missed the digital photography revolution. *The Journal of Strategic Information Systems*, 18(1), 46–55. DOI: 10.1016/j.jsis.2009.01.002

Luhar.H., (2021), An Analytical Study on Economy Upliftment through Virtual Tourism with special reference to Vasai Virar City, -- Palarch's Journal Of Archaeology Of Egypt/Egyptology 18(7). ISSN 1567-214x

Lui, T.-W., Piccoli, G., & Ives, B. (2007). Marketing strategies in virtual worlds. *The Data Base for Advances in Information Systems*, 38(4), 77–80. DOI: 10.1145/1314234.1314248

Lu, J., Xiao, X., Xu, Z., Wang, C., Zhang, M., & Zhou, Y. (2022). The potential of virtual tourism in the recovery of tourism industry during the COVID-19 pandemic. *Current Issues in Tourism*, 25(3), 441–457. DOI: 10.1080/13683500.2021.1959526

Lundberg, S. M., & Lee, S. I. (2017). A unified approach to interpreting model predictions. *Advances in Neural Information Processing Systems*, •••, 30.

Luo, C. L., Rong, Y., Chen, H., Zhang, W. W., Wu, L., Wei, D., Wei, X.-Q., Mei, L.-J., & Wang, F. B. (2019). A logistic regression model for noninvasive prediction of AFP-negative hepatocellular carcinoma. *Technology in Cancer Research & Treatment*, 18, 1533033819846632. DOI: 10.1177/1533033819846632 PMID: 31106685

Lynn, T., Rosati, P., Conway, E., Curran, D., Fox, G., & O'Gorman, C. (2022). *Digital Technologies and Civil Society*. Essay, Palgrave Macmillan Cham. DOI: 10.1007/978-3-030-91247-5_5

Mamdouh Farghaly, H., Shams, M. Y., & Abd El-Hafeez, T. (2023). Hepatitis C Virus prediction based on machine learning framework: A real-world case study in Egypt. *Knowledge and Information Systems*, 65(6), 2595–2617. DOI: 10.1007/s10115-023-01851-4

Mancini, J. (2019). E-health Records: A slow marathon to get online. *Health Science Inquiry*, 5(1). Advance online publication. DOI: 10.29173/hsi179

Manyika, J., Chui, M., Bughin, J., Dobbs, R., Bisson, P., & Marrs, A. (2017). *Disruptive technologies: Advances that will transform life, business, and the global economy*. McKinsey Global Institute.

Manzano-León, A., Camacho-Lazarraga, P., Guerrero, M. A., Guerrero-Puerta, L., Aguilar-Parra, J. M., Trigueros, R., & Alias, A. (2021). Between level up and game over: A systematic literature review of gamification in education. *Sustainability (Basel)*, 13(4), 1–14. DOI: 10.3390/su13042247

Marchiori, E., Niforatos, E., & Preto, L. (2017). Measuring the media effects of a tourism-related virtual reality experience using biophysical data. In Schegg, R., & Stangl, B. (Eds.), *Information and communication technologies in tourism 2017* (pp. 203–215). Springer. DOI: 10.1007/978-3-319-51168-9_15

Mardanov, I. (2020). Intrinsic and extrinsic motivation, organizational context, employee contentment, job satisfaction, performance and intention to stay. *Evidence-Based HRM*, 9(3), 223–240. DOI: 10.1108/EBHRM-02-2020-0018

Martin, F. (2018, October 29). *Unlocking success in digital transformations*. McKinsey & Company. https://www.mckinsey.com/capabilities/people-and-organizational-performance/our-insights/unlocking-success-in-digital-transformations#/

Mathenge, V. W., Sambaiga, R., Bitebo, A., Komba, M., Mushashu, A., Sang'anya, M., Shaweji, F., Mbwilo, G., Mayao, V., Mlembwa, J., Rangi, J., Onuekwe, C., Kileo, N., Mwengee, W., & Sagoe-Moses, C. (2023). The development of a digital NGO mapping platform in Tanzania, as a tool to strengthen civil society engagement, in pursuit of Universal Health Coverage. *The Pan African Medical Journal*, 46. Advance online publication. DOI: 10.11604/pamj.2023.46.115.42333 PMID: 38465003

Matt, D. T., & Rauch, E. (2020). SME 4.0: The role of small- and medium-sized enterprises in the digital transformation. *Industry 4.0 for SMEs*, 3–36. https://doi.org/DOI: 10.1007/978-3-030-25425-4_1

Maurice, J., & Manousou, P. (2018). Non-alcoholic fatty liver disease. *Clinical Medicine*, 18(3), 245–250. DOI: 10.7861/clinmedicine.18-3-245 PMID: 29858436

McKenzie, L. (2019, November 17). *Digital Learning*. Retrieved from Insidehighered.com: https://www.insidehighered.com/digital-learning/article/2019/11/18/pearson-introduces-aida-ai-powered-calculus-tutor

Mei, L., Zhang, T., & Chen, J. (2019). Survey data on inter-firm linkages and innovation activities of Chinese manufacturing SMEs. *Data in Brief*, 27, 104671. DOI: 10.1016/j.dib.2019.104671 PMID: 31720326

Menna, A., Walsh, P. R., & Ekhtari, H. (2019). Identifying enablers of innovation in developed economies: A National Innovation Systems approach. *Journal of Innovation Management*, 7(1), 108–128. DOI: 10.24840/2183-0606_007.001_0007

Mergel, I. (2016). Agile innovation management in government: A research agenda. *Government Information Quarterly*, 33(3), 516–523. DOI: 10.1016/j.giq.2016.07.004

Messner, W. (2023, February 2). Hypothesis Testing and Machine Learning: Interpreting Variable Effects in Deep Artificial Neural Networks using Cohen's f 2. Retrieved from https://arxiv.org/ftp/arxiv/papers/2302/2302.01407.pdf

Meyerhoff Nielsen, M., & Mahmoud Ali, W. K. (2021). Governing and monitoring the Digital Transformation: Assessing the Qatari experience since 2003. *14th International Conference on Theory and Practice of Electronic Governance*. https://doi.org/DOI: 10.1145/3494193.3494227

Miller, R. A., McNeil, M. A., Challinor, S. M., Masarie, F. E., & Myers, J. D. (1986). The INTERNIST-1/QUICK MEDICAL REFERENCE Project - Status report. *The Western Journal of Medicine*, 145(6). PMID: 3544509

Ministry of Health & Family Welfare. India (2013). National Programme for Prevention and Control of Deafness. https://main.mohfw.gov.in/sites/default/files/51892751619025258383.pdf

Ministry of Health and Welfare. PRC (2020). Hearing Specialists Act. https://law.moj.gov.tw/ENG/LawClass/LawAll.aspx?pcode=L0020149

Ministry of Health. Canada (2023). Hearing Devices Policy and Administration Manual https://files.ontario.ca/moh-adp-policy-and-administration-manual-hearing-devices-2023-07-01.pdf

Miroshnychenko, I., Strobl, A., Matzler, K., & De Massis, A. (2021a). Absorptive capacity, strategic flexibility, and business model innovation: Empirical evidence from Italian SMEs. *Journal of Business Research*, 130, 670–682. DOI: 10.1016/j.jbusres.2020.02.015

Mitchell, C. (2020). Virtual reality. Investopedia. Retrieved from https://www.investopedia.com/

Mohit, M., & Madhurima, D. (2021). Effects of multisensory virtual reality on virtual and physical tourism during the COVID-19 pandemic. *Current Issues in Tourism*. Advance online publication. DOI: 10.1080/13683500.2021.1978953

Mohsen, M. (2018). *What are NGOs doing in Egypt and what can you do to help? | Egyptian Streets*. Egyptian Streets. https://egyptianstreets.com/2018/07/31/what-are-ngos-doing-in-egypt-and-what-can-you-do-to-help/

Moran, M. E. (2007). Evolution of robotic arms. *Journal of Robotic Surgery*, 1(2), 103–111. Advance online publication. DOI: 10.1007/s11701-006-0002-x PMID: 25484945

Murphy, P., Pritchard, M. P., & Smith, B. (2000). The destination product and its impact on traveller perceptions. *Tourism Management*, 21(1), 43–52. DOI: 10.1016/S0261-5177(99)00080-1

Mustapha, S. (2023, September 29). Predictive Analysis of Students' Learning Performance Using Data Mining Techniques: A Comparative Study of Feature Selection Methods. *Applied System Innovation*, 6(5), 86. Advance online publication. DOI: 10.3390/asi6050086

Nafisa Zarrin Tasnim, Aoxin Ni, Edward Lobarinas, Nasser Kehtarnavaz, (2023). A Review of Machine Learning Approaches for the Personalization of Amplification in Hearing Aids DOI: 10.3390/s24051546

Nah, F. F. H., Eschenbrenner, B., Claybaugh, C. C., & Koob, P. B. (2019). Gamification of enterprise systems. *Systems*, 7(1), 1–21. DOI: 10.3390/systems7010013

Nam, D., Chapiro, J., Paradis, V., Seraphin, T. P., & Kather, J. N. (2022). Artificial intelligence in liver diseases: Improving diagnostics, prognostics and response prediction. *JHEP Reports : Innovation in Hepatology*, 4(4), 100443. DOI: 10.1016/j.jhepr.2022.100443 PMID: 35243281

Nandipati, S. C., XinYing, C., & Wah, K. K. (2020). Hepatitis C virus (HCV) prediction by machine learning techniques. *Applications of modelling and simulation, 4*, 89-100.

NANDIPATI. (2020). Satish CR; XINYING, Chew; WAH, Khaw Khai. Hepatitis C virus (HCV) prediction by machine learning techniques. *Applications of Modelling and Simulation*, 4, 89–100.

Nassir, U., Cheng, J., & Wang, W. (2023). Explainable clinical models for non-alcoholic fatty liver disease prediction: A Bayesian network approach. *Journal of Medical Imaging and Health Informatics*, 12(2), 1047–1054.

National Council on Aging (NCOA). USA (2023). FDA Hearing Aids Guidance: 2024 Compliance https://www.ncoa.org/adviser/hearing-aids/fda-hearing-aids-guidance

National Health Service (NHS). Hearing aids and implants https://www.nhs.uk/conditions/hearing-aids-and-implants

National Health Service. UK. Hearing aid dispensers https://www.healthcareers.nhs.uk/explore-roles/healthcare-science/roles-healthcare-science/physiological-sciences/hearing-aid-dispenser/hearing-aid-dispenser

National Institute on Deafness and Other Communication Disorders (NIDCD), National Institute of Health (NIH), USA. (2015). How do we hear? https://www.nidcd.nih.gov/health/how-do-we-hear

Nel, F. (2010). Where else is the money?: A study of innovation in online business models at newspapers in britain's 66 cities. *Journalism Practice*, 4(3), 360–372. DOI: 10.1080/17512781003642964

Neshat, M., Yaghobi, M., Naghibi, M. B., & Esmaelzadeh, A. (2008, December). Fuzzy expert system design for diagnosis of liver disorders. In *2008 International Symposium on Knowledge Acquisition and Modeling* (pp. 252-256). IEEE. DOI: 10.1109/KAM.2008.43

Neuberger, J., & Cain, O. (2021). The need for alternatives to liver biopsies: Non-invasive analytics and diagnostics. *Hepatic Medicine : Evidence and Research*, 13, 59–69. DOI: 10.2147/HMER.S278076 PMID: 34163263

Newman, N., Richard Fletcher, with, Kalogeropoulos, A., & Kleis Nielsen, R. (2019). *Reuters Institute Digital News Report 2019.*

Nguyen, N. T. D., & Aoyama, A. (2014). Achieving efficient technology transfer through a specific corporate culture facilitated by management practices. *The Journal of High Technology Management Research*, 25(2), 108–122. DOI: 10.1016/j.hitech.2014.07.001

O'Brien, C. (2023, September 14). *8-universities-leveraging-ai-to-drive-student-success*. Retrieved from Digital Marketing Institute: https://digitalmarketinginstitute.com/blog/8-universities-leveraging-ai-to-drive-student-success

O'Donoghue, G. (2013). Cochlear Implants – Science, Serendipity, and Success. *The New England Journal of Medicine*, 369(13), 1190–1193. Advance online publication. DOI: 10.1056/NEJMp1310111 PMID: 24015974

OECD. (2010). *Handbook on deriving capital measures of intellectual property products*. OECD Publishing.

OECD. (2010). *Measuring innovation: A new perspective*. OECD Publishing., DOI: 10.1787/9789264059474-

Oladimeji, O. O., Oladimeji, A., & Olayanju, O. (2021). Machine Learning Models for Diagnostic Classification of Hepatitis C Tests. *Frontiers in Health Informatics*, 10(1), 70. DOI: 10.30699/fhi.v10i1.274

Omdena. (2024). *How can NGOs develop an effective data strategy for digital transformation*. https://www.omdena.com/blog/ngos-digital-transformation

Open, A. I. GPT 4 model. (2024). Retrieved from www.chatgpt.com

Orhanbulucu, F., Acer, İ., Latifoğlu, F., & İçer, S. (2022). Predicting liver disease using decision tree ensemble methods. *Erciyes Üniversitesi Fen Bilimleri Enstitüsü Fen Bilimleri Dergisi*, 38(2), 261–267.

Ozyilmaz, L., & Yildirim, T. (2003, July). Artificial neural networks for diagnosis of hepatitis disease. In *Proceedings of the International Joint Conference on Neural Networks,* 2003. (Vol. 1, pp. 586-589). IEEE. DOI: 10.1109/IJCNN.2003.1223422

Pacolli, M. (2022). Importance of Change Management in Digital Transformation Sustainability. *IFAC-PapersOnLine*, 55(39), 276–280. DOI: 10.1016/j.ifacol.2022.12.034

Pane, J. F., Steiner, E. D., Baird, M. D., Hamilton, L. S., & Pane, J. D. (2017). *Insights on Personalized Learning Implementation and Effects.* RAND Corporation. Retrieved from https://www.rand.org/pubs/research_reports/RR2042.html

Pantano, E., & Servidio, R. (2009). An exploratory study of the role of pervasive environments for promotion of tourism destinations. *Journal of Hospitality and Tourism Technology*, 2(1), 50–65. DOI: 10.1108/17579881111112412

Papsin, B. C., & Gordon, K. A. (2007). Cochlear implants for children with severe-to-profound hearing loss. *The New England Journal of Medicine*, 357(23), 2380–2387.

Patel, R., Jain, A., Patel, Z., Kavani, H., Patel, M., Gadhiya, D. K., ... & Yang, C. (2024). Artificial Intelligence and Machine Learning in Hepatocellular Carcinoma Screening, Diagnosis and Treatment-A Comprehensive Systematic Review. *Glob Acad J Med Sci, 6.*

Pathiranage, Y. L., Jayatilake, L. V. K., & Abeysekera, R. (2020). A literature review on organizational culture towards corporate performance. *International Journal of Management. Accounting & Economics*, 7(9), 522–544.

Pearson. (2024, January). *Explore Pearson*. Retrieved from Pearson: https://www.pearson.com/explore-pearson.html

Peng, H., Ma, S., & Spector, J. M. (2019). *Personalized adaptive learning: an emerging pedagogical approach enabled by a smart learning environment*. Smart Learning Environments., DOI: 10.1186/s40561-019-0089-y

Perryer, C., Celestine, N. A., Scott-Ladd, B., & Leighton, C. (2016). Enhancing workplace motivation through gamification: Transferrable lessons from pedagogy. *International Journal of Management Education*, 14(3), 327–335. DOI: 10.1016/j.ijme.2016.07.001

Phobun, P., & Vicheanpanya, J. (2010). Adaptive intelligent tutoring systems for e-learning systems. *Procedia: Social and Behavioral Sciences*, 4065-4068(2), 4064–4069. Advance online publication. DOI: 10.1016/j.sbspro.2010.03.641

Powerschool. (2024). *Clasroom*. Retrieved from www.powerschool.com: https://www.powerschool.com/classroom/schoology-learning/

Priyanka Saha, Minhaj Uddin Monir, Shaheen M. Sarkar, Arzina Abd Aziz (2023). Crises in Oil, Gas and Petrochemical Industries, Elsevier. 18 – Noise pollution from oil, gas, and petrochemical industries, 2.1 DOI: 10.1016/B978-0-323-95154-8.00012-8

Province of British Columbia. Canada (2024). Medical Equipment – Hearing Instruments https://www2.gov.bc.ca/gov/content/governments/policies-for-government/bcea-policy-and-procedure-manual/health-supplements-and-programs/medical-equipment-hearing-instruments

Qadir, J. (2022, December 30). Engineering Education in the Era of ChatGPT:Promise and Pitfalls of Generative AI for Education. *Techrxiv*. DOI: 10.36227/techrxiv.21789434.v1

Rahoma, K. H., & Ali, A. A. (2022, April). Applying Machine Learning Technology to Perform Automatic Provisioning of the Optical Transport Network. In *International Conference on Advanced Machine Learning Technologies and Applications* (pp. 169-177). Cham: Springer International Publishing. DOI: 10.1007/978-3-031-03918-8_16

Rahouma, K. H., & Ali, A. A. (2020, October). Applying Smart Security Model to Protect the Client Services from the Threats of the Optical Network. In *2020 2nd Novel Intelligent and Leading Emerging Sciences Conference (NILES)* (pp. 27-34). IEEE. DOI: 10.1109/NILES50944.2020.9257917

Rahouma, K. H., & Ali, A. A. (2021). Applying Smart Security Model to Protect Client Services From the Threats of the Optical Network. In *Advancements in Security and Privacy Initiatives for Multimedia Images* (pp. 80–113). IGI Global. DOI: 10.4018/978-1-7998-2795-5.ch004

Rahouma, K., & Ali, A. (2019). Applying intrusion detection and response systems for securing the client data signals in the Egyptian optical network. *Procedia Computer Science*, 163, 538–549. DOI: 10.1016/j.procs.2019.12.136

Ramesh, N., & Delen, D. (2021). Digital Transformation: How to Beat the 90% Failure Rate? *IEEE Engineering Management Review*, 49(3), 22–25. DOI: 10.1109/EMR.2021.3070139

Ramyap, P., Abrahamp, S., Pattasserilp, M. B., Mp, M. M., Mohan Kp, N., & Kp, S. (2022). Role of Artificial Intelligence in the Pharmaceutical Industry. In *IJISET-International Journal of Innovative Science* (Vol. 09). Engineering & Technology.

Ranchon, F., Chanoine, S., Lambert-Lacroix, S., Bosson, J. L., Moreau-Gaudry, A., & Bedouch, P. (2023). Development of artificial intelligence powered apps and tools for clinical pharmacy services: A systematic review. In *International Journal of Medical Informatics* (Vol. 172). DOI: 10.1016/j.ijmedinf.2022.104983

Regmi, P. R., Waithaka, E., Paudyal, A., Simkhada, P., & Van Teijlingen, E. (2017). Guide to the design and application of online questionnaire surveys. *Nepal Journal of Epidemiology*, 6(4), 640–644. DOI: 10.3126/nje.v6i4.17258 PMID: 28804676

Reis, D., & Afonso, R. (2011). PDSS: Pharmacy Decision Support System - Concept approach. *Proceedings of the IADIS International Conference E-Health 2011, Part of the IADIS Multi Conference on Computer Science and Information Systems 2011, MCCSIS 2011.*

Ribeiro, M. T., Singh, S., & Guestrin, C. (2018). Anchors: What makes a good explanation. In *AAAI Conference on Artificial Intelligence* (pp. 804-810).

Rix, A. W., Beerends, J. G., Hollier, M. P., & Hekstra, A. P. (2001, May). Perceptual evaluation of speech quality (PESQ)-a new method for speech quality assessment of telephone networks and codecs. In *2001 IEEE international conference on acoustics, speech, and signal processing. Proceedings (Cat. No. 01CH37221)* (Vol. 2, pp. 749-752). IEEE.

Roberts, R., Flin, R., Millar, D., & Corradi, L. (2021). *Psychological factors influencing technology adoption: A case study from the oil and gas industry*. Elsevier., DOI: 10.1016/j.technovation.2020.102219

Robson, K., Plangger, K., Kietzmann, J. H., McCarthy, I., & Pitt, L. (2015). Is it all a game? Understanding the principles of gamification. *Business Horizons*, 58(4), 411–420. DOI: 10.1016/j.bushor.2015.03.006

Rosanna Cole, Claire Frances Lindsay & Fiona Barker, (2018). Reverse exchange of healthcare devices: the case of hearing aid equipment in the UK DOI: 10.1080/09537287.2018.1506892

Rus, V. (2023). *The University of Memphis*. Retrieved from https://www.memphis.edu/iis/projects/deeptutor.php

Rus, V., Niraula, N., & Banjade, R. (2015). DeepTutor: An Effective, Online Intelligent Tutoring System That Promotes Deep Learning., *37*. Retrieved from https://ojs.aaai.org/index.php/AAAI/article/view/9269

Ryan, C. (2002). Equity, Management, Power Sharing and Sustainability: Issues of the 'New Tourism.'. *Tourism Management*, 23(1), 17–26. DOI: 10.1016/S0261-5177(01)00064-4

Saarikko, T., Westergren, U. H., & Blomquist, T. (2020). Digital Transformation: Five recommendations for the digitally conscious firm. *Business Horizons*, 63(6), 825–839. DOI: 10.1016/j.bushor.2020.07.005

Saeed, S., Altamimi, S. A., Alkayyal, N. A., Alshehri, E., & Alabbad, D. A. (2023). Digital transformation and cybersecurity challenges for businesses resilience: Issues and recommendations. *Sensors (Basel)*, 23(15), 6666. DOI: 10.3390/s23156666 PMID: 37571451

Safdari, R., Deghatipour, A., Gholamzadeh, M., & Maghooli, K. (2022). Applying data mining techniques to classify patients with suspected hepatitis C virus infection. *Intelligent Medicine*, 2(04), 193–198. DOI: 10.1016/j.imed.2021.12.003

Sakamoto, H., & Yasukawa, F. (2022). A Consideration of the Hearing Aid Business in Japan with the Comparison of the European Union Countries and the U.S. *Iryo To Shakai = Academic Studies*, 32(3), 405–418. Advance online publication. DOI: 10.4091/iken.2022.005

Salah El Din, P. (2023). The Role of Digital Organizational Culture in Organizational Development In light of Digital Transformation adoption in Egypt. *Arab Journal of Administration*, 45(4), 1 20. DOI: 10.21608/aja.2023.173954.1353

Schmidhuber, J. (2015). Deep Learning in neural networks: An overview. In *Neural Networks* (Vol. 61). DOI: 10.1016/j.neunet.2014.09.003

Schneider, S., & Kokshagina, O. (2021). Digital Transformation: What we have learned (thus far) and what is Next. *Creativity and Innovation Management*, 30(2), 384–411. DOI: 10.1111/caim.12414

Schreiber, B. E., Agrup, C., Haskard, D. O., & Luxon, L. M. (2010). Sudden sensorineural hearing loss. *Lancet*, 375(9721), 1203–1211.

Schussel, K. E., Forbes, S., Taylor, A. M., & Cooley, J. H. (2019). Implementation of an interprofessional medication therapy management experience. *American Journal of Pharmaceutical Education*, 83(3), 6584. Advance online publication. DOI: 10.5688/ajpe6584 PMID: 31065160

Schutz, N., Olsen, C. A., Mclaughlin, A. J., Yi, W. M., Nelson, S. D., Kalichira, A. L., Smith, A. H., Miller, K. A., Le, T., Chaffee, B. W., Worthy Woodbury, C. D. R. K., & Patel, H. (2020). ASHP Statement on the Use of Artificial Intelligence in Pharmacy. In *American Journal of Health-System Pharmacy* (Vol. 77, Issue 23). DOI: 10.1093/ajhp/zxaa249

SciSpace. (2023). Retrieved from https://typeset.io/resources/scispace-literature-review-workspace/

Searchfield, G. D., Sanders, P. J., Doborjeh, Z., Doborjeh, M., Boldu, R., Sun, K., & Barde, A. (2021). A State-of-Art Review of Digital Technologies for the Next Generation of Tinnitus Therapeutics. In *Frontiers in Digital Health* (Vol. 3). DOI: 10.3389/fdgth.2021.724370

Secinaro, S., Calandra, D., Secinaro, A., Muthurangu, V., & Biancone, P. (2021). The role of artificial intelligence in healthcare: A structured literature review. *BMC Medical Informatics and Decision Making*, 21(1), 125. Advance online publication. DOI: 10.1186/s12911-021-01488-9 PMID: 33836752

Seo, K., Tang, J., Roll, I., Fels, S., & Yoon, D. (2021, October 26). The impact of artificial intelligence on learner–instructor interaction in online learning. *International Journal of Educational Technology in Higher Education*. Retrieved from https://educationaltechnologyjournal.springeropen.com/articles/10.1186/s41239-021-00292-9

Shaikh, S., & Bokde, K., INGALE, A., & Tekwani, B. (2018). Virtual Tourism. *International Research Journal of Engineering and Technology*, 5(4), 2044–2046.

Sheombar, A., & Ravesteijn, P. (2022). Digital Transformation of development NGOs: The case of transitioning northern-based development NGOs. *Lecture Notes in Networks and Systems*, 600–613. https://doi.org/DOI: 10.1007/978-3-030-96293-7_50

Shieber, S. M. (2007). The turing test as interactive proof. *Noûs (Detroit, Mich.)*, 41(4), 686–713. Advance online publication. DOI: 10.1111/j.1468-0068.2007.00636.x

Shortliffe, E. H. (1977). MYCIN: A KNOWLEDGE-BASED COMPUTER PROGRAM APPLIED TO. *Infectious Diseases*.

Shukla, A., Tiwari, R., Kaur, P., & Janghel, R. R. (2009, March). Diagnosis of thyroid disorders using artificial neural networks. In *2009 IEEE International Advance Computing Conference* (pp. 1016-1020). IEEE. DOI: 10.1109/IADCC.2009.4809154

Sifo, K. (2019). Mediebild, opinion och faktakoll i klimatfrågan. Retrieved from: https://www.kantarsifo.se/rapporter-undersokningar /mediebild-opinion-och-faktakoll-i-klimatfragan

Siminto, Akib, Hasmirati, & Widianto, D. S. (2023). Educational Management Innovation by Utilizing Artificial Intelligence in Higher Education. *Jurnal Manajemen Pendidikan*, 284-296. DOI: 10.31958/jaf.v10i1.6007

Singh, P., Chahal, D., & Kharb, L. (2020). Predictive strength of selected classification algorithms for diagnosis of liver disease. In *Proceedings of ICRIC 2019: Recent innovations in computing* (pp. 239-255). Springer International Publishing. DOI: 10.1007/978-3-030-29407-6_19

Singh, R., Saurav, S., Kumar, T., Saini, R., Vohra, A., & Singh, S. (2023). Facial expression recognition in videos using hybrid CNN & ConvLSTM. *International Journal of Information Technology : an Official Journal of Bharati Vidyapeeth's Institute of Computer Applications and Management*, 15(4), 1819–1830. DOI: 10.1007/s41870-023-01183-0 PMID: 37256027

Slater, M., & Sanchez-Vives, M. V. (2016). Enhancing Our Lives with Immersive Virtual Reality. *Frontiers in Robotics and AI*, 3, 1–47. DOI: 10.3389/frobt.2016.00074

Snyder, H. (2019). Literature review as a research methodology: An overview and guidelines. *Journal of Business Research*, 104, 333–339. DOI: 10.1016/j.jbusres.2019.07.039

Society for Learning Analytics Research. (2023). Retrieved from https://www.solaresearch.org/about/what-is-learning-analytics/

Sparrow, S. (2023). Retrieved from https://www.smartsparrow.com/

Stokel-Walker, C., & Van Noorden, R. (2023, February 9). THE PROMISE AND PERIL OF GENERATIVE AI. *Springer*. Retrieved from https://www.nature.com/articles/d41586-023-00340-6

Straus, S. E., Glasziou, P., Richardson, W. S., & Haynes, R. B. (2018). *Evidence-Based Medicine: Evidence-Based Medicine E-Book*. Elsevier Health Sciences.

Streimikiene, D., Svagzdiene, B., Jasinskas, E., & Simanavicius, A. (2021). Sustainable tourism development and competitiveness: The systematic literature review. *Sustainable Development (Bradford)*, 29(1), 259–271. DOI: 10.1002/sd.2133

Subai, Ada & Konya, K & Adim, Victor. (2020). Technological Innovation of Hotel Services: A COVID-19 Strategic Response. British Journal of Economics, Finance and Management Sciences 19, August 2020, Vol. 17 (2)

Subawa, N. S., Widhiasthini, N. W., Astawa, I. P., Dwiatmadja, C., & Ni, P. I. P. (2021). The practices of virtual reality marketing in the tourism sector, a case study of Bali, Indonesia. *Current Issues in Tourism*, 24(23), 3284–3295. Advance online publication. DOI: 10.1080/13683500.2020.1870940

Sundararajan, M., Taly, A., & Yan, Q. (2017, July). Axiomatic attribution for deep networks. In *International conference on machine learning* (pp. 3319-3328). PMLR.

Support to implementation of strategies to foster Micro, Small and Medium Enterprises (MiSMESIS) development in Egypt. (n.d.). EEAS. https://www.eeas.europa.eu/node/43122_en

Svahn, F., Mathiassen, L., & Lindgren, R. (2017). Embracing digital innovation in incumbent firms: How Volvo Cars managed competing concerns. *Management Information Systems Quarterly*, 41(1), 239–253. DOI: 10.25300/MISQ/2017/41.1.12

Tabrizi, B., Lam, E., Girard, K., & Irvin, V. (2019, March 13). *Digital transformation is not about technology*. Harvard Business Review. https://hbr.org/2019/03/digital-transformation-is-not-about-technology

The definition of clinical pharmacy. (2008). In *Pharmacotherapy* (Vol. 28, Issue 6). DOI: 10.1592/phco.28.6.816

The National Alliance for Developmental Civil Work. SIS. (2024, January). https://shorturl.at/dveoT

Ting, H. W., Chung, S. L., Chen, C. F., Chiu, H. Y., & Hsieh, Y. W. (2020). A drug identification model developed using deep learning technologies: Experience of a medical center in Taiwan. *BMC Health Services Research*, 20(1), 312. Advance online publication. DOI: 10.1186/s12913-020-05166-w PMID: 32293426

Torlaković, A., Veselinović, L., & Bičo Ćar, M. (2022). The Application of Gamification in a Business Environment: The Most Common Themes and Elements. *Lecture Notes in Networks and Systems, 472 LNNS*, 996–1002. DOI: 10.1007/978-3-031-05230-9_117

Torrijos, J. L. R. (2020). Gamification of sports media coverage: An infotainment approach to olympics and football world cups. *Communicatio Socialis*, 33(1), 29–44. DOI: 10.15581/003.33.1.29-44

Treceño-Fernández, D., Calabia-Del-Campo, J., Bote-Lorenzo, M. L., Gómez-Sánchez, E., Luis-García, R., & Alberola-López, C. (2020, October). Integration of an intelligent tutoring system in a magnetic resonance simulator for education: Technical feasibility and user experience. *Computer Methods and Programs in Biomedicine*, 195, 105634. Advance online publication. DOI: 10.1016/j.cmpb.2020.105634 PMID: 32645627

Truyols, M. (2023, February, 24). The 6 phases of a tourist destination's life cycle. Mize. https://mize.tech/blog/the-6-phases-of-a-tourist-destinations-life-cycle

Tsvetkov, V., Tokin, I., & Lioznov, D. (2021). Machine learning model for diagnosing the stage of liver fibrosis in patients with chronic viral hepatitis C.

Tu, R., Hsieh, P., & Feng, W. (2019). Walking for fun or for "likes"? The impacts of different gamification orientations of fitness apps on consumers' physical activities. *Sport Management Review*, 22(5), 682–693. DOI: 10.1016/j.smr.2018.10.005

Tussyadiah, I. P., Wang, D., & Jia, C. H. (2017). Virtual reality and attitudes toward tourism destinations. In Schegg, R., & Stangl, B. (Eds.), *Information and Communication Technologies in Tourism 2017* [Electronic Resource]. (pp. 229–239)., DOI: 10.1007/978-3-319-51168-9_17

Tutt, A. (2017). An fda for algorithms. In *Administrative Law Review* (Vol. 69, Issue 1). DOI: 10.2139/ssrn.2747994

U.S Department of Education. (2016, June). Retrieved from https://ies.ed.gov/ncee/wwc/Docs/InterventionReports/wwc_cognitivetutor_062116.pdf

UNESCO. (2023). *Guidance for Generative AI in education and research.* Retrieved from https://unesdoc.unesco.org/ark:/48223/pf0000386693

UNESCO. (2023, April). *ChatGPT and Artifical Intelligence in Higher Education.* United Nations Educational, Scientific and Cultural Organization. Retrieved from https://www.iesalc.unesco.org/wp-content/uploads/2023/04/ChatGPT-and-Artificial-Intelligence-in-higher-education-Quick-Start-guide_EN_FINAL.pdf

United Nations Environment Program. UNEP (2016b) Negative socio-cultural impacts from tourism,https://www.unep.org/resourceefficiency/Business/SectoralActivities/Tourism/FactsandFiguresaboutTourism/ImpactsofTourism/SocioCulturalImpacts/NegativeSocio-CulturalImpactsFromTourism/tabid/78781/Default.aspx, accessed July 29, 2016

Valenzuela-Fernández, L., Garro-Aburto, L., & Ocaña-Fernández, Y. (2019). Artificial Intelligence and Its Implications in Higher Education. Retrieved from https://eric.ed.gov/?id=EJ1220536

van Leeuwen, K. G., Schalekamp, S., Rutten, M. J. C. M., van Ginneken, B., & de Rooij, M. (2021). Artificial intelligence in radiology: 100 commercially available products and their scientific evidence. *European Radiology*, 31(6), 3797–3804. Advance online publication. DOI: 10.1007/s00330-021-07892-z PMID: 33856519

Van Veldhoven, Z., & Vanthienen, J. (2023). Best practices for digital transformation based on a systematic literature review. *Digital Transformation and Society*, 2(2), 104–128. DOI: 10.1108/DTS-11-2022-0057

Vandenbroucke, J. P., Elm, E. V., Altman, D. G., Gøtzsche, P. C., Mulrow, C. D., & Pocock, S. J.. (2007). Strengthening the Reporting of Observational Studies in Epidemiology (STROBE): Explanation and elaboration. *Annals of Internal Medicine*, 147(8), W-163. DOI: 10.7326/0003-4819-147-8-200710160-00010-w1 PMID: 17938389

Vashkevich, M., Azarov, E., Petrovsky, N., & Petrovsky, A. (2017). *"Petralex: A smartphone-based real-time digital hearing aid with combined noise reduction and acoustic feedback suppres-sion," 2017 Signal Processing: Algorithms, Architectures, Arrangements, and Applications*. SPA., DOI: 10.23919/SPA.2017.8166873

Veerappa, M., & Rinzivillo, S. (2023, July). Explainable AI. *Ercim News*. Retrieved from https://ercim-news.ercim.eu/images/stories/EN134/EN134-web.pdf

Vial, G. (2021). Understanding digital transformation. *Managing Digital Transformation*, 13–66. DOI: 10.4324/9781003008637-4

Vijay M. Mane, Seema S. Arote, Shakil A Shaikh (2024). An Integrated Deep Learning Model for Concurrent Speech Dereverberation and Denoising DOI: 10.12720/jait.15.2.281-287

Virtual reality experiences for sustainability & environmental awareness. (n.d.). *steamedegg*.https://steamedegg.co.uk/environmental-awareness-virtual-reality/

Von Leipzig, T., Gamp, M., Manz, D., Schöttle, K., Ohlhausen, P., Oosthuizen, G., Palm, D., & von Leipzig, K. (2017). Initializing customer-orientated digital transformation in enterprises. *Procedia Manufacturing*, 8, 517–524. DOI: 10.1016/j.promfg.2017.02.066

Voronkova, L. (2018). Virtual Tourism: On the Way To the Digital Economy. *IOP Conference Series. Materials Science and Engineering*, 463, 042096. DOI: 10.1088/1757-899X/463/4/042096

Vos, T. P., & Perreault, G. P. (2020). The discursive construction of the gamification of journalism. *Convergence (London)*, 26(3), 470–485. DOI: 10.1177/1354856520909542

Wade, M., & Shan, J. (2020). Covid-19 has accelerated digital transformation, but may have made it harder not easier. *MIS Quarterly Executive*, 19(3), 213–220. DOI: 10.17705/2msqe.00034

Wang, F., Liu, Y., Luo, J., & Yang, J. (NA). Explainable machine learning for predicting hepatocellular carcinoma recurrence in hepatitis B patients with digital pathology.

Wang, H., Tlili, A., Huang, R., Cai, Z., Li, M., Cheng, Z., & Zhu, X. (2023). Examining the applications of intelligent tutoring systems in real educational contexts: A systematic literature review from the social experiment perspective. *Education and Information Technologies*, 28(7), 9113–9148. DOI: 10.1007/s10639-022-11555-x PMID: 36643383

Wang, Z., Li, M., Lu, J., & Cheng, X. (2022). Business Innovation based on artificial intelligence and Blockchain technology. *Information Processing & Management*, 59(1), 102759. Advance online publication. DOI: 10.1016/j.ipm.2021.102759

Weizenbaum, J. (1966). ELIZA-A computer program for the study of natural language communication between man and machine. *Communications of the ACM*, 9(1), 36–45. Advance online publication. DOI: 10.1145/365153.365168

Wen, Z., & Huang, H. (2022). The potential for artificial intelligence in healthcare. *Journal of Commercial Biotechnology*, 27(4). Advance online publication. DOI: 10.5912/jcb1327

What is Sustainable Tourism? (n.d.).*GSTC*. https://www.gstcouncil.org/what-is-sustainable-tourism/

Wiest, R., Albillos, A., Trauner, M., Bajaj, J. S., & Jalan, R. (2018). Corrigendum to "Targeting the gut-liver axis in liver disease" [J Hepatol 67 (2017) 1084–1103]. *Journal of Hepatology*, 68(6), 1336. DOI: 10.1016/j.jhep.2018.03.001 PMID: 29655855

Wiltshier, P., & Clarke, A. (2016). Virtual cultural tourism: Six pillars of VCT using co-creation, value exchange and exchange value. In *Tourism and Hospitality Research* (pp. 1–12). SAGE Publications.

Winter. (2008). Demonstration effect. *UNBC Outdoor Recreation & Tourism Management.*https://www2.unbc.ca/sites/default/files/sections/outdoor-recreation-tourism-management/demonstration_effect.pdf

Woodall, T. (2023, December 15). Retrieved from https://news.osu.edu/using-ai-to-pinpoint-hidden-sources-of-clean-energy-underground/

World Health Organization (WHO). (2021). World Report on Hearing. https://www.who.int/publications/i/item/9789240020481

World Health Organization. (2024). Hearing aid service delivery approaches for low- and middle-income settings https://www.who.int/publications/i/item/9789240087927

WTO. (2005).Making Tourism More Sustainable - A Guide for Policy Makers. *UNWTO*.https://www.e-unwto.org/doi/book/10.18111/9789284408214

Wu, S., Li, J., Tian, Z., & Zhang, H. (2023). Explainable machine learning for non-invasive diagnosis of fatty liver disease.

Wu, J., Li, Q., Yang, G., Li, L., Senhadji, L., & Shu, H. (2023). Self-supervised speech denoising using only noisy audio signals. *Speech Communication*, 149, 63–73. Advance online publication. DOI: 10.1016/j.specom.2023.03.009

Xiang, K., Jiang, B., & Shang, D. (2021). The overview of the deep learning integrated into the medical imaging of liver: A review. *Hepatology International*, 15(4), 868–880. DOI: 10.1007/s12072-021-10229-z PMID: 34264509

Yi, H., Li, Z., Xu, J., Zhang, Y., & Yang, Y. (2020). Topological data analysis combined with machine learning for classification of liver tissue images. *IEEE Access : Practical Innovations, Open Solutions*, 8, 138132–138142.

Ying, T., Tang, J., Ye, S., Tan, X., & Wei, W. (2021). Virtual Reality in Destination Marketing: Telepresence, Social Presence, and Tourists' Visit Intentions. *Journal of Travel Research*. Advance online publication. DOI: 10.1177/00472875211047273

Yip, T. F., Ma, A. J., Wong, V. S., Tse, Y. K., Chan, H. Y., Yuen, P. C., & Wong, G. H. (2017). Laboratory parameter-based machine learning model for excluding non-alcoholic fatty liver disease (NAFLD) in the general population. *Alimentary Pharmacology & Therapeutics*, 46(4), 447–456. DOI: 10.1111/apt.14172 PMID: 28585725

Yon, M., Willing, A., McMahon, C., McPherson, B., Nieman, C. L., Reed, N. S., & Lin, F. R. (2019). Bulletin of the World Health Organization Vol. 101: Access to adults' hearing aids: policies and technologies used in eight countries. https://www.ncbi.nlm.nih.gov/pmc/articles/PMC6796668

Yu, L., Kim, S., Moon, S., & Kim, D. S. (2018). Deep learning in non-invasive liver disease diagnosis using medical imaging: A review. *Journal of Medical Imaging and Health Informatics*, 10(7), 1422–1430.

Yung, R., & Khoo-Lattimore, C. (2019). New realities: A systematic literature review on virtual reality and augmented reality in tourism research. *Current Issues in Tourism*, 22(17), 2056–2081. DOI: 10.1080/13683500.2017.1417359

Yung, R., Khoo-Lattimore, C., & Potter, L. E. (2021). Virtual reality and tourism marketing: Conceptualizing a framework on presence, emotion, and intention. *Current Issues in Tourism*, 24(11), 1505–1525. DOI: 10.1080/13683500.2020.1820454

Zhang, H., Sun, C., Li, X., & Zheng, L. (2022). Machine learning-based prediction of liver fibrosis in chronic hepatitis B patients using ultrasound radiomics. *Ultrasound in Medicine & Biology*, 48(10), 2747–2757.

Zhao, M., Hoti, K., Wang, H., Raghu, A., & Katabi, D. (2021). Assessment of medication self-administration using artificial intelligence. *Nature Medicine*, 27(4), 727–735. Advance online publication. DOI: 10.1038/s41591-021-01273-1 PMID: 33737750

Zhen, S. H., Cheng, M., Tao, Y. B., Wang, Y. F., Juengpanich, S., Jiang, Z. Y., Jiang, Y., Yan, Y., Lu, W., Lue, J., Qian, J., Wu, Z., Sun, J., Lin, H., & Cai, X. J. (2020). Deep learning for accurate diagnosis of liver tumor based on magnetic resonance imaging and clinical data. *Frontiers in Oncology*, 10, 680. DOI: 10.3389/fonc.2020.00680 PMID: 32547939

Zhifeng Kong, Wei Ping, Ambrish Dantrey, Bryan Catanzaro (2023). CleanUNet 2: A Hybrid Speech Denoising Model on Waveform and Spectrogram DOI: 10.21437/Interspeech.2023-1287

Zhou, L. Q., Wang, J. Y., Yu, S. Y., Wu, G. G., Wei, Q., Deng, Y. B., Wu, X.-L., Cui, X.-W., & Dietrich, C. F. (2019). Artificial intelligence in medical imaging of the liver. *World Journal of Gastroenterology*, 25(6), 672–682. DOI: 10.3748/wjg.v25.i6.672 PMID: 30783371

Zhu, B. (2020). How the tourism industry in China is finding new ways to engage customers.Capgemini.https://www.capgemini.com/2020/05/tourism-industry-engaging-customers/

About the Contributors

Ahmed G. Radwan (SMIEEE, Fellow-AAS) is the Vice-President for Research, Nile University, Egypt. He is the Founder and general co-Chair of NILES International Conference, and also the Founder of the Undergraduate Research Forum (UGRF) at Nile University . He has a total number of indexed publications of 450, 9600 citations and h-index: 51 according to Scopus. He is Co-inventor of 6 US patents, author/co-author of 10 international books as well as 26 book chapters in the highly ranked publishers. Prof. Radwan is leading the fractional-order and chaotic circuits tracks worldwide. He is an associate editor in 4 prestigious Scopus-indexed journals. Some of his prestigious Awards: The State Encouragement Award, The State Excellence Award, Abdul Hameed Shoman Award, Scopus Award in Engineering and Technology, Prof. Tarek Khalil Award for Distinguished Leadership from Nile university and many more. He managed funded projects with budgets over 1M$ and supervised more than 40 (Master's & Ph.Ds). He received many research grants as Principal Investigator (PI), CO-PI, and Consultant from different national/ international organizations. He was Invited to be Lead/Guest Editors in Journal of Circuits, Systems and Signal Processing, and Journal of Mathematical Problems in Engineering, and Complexity. He organized many special sessions and participated as Technical Program Committee (TPC) in various international conferences. He was selected as a member of the first scientific council of Egyptian Young Academy of Sciences (EYAS) as well as in the first scientific council of the Egyptian Center for the Advancement of Science, Technology, and Innovation (ECASTI) to empower and encourage Egyptian young scientists in science and technology and build knowledge-based societies.

Salwa K. Abd-El-Hafiz received the B.Sc. degree in Electronics and Electrical Communications Engineering from Cairo University, Egypt, in 1986 and the M.Sc. and Ph.D. degrees in Computer Science from the University of Maryland, College Park, Maryland, USA, in 1990 and 1994, respectively. Since 1994, she has been working as a Faculty Member in the Engineering Mathematics and Physics Department, Faculty of Engineering, Cairo University, and has been promoted to a Full Professor in the same department in 2004. She was the Director of the Technical Center for Career Development, Cairo University, Egypt, from August 2014 to July 2016. She served as Chair of the Engineering Mathematics and Physics Department from August 2016 to July 2017. From August 2017 to October 2020, she served as the Vice Dean for Graduate Studies and Research, Faculty of Engineering, Cairo University. Prof. Abd-El-Hafiz co-authored two books, contributed different chapters to four books and published more than 110 refereed papers. Her research interests include information security, chaos theory, number theory, computational intelligence, software engineering, and numerical analysis. She is a recipient of the 2001 Egyptian State Encouragement Prize in Engineering Sciences, recipient of the 2012 National Publications Excellence Award from the Egyptian Ministry of Higher Education, recipient of the 2014 African Union Kwame Nkrumah Regional Scientific Award for Women in basic science, technology and innovation, recipient of several international publications' awards from Cairo University, and an IEEE Senior Member.

Islam Tharwat Abdel Halim received his B.Sc. and M.Sc. degrees and his Ph.D. degree in computer and systems engineering from the Faculty of Engineering, Ain Shams University, Cairo, Egypt. His Ph.D. thesis was granted the best thesis award. He is currently an Assistant Professor in the School of Information Technology and Computer Science, Nile University (NU), Giza, Egypt. He has authored many research articles in various refereed conferences and journals, including IEEE Intelligent Transportation Systems Magazine, Computer Networks (Elsevier), and Wireless Networks (Springer). His current research interests include IoT, Fog/Edge computing, mobile and wireless networks, with emphasis on security and performance evaluation in vehicular communication, and 5G networks. He is a senior member of the IEEE and is currently appointed as IEEE Computer Society Coordinator for Region 8 (Europe, Africa, and the Middle East) as well as the chair of the awareness and customer advocacy committee in the Distinguished Visitor Program (DVP). Also, he is the founder and currently recorded as the chair of the IEEE Egypt Section Consultants Network and the vice-chair of ACM Cairo Chapter Besides, he has over 20 years of experience in the field of information technology and digital transformation. He is a member of the Digital Transformation Committee at Nile University and participated in the development of the digital transformation training plan for the Egyptian government. He has also participated in numerous national projects in the Ministry of Finance, the Tax and Customs authorities, and the government certificate authority. Furthermore, he has obtained several accredited certifications from major global IT companies such as Cisco, Microsoft, VMware, and IBM. He contributed to the establishment and organization of several local and international conferences in the field of computer engineering and sciences, serving as a session chair, invited speaker, program committee member, and editor for various international conferences and journals. In addition, he is a member of several specialized technical committees, including the IEEE computer Society special technical community on Blockchain.

Yuhong Liu, Associate Professor at the Department of Computer Science and Engineering, Santa Clara University. Her research interests include responsible AI, trustworthy computing and cyber security of emerging applications, such as online social media, Internet-of-things and Blockchain. She has published over 100 papers on prestigious journals and peer reviewed conferences. She is currently serving as an Associate Editor of IEEE Transactions of Service Computing, Multimedia Tools and Applications, APSIPA Transactions on Signal and Information Processing, and a Youth EBM Editor of the Blockchain: Research and Applications Journal. She serves as an IEEE Computer Society Distinguished Visitor (2022-2024) and APSIPA Distinguished Lecturer (2021-2022). She is the recipient of the 2023 Rising Star Service Award, IEEE CS Technical & Conference Activities Board, and the 2019 Researcher of the Year Award at School of Engineering, Santa Clara University.

Meikang Qiu received the BE and ME degrees from Shanghai Jiao Tong University and received Ph.D. degree of Computer Science from University of Texas at Dallas in 2007. Dr. Qiu comes to us with a background in Cyber Security, Machine Learning, Artificial Intelligence, Big Data, and Cloud Computing. Gaining expert-level knowledge of theory and practice in many fields of Cyber Security, Machine Learning, and Artificial Intelligence, he had produced many novel results, which have been published in 500+ peer-reviewed journal/conference papers, 20+ books, and 3 patents. He is the winner of 10+ Best Paper Awards in recent years on IEEE/ACM international journals and conferences and is or was the associate editor of more than 10 international journals, including IEEE Transactions on Computers and IEEE Transactions on Cloud Computing. He is the Chair of IEEE Smart Computing Technical Committee. His paper on cloud systems has won IEEE Systems Journal 2018 Best Paper Award. He has received 2018 Middle Career Researcher Award from TCSC of IEEE Computer Society. His research has been funded by NSF, Air Force, Navy, and many companies. He is the General Chair/Program Chair of a dozen of IEEE/ACM international conferences, such as IEEE TrustCom, IEEE BigDataSecurity, IEEE CSCloud, and IEEE HPCC. He has won Navy Summer Faculty Award in 2012 and Air Force Summer Faculty Award in 2009.

* * *

Alaa-Eldin Adris is a Professor at The Graduate School of Management of Technology, Nile University. Throughout his 38-year career in science, technology, and innovation, I have been at the interaction point between industry and academia, serving on a different side at different times. As a result, he has been involved with all elements of the STI value chain, including policy, strategy, research, development, demonstration, commercialization, deep and science-based technology entrepreneurship, and licensing (playing the roles of both a licensor and a licensee).

Ayman Ali is a communication engineer, researcher, and author with more than 23 years of experience in optical transmission networks. He holds a PhD in communication engineering from Minia University, a Master of Communication Engineering from Ain Shams University, a BSc of communication engineering from Banha University, and a Master of Business Administration from Arab Academy of Science. He has published many papers in the field of artificial intelligence in optical transport networks, and he is an expert in the principles, techniques, and standards of digital transmission systems. He works as an O&M engineer in one of the largest communication operators in the Middle East, where he is responsible for the design, implementation, operation, and maintenance of optical transmission networks. He was born in 1976 in Egypt, and he lives with his wife and sons in Cairo, Egypt. He is passionate about digital communication, and he enjoys sharing his knowledge and experience with his students, colleagues, and readers.

Mohamed Mamdouh Awny has a well-balanced blend of in-depth experiences in academia, applied research, and industry. He obtained his Ph.D. from Queen Mary College, University of London, UK, in "Computer Control of Engine Testing". He holds B.Sc. and M.Sc. degrees in Aeronautical Engineering from Cairo University. Prof. Awny has a rich background in the aerospace industry, including R&D of ground-to-ground missiles and consultancy for the Ford Foundation in the UK. He played a key role in the Arab Organization for Industrialization, including the establishment of the anti-tank missile factory and technology transfer negotiations with British partners. Prof. Awny has served as Deputy Manager of the Arab Institute for Aerospace Technology and Director-General of the Information Technology Center at the Arab Organization for Industrialization. His leadership roles also include Vice President and Director of Consulting and Information Technology at Engineering and Management Consultants Company (TEAM International) and Director-General for Small Enterprise Development Organization at the Social Fund for Development (World Bank/The Cabinet of Egypt). Before joining Nile University, Prof. Awny directed the Technology Management Program at the College of Graduate Studies, Arabian Gulf University, Kingdom of Bahrain (2002-2009). He chaired the UN International Task Force for developing the Informal Sector in

Africa and was a member of the UN Expert Group in Public Administration and Finance. He has been involved in various national and international committees and associations, including the Economic Committee of the National Council of Women in Egypt, Egyptian Incubator Association, and the International Association of Management of Technology (USA). Prof. Awny is an internationally recognized expert with numerous publications and a wide spectrum of experience.

Omar M. Badreldin, M.Sc. is a Technology Director, independent TMT (Technology,Media,Telecommunication) consultant, and legal technology expert with over 23 years' experience. His career spans digital transformation, AI, software development, telecommunication, TV/DVB broadcasting, and multi-modal applications. He has led cross-functional teams, managed complex initiatives, driven successful digital transformations, and delivered innovative solutions to critical business challenges. He has taken on diverse roles in dozens of technology projects and was appointed by the Egyptian Economic Courts as one of only four accredited technology and telecommunications experts. In this capacity, he provides expert testimony and strategic counsel on technology-related legal disputes, telecommunications regulations, and digital forensics. His blend of technical acumen, operational leadership, and legal expertise positions him as a leading authority in navigating the evolving TMT landscape.

ElHassan ElSabry is an Assistant Professor at the Graduate School of Management of Technology in Nile University Egypt. He earned his doctoral degree in Science, Technology and Innovation Policy from Japan's leading public policy school, National Graduate Institute for Policy Studies (GRIPS). His research interests cover the topics of research funding, science philanthropy and strategic innovation management.

Hebatalla Kaoud is a PhD holder in Management Science from Nantes University in France, Dr. Hebatalla Kaoud is an assistant professor at the Graduate School of Management of Technology at Nile University and the Director of Intellectual Property and Technology Transfer Office (IPTTO).

Laura Labib is a professional with 16 years of experience in the Information and Communication Technology (ICT) sector spanning both private and public sectors. She holds a Bachelor of Science in Economics, and in 2019 she enrolled for a Master's in Management of Technology which she is currently concluding. Additionally, Laura embarked on a journey to study psychology and philosophy and she started a master's in Psychology in 2022. In her professional journey, Laura has contributed 13 years to the private sector, taking on various roles in Operations and Support at Oracle Corporation, and 3 years of experience in the public sector as she currently serves as an Institutional Development Consultant at Egypt's Ministry of Communications and Information Technology, focusing on institutional development, digital transformation and enhancing digital governance. An advocate for continuous learning and naturally curious, Laura has explored various disciplines and earned diverse certifications. She studied neurolinguistic programming and is a certified NLP Master Practitioner, Project Management Professional (PMP), and ITIL practician but also has courses in Artificial Intelligence, Blockchain.

Nahed Salem is an MSc student at The Graduate School of Management of Technology, Nile University.

Mallika Sankar is an assistant professor in the School of Business and Management at the Christ University in Bangalore, India. She received her Ph.D. in management from Bharathiar University, MBA from Pondicherry University and undergraduate studies at Mahatma Gandhi University. She has collaborated actively with researchers in several other disciplines of business management, published articles and co-authored a book in research methodology. She is preferred as a resource person in several faculty development programs, conferences, workshops. She has also served as trainer in the programs organized by AICTE and an invited speaker in many public events and academic institutions. She has also served in industry at various administrative positions for over 12 years. Her areas of teaching include research methodology, business analytics and strategic management at the post graduate, graduate and undergraduate levels.

Al-Hassan Wadan is a dentist who obtained his Bachelor's degree in Dentistry from Sinai University in Egypt. Soliman is interested in researching regenerative medicine, tissue therapy, and stem cell biology. He has published several review articles in scientific journals such as Frontiers in Cell and Developmental Biology, Life Sciences, IBRO Neuroscience Reports, etc.

Asmaa Youssef is the Deputy Director for the Innovation, Entrepreneurship, and Competitiveness Center (IECC) at Nile University, where she also serves as Associate Director for the University-Industry Collaboration Office. With over 15 years of experience, Asmaa has developed a strong expertise in innovation management, business development, and fostering strategic collaborations between academia and industry. Her work focuses on helping startups and SMEs improve their innovation capabilities, product development strategies, and market entry approaches. As part of her role at IECC, Asmaa has mentored and coached over 200 startups and SMEs, guiding them in innovation management, new product development, and market entry strategies. Her hands-on approach has contributed to the successful launch of numerous products across various sectors, including technology, manufacturing, and service industries. She is committed to driving strategic initiatives that support the growth of SMEs and commercialization of research outputs, which has resulted in fostering stronger ties between the university and industry partners. Asmaa is responsible for overseeing and executing a wide range of innovative projects within IECC. These projects aim to promote a culture of creativity and forward-thinking, enabling startups, SMEs, and business owners to remain competitive in their markets. She works closely with multidisciplinary teams to identify opportunities for innovation, monitor emerging technologies, and establish partnerships that drive the commercialization of innovative solutions. Additionally, Asmaa ensures that projects align with organizational goals and oversees the development of Monitoring and Evaluation (M&E) systems to track progress and outcomes. Prior to joining Nile University in 2018, Asmaa had an extensive background in project management, particularly within the engineering and construction industries, where she led multiple large-scale projects and managed operations across various units. Her career shift toward innovation management was driven by her passion for applying engineering principles to solve business challenges and foster innovation in high-growth sectors. Asmaa holds a BSc in Civil Engineering from Ain Shams University and is currently pursuing her MSc in Management of Technology at Nile University. Her research focuses on creating a holistic framework for measuring innovation capabilities within organizations, specifically targeting the needs of SMEs in developing countries. Her work aims to fill the gap in customizable innovation measurement tools and frameworks that can be adapted across various sectors and organizational contexts. Asmaa's leadership at Nile University's IECC has made a significant impact on advancing the innovation ecosystem in Egypt and the broader MENA region. She has led multi-stakeholder, multi-country projects in collaboration with university leadership and industry partners, focusing on promoting innovation-driven growth and entrepreneurship. Her ability to blend academic insights with practical applications makes her a valuable contributor to both academic research and the business community.

Index

A

B

C

D

E

G

H

V

www.ingramcontent.com/pod-product-compliance
Lightning Source LLC
LaVergne TN
LVHW080557300926
849162LV00005B/983

* 9 7 9 8 3 3 7 3 1 1 3 2 6 *